Breakwaters and Closure Dams

Breakwaters and Closure Dams

K. d'Angremond

F.C. van Roode

CRC Press
Taylor & Francis Group
Boca Raton London New York

CRC Press is an imprint of the
Taylor & Francis Group, an **informa** business

CRC Press
Taylor & Francis Group
6000 Broken Sound Parkway NW, Suite 300
Boca Raton, FL 33487-2742

© 2004 by Taylor & Francis Group, LLC
CRC Press is an imprint of Taylor & Francis Group, an Informa business

No claim to original U.S. Government works

Visit the Taylor & Francis Web site at
http://www.taylorandfrancis.com

and the CRC Press Web site at
http://www.crcpress.com

Contents

Preface

This book is primarily a study book for graduate students in Civil Engineering at Delft University of Technology. The consequence is that in addition to treating the latest insights into the subject matter, it places the developments in their historic perspective, at least when this contributes to better understanding. It also means that this book cannot replace comprehensive textbooks or original scientific publications. The reader is strongly advised to consult the original references rather than blindly following this textbook.

In the curriculum of Delft University, the present course on breakwaters and closure dams is preceded by a variety of courses on subjects such as fluid mechanics, hydraulic engineering, coastal engineering and bed, bank and shore protection, design process, and probabilistic design. Many subjects dealt with in these courses will be refreshed briefly in the present, more specialized, course.

Previously, the designing of breakwaters and of closure dams were treated in separate courses, but since 1998, these courses have been combined. At first sight it seems strange to combine the design of two rather dedicated types of structures with distinctly different purposes, however from an educational point of view this is not so.

In both cases the design process requires that due attention should be paid to:

- the functional requirements
- the various limit states to which a structure will be exposed in relation to the requirements
- the various limit states that occur during construction phases
- the relation between these limit states and the occurrence of certain natural conditions

The differences between closure dams and breakwaters will enable us to focus attention on the above mentioned considerations.

In addition to this, there are also quite a number of similarities. In this respect, we refer to the construction materials, such as quarry stone, concrete blocks and caissons, which are widely used in both types of structures. The same applies to a wide range of construction equipment, both floating and rolling, and, last but not least, the interdependence between design and construction.

It is good to mention here that the design of closure dams, and more specifically closure dams in estuaries, has undergone a major development in the period between 1960 and 1985, when the Delta Project in the Netherlands was being executed. No projects on a similar scale have been started since, but it is safe to say that similar projects will come up in the future when low lying and densely populated areas are threatened by a rising sea level. Breakwaters, and specifically various kinds of rubble mound breakwaters, underwent a tremendous development in the past decade (1985-1995). Now the pace of innovation seems to be slowing down, although monolithic breakwaters are gaining attention and it is expected that the state-of-the-art with respect to monolithic breakwaters will change rapidly in the near future. Therefore, the present study book does not represent a static subject. This necessitates that both the teacher and the student should continuously observe the latest developments.

The authors very much appreciate the contribution of H.J. Verhagen of Delft University of Technlogy, who added valuable sections on the determination of design conditions and on probablistic design.

K. d'Angremond Delft   December 2003
F.C. van Roode

1 INTRODUCTION

1.1 Scope

In the Preface, it was mentioned that closure dams and breakwaters were previously treated in separate lectures, each with their own study material. Even with the two subjects combined in one series of lectures, it would have been possible to treat them as separate entities. However, we have deliberately chosen that the text of this book should follow a more or less logical design procedure. This means that in each step of the procedure attention is paid to both subjects and that every time the two types of structures are compared the similarities and differences are emphasized.

With respect to breakwaters, all existing types are discussed briefly but only the types that are frequently used all over the world (i.e. rubble mound breakwaters, berm breakwaters and monolithic breakwaters) are treated in detail.

With regard to closing dams, it is emphasized that only the actual closing dams are considered in this book. This means that only the closing operation itself is treated; the transformation of the closing dam into a permanent structure like an embankment is beyond the scope of this book.

It is expected that the reader will possess some basic knowledge and duplication of theories treated in the preceding lectures has been avoided. Here and there, where they are deemed useful for a proper understanding of the actual design process, the results of theoretical considerations and derivations are given. Sources of basic knowledge are listed in a separate section of the list of references.

1.2 Authors

This book has been compiled by a great number of people on the staff of or attached to the Hydraulic Engineering Section of the Faculty of Civil Engineering and Geo Sciences of Delft University of Technology.

The main authors are:

- Prof. Ir. K. d'Angremond, Professor of Coastal Engineering, Delft University of Technology
- Ir. F.C. Van Roode, Associate Professor, Delft University of Technology

Valuable contributions in the form of comments and/or text were received from:
- Dr Ir. M.R.A. van Gent, WL | Delft Hydraulics
- Dr Ir. J. van der Meer, INFRAM b.v.
- Ir. G.J. Schiereck, Associate Professor, Delft University of Technology
- Ir. H.J. Verhagen, Delft University of Technlogy.

Many others contributed in a variety of ways, including correcting text and preparing figures. In this respect the contributions of the following are greatly acknowledged:
- V.L. van Dam - Foley
- Ing. M.Z. Voorendt

1.3 References

Although a study book has its own right to existence, there are some outstanding reference books in the field treated by this textbook and these are often far more comprehensive than any study book can be. Therefore a number of books and periodicals that should be available to anybody who will ever be in charge of design or construction of breakwaters and closure dams are mentioned here.

For *breakwaters* such books include: Shore Protection Manual (Coastal Engineering Research Center [1984]), Manual on the use of Rock in Hydraulic Engineering (CUR/RWS [1995]), PIANC Working Group reports (Anonymous [1976]), Report of PIANC Working Group no. 12 [1993]). For *closure dams reference may be made to:* The Closure of Tidal Basins (Huis in 't Veld, Stuip, Walther, van Westen [1984]) and the Manuals of the Advisory Commission on Sea Defences (TAW, in Dutch). Useful periodicals include the Journals of the ASCE, the magazine "Coastal Engineering" and the proceedings of the international conferences on Coastal Engineering.

1.4 Miscellaneous

The present text has been written in English by authors from the Netherlands and some of the techniques discussed have been developed in the Netherlands, both recently and centuries ago. The English therefore carries a Dutch flavour. British spelling has been used except where reference is made to American literature.

To avoid misunderstandings, a glossary of the terms used in this book in both English and Dutch is added as Appendix 1. The reader is also referred to a more general vocabulary on coastal engineering (The Liverpool Thessaloniki Network [1996]).

In this book, the metric (mks) system (based on the definition of mass [kg], length [m], and time [s] has been used, except for some widely accepted nautical and hydrographic terms such as knots, fathoms and miles.
Where applicable,

- the X-co-ordinate is used in the direction of current or wave propagation
- the Y-co-ordinate is used for the horizontal direction, normal to the X- co-ordinate
- the Z-co-ordinate is defined in the vertical direction, positive upward, with the origin either at the bottom or at the surface level.

Since data from existing literature, often from different disciplines, are used and copied, the reader cannot expect a 100% systematic use of symbols throughout the book. For instance, the water depth may be indicated by the letters h, d or D, and stone size by d or D. The Greek letter σ may thus be used for the stress in a material, or as a symbol for the standard deviation. Therefore, since it is not unambiguous, the list of symbols must be used with care. In cases where serious confusion may arise, the symbols are defined and explained as and where they are used. The authors feel that students should be trained to adapt to different notations when they read literature from different sources.

2 POSITIONING THE SUBJECT

2.1 General

Breakwaters are widely used throughout the world, mainly to provide shelter from wave action. This protection is primarily designed for vessels in port and for port facilities, but sometimes breakwaters are also used to protect valuable habitats that are threatened by the destructive forces of the sea or to protect beaches from erosion. Although the threat is usually caused by wave action, protection against currents is also important. Additionally, breakwaters can prevent or reduce the siltation of navigation channels. In some cases breakwaters also accommodate loading facilities for cargo or passengers.

Closure dams are constructed for a variety of very different purposes, such as the creation of a separate tidal basin for power generation or as sea defence structures to increase safety.

Compared to closure works, few engineering works have such an extensive impact on the environment in all its aspects. For instance, the main purposes of the construction of the Afsluitdijk in the Netherlands, which changed part of the Zuiderzee into IJsselmeer, were to provide protection against high storm surge levels and to facilitate land reclamation. Additional advantages were fresh water conservation and the road connection (a railway was considered but never realized). The purpose of the closure may be one or more of such objectives, but these are automatically accompanied by other side effects. A thorough study of these impacts is part of the design process. A feasibility study that does not mention and estimate the negative aspects of the closing work is incomplete and valueless. For example, the negative effects for the Afsluitdijk include: the drastic change in tidal amplitude in the Waddenzee, with its consequential impact on the morphological equilibrium of the tidal flats and channel system, the social impact on life and employment in the bordering cities, the influence on drainage and ground water table in the surrounding land areas, the changes to the fisheries, and changes to flora and fauna.

Several aspects are non-technical in nature and some, like environmental, social and cultural values, cannot be expressed in financial terms. The evaluation of such considerations is not within the scope of this book. Nevertheless, the engineer must identify the consequential effects to the best of his ability and present them in such a way that they are understood by decision-makers.

This book focuses on the technical aspects of the construction of a closure dam in a variety of circumstances. Every closure operation is a struggle with nature. Flowing water on an erodable bed has to be controlled. Every action taken to obstruct the flow will immediately be counteracted in some way or another by nature itself. Of course this happens within the laws of nature, many (but not all) of which are known. The knowledge gained from both good and bad experience is supplemented by the results of advanced research and experiment. Nevertheless, the changes in conditions during the progression of the closure are sometimes difficult to predict. Flexibility in operations that is incorporated in the design provides an important tool.

For a design to be made, the hydrology of the water body or watercourse to be closed has to be fully understood. The main distinction is made between tidal and riverine regimes. Tides are characterized by short-term variations in water level and in flow direction. The design must cater for quick action during high or, more typically, low water periods and during the daily occurring slack water periods. River flows are steadier in the short term, generally one-directional and never cease. Damming rivers is therefore a completely different process.

Comparison of the designs for breakwaters and closure dams shows some identical aspects but other aspects require a completely different approach. For instance:

Comparable:
- Many construction materials used are similar: bottom protection, quarry stone, concrete blocks, specially designed concrete structures (caissons).
- In both cases the equipment used is either land based or water-borne: for example hydraulic excavators and cranes, dump trucks and dump-vessels, barges and bulldozers.

Differences:
- The main determining parameter for breakwater design is wave-action, while for closure design it is flow velocity.
- The estimated design wave is unlikely to occur during the construction of a breakwater, but may occur in its lifetime. The estimated maximum flow during closure will occur during construction and will never occur after closure.

- The breakwater construction is the final design intended to withstand all future attack. The closure dam is a temporary construction that halts the flow, after which, for future safety, the desired definite dam profile can be made. This is based on construction in no-flow conditions.

2.2 Types of breakwaters

There are many different types of breakwaters that can be divided into categories according to their *structural features*:

2.2.1 Mound types

Mound types of breakwaters are actually no more than large heaps of loose elements, such as gravel and quarry stone or concrete blocks. The stability of the exposed slope of the mound depends on the ratio between load and strength i.e. wave height (H) on one hand and size and the relative density of the elements (ΔD) on the other hand. One extreme example is a gravel beach that is subject to changes in the equilibrium profile as the wave characteristics change and also to longshore transport phenomena. Another extreme is the 'statically stable breakwater', where the weight of the elements in the outer layer (armour) is sufficient to withstand the wave forces. Between these is the 'berm breakwater', where the size of the armour is not sufficient to guarantee stability under all conditions, but where some extra quantity of material is provided so that the slope of the structure can reshape between given limits. Typical values of $H/\Delta D$ for the three types of structures are given in Table 2-1.

2.2.2 Monolithic types

Monolithic types of breakwater have a cross-section designed in such a way that the structure acts as one solid block. In practice, one may think of a caisson, a block wall, or a masonry structure. This type of structure can be categorized by a typical value of $H/\Delta D$ that is given (as caisson) in Table 2-1.

Type of structure	$H/(\Delta D)$
Sandy Beach	> 500
Gravel Beach	20 – 500
Rock slope	6 – 20
Berm Breakwater	3 – 6
(Stable) Rubble Mound Breakwater	1 – 4
Caisson	< 1

Table 2-1 Characteristic values of $H/(\Delta D)$

The main differences between the mound type and the monolithic type of breakwater are caused by the interaction between the structure and the subsoil and by the behaviour at failure. The mound-type structures are more or less flexible; they can follow uneven settlement of the foundation layers, whereas the monolithic structures require a solid foundation that can cope with high and often dynamic loads. The behaviour of the structures close to failure is also quite different. When a critical load value is exceeded a monolithic structure will lose stability at once, whereas a mound type of structure will fail more gradually as elements from the armour layer disappear one after another.

2.2.3 Composite types

Composite types of breakwaters combine a monolithic element with a berm composed of loose elements. In fact, there is an abundance of alternatives that combine a rigid element and a flexible structure.

2.2.4 Special (unconventional) types

Many methods can be used to break the wave action other than the traditional types defined above. These include:

- Floating breakwaters
- Pneumatic breakwaters
- Hydraulic breakwaters
- Pile breakwaters
- Horizontal plate breakwaters

All these unconventional breakwaters are used, or their use has been proposed, in exceptional cases under exceptional conditions. Under standard conditions their use usually appears to be either unfeasible or uneconomic. Floating, pneumatic and hydraulic breakwaters require either large dimensions or a lot of energy to damp longer waves that occur at sea. Pile breakwaters and horizontal plate breakwaters require very high structural strength to survive wave loads under extreme conditions. Apart from a distinction between the categories described so far, there is also a distinction in terms of the *freeboard* of the crest above the still water level (SWL)[1]. Traditional structures usually have a crest level that is only overtopped occasionally. It is also possible to choose a lower crest level that is overtopped more frequently, or even a submerged crest. When a low crest level is combined with the design philosophy of a berm breakwater, (i.e. a reshaping mound) it is termed a reef-type breakwater. Examples of all types of breakwater are shown in Figure 2-1 to Figure 2-4.

---

[1] SWL is the water level that would exist in the absence of sea and swell (instantaneous mean water level in the absence of waves).

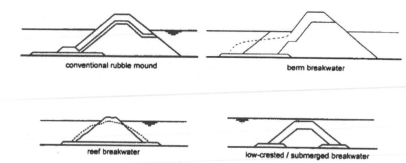

Figure 2-1 Mound breakwater types

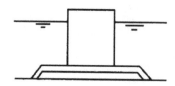

Figure 2-2 Monolithic breakwater type

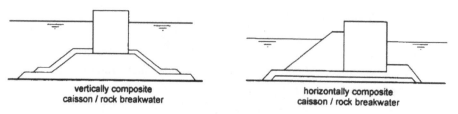

Figure 2-3 Composite breakwater types

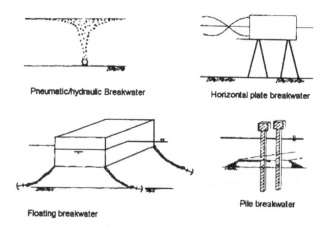

Figure 2-4 Special breakwater types

In the present book, attention will be mainly focused on the traditional types of breakwater, i.e. the mound type and the monolithic type.

2.3 Types of closure dams

Several names have been adopted to distinguish various types of closure operations. The names used may refer to different aspects. However, the adoption of names has been random rather than systematic. Some names are typically Dutch and there may be no literal English translation.

A main distinction can be made according to the construction method. This is illustrated in Figure 2-5.

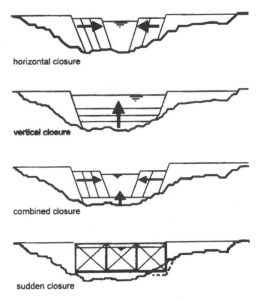

horizontal closure

vertical closure

combined closure

sudden closure

Figure 2-5 Basic methods of closure

The construction method is related to the equipment used, which is either land-based or water-borne. This leads to a distinction between horizontal or vertical closure and the possible combination of these two methods. Using large structures (caissons) gives a type of horizontal closure with very large units. Figure 2-5 illustrates these methods.

Basic methods of closure:

A. Gradual closure:

Relatively small sized, flow resistant material is progressively deposited in small quantities into the flow until complete blockage is attained. This can be used for either a vertical or a horizontal or a combined closure:

- Horizontal (gradual) closure: sideways narrowing of the closure gap.
- Vertical (gradual) closure: layer by layer upward closing of the gap.
- Combined vertical and horizontal closure: a sill is first constructed, on which sideways narrowing takes place.

B. Sudden closure:

Blocking of the flow in a single operation by using pre-installed flap gates or sliding gates, or by the placing of a caisson or vessel.

Methods of closure may also be distinguished according to:
The *topography* of the gap to be closed, as is illustrated in **Figure 2-6**:
- Tidal gully closure [stroomgat-sluiting]: closure of a deeply scoured channel in which high flow-velocities may occur.
- Tidal-flat closure [maaiveld-sluiting]: closure across a shallow area that is generally dry at low water. This is characterized by critical flow at certain tide-levels.
- Reservoir dam (beyond the scope of this book): used in mountainous areas; this requires temporary diversion of the flow in order to obtain solid foundation in the riverbed at bedrock level.

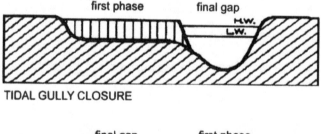

TIDAL GULLY CLOSURE

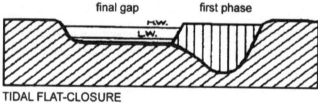

TIDAL FLAT-CLOSURE

Figure 2-6 Closure named after topography

The hydrologic conditions that determine the type of closure (see Figure 2-7):
- Tidal-basin closure: characterized by regularly changing flow directions and still water in between; mainly determined by the tidal volumes and the storage capacity of the enclosed basin.
- Partial tidal closure: a closure in a system of watercourses, such that after closure there is still a variation in water-level at both sides of the closure dam.

- River closure (non-tidal): closure determined by upland discharge characteristics and backwater curves.

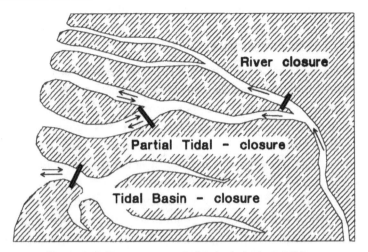

Figure 2-7 Closures named after hydrologic conditions

The *materials used*, which may vary according to the method of closure:
- Stacking-up willow mattresses [opzinken]: Closure realized by successively dropping mattresses (made of willow faggots, ballasted by clay or cobbles) onto each other.
- Sand closure: Closure realized by pumping sand at a very high rate of production.
- Clay or boulder-clay closure: Lumps of flow-resistant clay, worked up by grabs from floating cranes.
- Stone-dam closure: Closure realized by dumping rock, boulders or concrete blocks in the gap, either by using dump-barges and floating cranes, or by cableway.
- Caisson closure: Closure by using large concrete structures or vessels, floated into position and then sunken in the gap, (possibly provided with sluice gates).

The equipment used (typically used for vertical closure):
- Bridge closure: Closure realized by dumping material from a bridge, pre-installed across the gap.
- Cableway closure: Dropping materials from a pre-installed cableway.
- Helicopter closure: Dropping materials from a helicopter.

Special circumstances leading to typical closure types:
- Emergency closure, is characterized by improvisation; the basic idea is that quick closure, even at the high risk of failure, prevents escalation of conditions; mainly used for closing dike breaches; needs strengthening later.

- Temporary closure: used to influence the conditions elsewhere, for instance by stepwise reducing the dimensions of the basin; needs to be sufficiently strong during the required period but easily removable afterward.

2.4 Historical breakwaters

The first breakwaters that are described in traceable sources date back to the ancient Egyptian, Phoenician, Greek and Roman cultures. Some of them were simple mound structures, composed of locally found rock. As early as 2000 BC, mention was made of a stone masonry breakwater in Alexandria, Egypt (Takahashi [1996]). The Greeks also constructed breakwaters (mainly rubble mound) along some parts of the Mediterranean coast. The Romans also constructed true monolithic breakwaters, since they had mastered the technique of making concrete. The Roman emperor Trajan (AD 53 - 117) initiated the construction of a rubble mound breakwater in Civitavecchia, which still exists today (Figure 2-8). The very flat seaward slope and the complicated superstructure are proof of a history of trial and error, damage and repair (Vitruvius [27 BC]; Shaw [1974] Blackman [1982]; De la Pena, Prada and Redondo [1994]; Franco [1996])

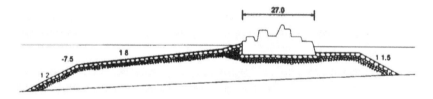

Figure 2-8 Rubble mound breakwater at Citavecchia

In modern times similar breakwaters were constructed at Cherbourg (1781/1789/1830), and at Plymouth (1812/1841). In both cases, the stability of the seaward slope was insufficient and during subsequent repair operations the final slopes were between 1:8 and 1:12 (See Figure 2-9 and Figure 2-10).

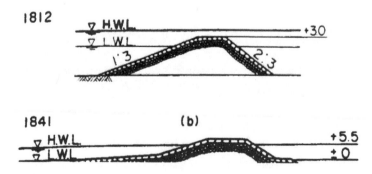

Figure 2-9 Breakwater at Cherbourg

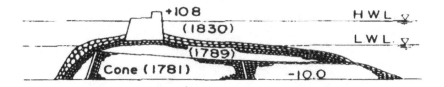

Figure 2-10 Breakwater at Plymouth

In view of the difficulties encountered in Cherbourg and Plymouth, in 1847 it was decided that a monolithic breakwater should be built at Dover. The construction posed a lot of problems, but the result was quite satisfactory since this breakwater has survived without major damage (Figure 2-11)

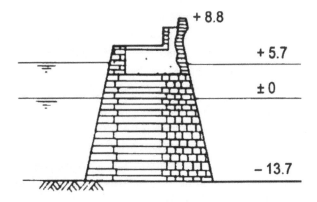

Figure 2-11 Monolithic breakwater at Dover

The rapidly increasing sea-borne trade in the 19th century led to a large number of breakwaters being built in Europe and in the emerging colonies. The British engineers in particular took the lessons from the Dover breakwater to heart. To avoid the problems of construction in deep water, rubble mound berms were used for the foundation of a monolithic superstructure, and thus the first real composite breakwaters came into existence. Here also, however, the process of trial and error took its toll. Many breakwaters had to be redesigned because the berms were originally erected to too high a level.

In France, engineers tried to solve the stability problems by designing flatter slopes above SWL, and by applying extremely heavy (cubic and parallelepiped) concrete blocks as the armour layer. They also started to use smaller-sized stone systematically in the core of the structure. The breakwater of Marseilles (1845) became a successful example for the French speaking world, just like the Dover breakwater in the English speaking world. However, it was recognized that the Marseilles type of solution required very heavy armour units and also a lot of material in the cross section, especially in deeper water (Figure 2-12).

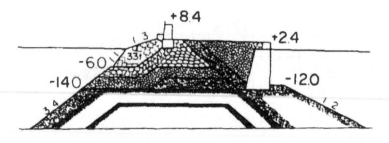

Figure 2-12 Breakwater at Marseilles

These developments made the composite breakwater the most widely used type in the early 20th century, especially in Italy, where a lot of breakwaters were constructed in relatively deep water along the Mediterranean coast. The logical solution therefore seemed to be a composite structure consisting of a berm to about the half water depth, with a vertical faced wall on top of it. The wall was built of extremely large (Cyclopean) blocks, sometimes interlocking to create the monolithic effect (Figure 2-13). However, these breakwaters were not a success, since the mound caused waves to break and to slam against the vertical wall, which subsequently tended to fail.

These failures worried the port engineers gathered in PIANC[2] so much that they decided to set up an international association for hydraulic research (IAHR). The failures of the vertical-wall breakwaters around the Mediterranean in the first half of the 20th century marked the end of this type of breakwater in W. Europe.

The French continued their efforts to optimize their rubble mound concept and to reduce the required weight of the armour blocks they developed the idea of interlocking them. Thus, in 1949, P. Danel [1953] of the Laboratoire Dauphinois d'Hydraulique (later SOGREAH) designed the Tetrapod armour unit, which was the start of a long series of similar blocks. To avoid the payment of royalties, Rijkswaterstaat and Delft Hydraulics developed the Akmon. The Dolos (South Africa) seemed to provide the ultimate solution, until the limited mechanical strength of this block triggered a new series of mishaps. One of the most spectacular mishaps was the failure of the breakwater of Sines (Portugal) in February 1978. The development of special shaped blocks went on, however, resulting in two other French blocks, which are still quite successful: the "Antifer cube" and the Accropode®.

---

[2] Permanent International Association of Navigation Congresses.

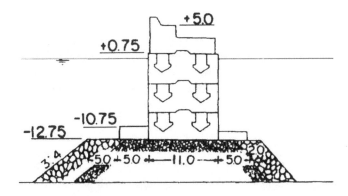

Figure 2-13 Typical breakwater along the mediterranean coast

In the meantime, the Japanese continued to build and develop the monolithic breakwater. There is no other country where so many monolithic and composite breakwaters have been built (with varying success). The principal contribution, however, was made by a French engineer, G.E. Jarlan [1961], who introduced the perforated front wall to reduce reflection and wave impact forces.

2.5 Historical closures

2.5.1 Introduction

Closure dams have probably been constructed since mankind started agriculture and needed water for irrigation. Another reason for their construction could be political strategy because of the need for road or navigational connections. There is little evidence of these activities in ancient times, but the irrigation projects that once existed in ancient Babylon and Egypt suggest the presence of such works. As such dams would have been constructed from locally available perishable materials, no ruins (like pyramids) are found, even though they might have been quite extensive, considering that the builders were able to construct pyramids.

2.5.2 The damming of the rivers Rhine and Meuse in the late Middle Ages

In the delta area of the rivers Rhine and Meuse, the damming of rivers and watercourses developed in the early Middle Ages. Because of the need for agricultural expansion, areas of marshland that are flooded only during extremely high tides or when rivers are in spate, were artificially drained. This caused the soil, mainly peat, to settle and as a result the incidence of flooding increased. Therefore, small earthen walls surround the areas and the natural drainage channels are dammed off. Many cities and villages in Holland are named after such dams (Rotterdam, Amsterdam). In the period 1100 to 1300, damming activities drastically changed even the courses of the two main rivers.

The damming of the river Rhine:

It is probable that to prevent the river that was choked by sediments from overflowing its banks, the ruler of Utrecht dammed the river Rhine at Wijk bij Duurstede around the year 1200. The flow was diverted via the Lek river-branch and the original river mouth near Katwijk shoaled and disappeared. For further details see Section 4.4.

The damming of the river Meuse (Maas):

In 1270 the river Meuse was diverted by damming it at Maasdam (near the city of Dordrecht) and upstream near Heusden, where the flow was directed towards the town of Woudrichem. For further details see Section 4.4.

2.5.3 From the Middle Ages to 1920

Historic sources give a fair idea about the old methods used. The dams had to be constructed from locally available materials that could be handled by hand and simple equipment. These materials were not stable under conditions of high flow velocities. Therefore the essence of the process followed was to limit the flow velocities during closure process in accordance with limitations on the size and weight of these materials. One way to achieve this was to split the basin area into separate small compartments and then to close these compartments successively. Experience indicated the maximum area that could be taken in relation to tidal rise. Furthermore, flow velocities were kept low by using the vertical closing method, as will be clarified in Section 5.2.3. Branches cut from willow trees (osiers), were the main construction materials. With these an interwoven structure (fascine mattress [zinkstuk]) was made. When ballasted with clay this could be sunk onto the bottom. For further details see Section 13.2 and Appendix 9.

The closure was realized by sinking these mattresses successively one on top of the other on every tide during the short period of slack water. In this way a stack of mattresses created a sill in the closure gap. This continued up to about low water level. Further sinking was then impossible, as the mattresses could not be floated above the sill. The closure was completed by using a different type of structure. This was again composed of willow (osier) and clay, but this time built out from the sides of the gap and directly positioned on the sill (for details see Appendix 9).

The closure of the Sloe between the isles of Walcheren and Z-Beveland in the year 1871 is a good example of this procedure. The gap was 365 m wide at low water level and had a maximum water depth of 10 m. The tidal range was about 4 m. By sinking mattresses a sill was constructed up to the level of about low water. This sill had side slopes of 1 in 1 and a crest width of 18 m. The next stage was to construct an osier revetment [rijspakwerk] on top of the sill. In consequence of the added weight, the sill settled 1. 80 m so in order to fabricate this wall up to the level of high water (at a height of 4 m above the original height of the sill), a 5.80 m high dam had to be made and that took a full month. Part of the final profile was made by adding a clay profile against the osier revetment.

In cases where the construction of an osier revetment failed, an attempt was made to position a vessel in the final gap and sink that onto the sill. This was not a simple operation, as transport was done by sailing or by rowing, and winching by hand was the only driving force. Timely ballasting and the prevention of the escalation of piping under and around the vessel were very critical. This method can be seen as the precursor of the caisson closure.

A historic example is found in the closure of the "Bottschlottertief" near Dagebuell (NW-Germany) in 1633. Clay had to be transported over a long distance by sailing vessels and it took 5500 labourers to execute the job. The closure was done by sinking a vessel in the gap. This was then ballasted and surrounded by clay, for the transport of which some 350 carts were used.

2.5.4 1920 until 1952

Gradually mechanization started to influence the work methods. The steam engine had already been in use for decades but the equipment was voluminous and heavy, both of which were troublesome in swift water and on soft ground. However, steam power could be used to drive winches, to drive sheet-piles and poles, to power the cranes used to transfer materials and for ship propulsion. Transport across the foreshore and newly constructed dam bodies was easier when locomotive engines, for which a stable railway had to be constructed, were used. Therefore initially the only change was the substitution of hard manual labour by engine work. However, better foundations for the transport roads and rails were needed since these were vulnerable to settlement in freshly worked ground and transport over water required greater water depth.

The difficulties encountered in building such closure dams are illustrated by the closure of the Hindenburgdam. This connection between the Isle of Sylt and the mainland of NW-Germany was completed between 1923 and 1927. The area was very shallow and sailing was impossible. The average tidal range was 1.70 m, but local wind effects much influenced the tides. The selected working method was to extend a wooden sheet-pile wall into the gap. The piling process was followed by the tipping of quarry stone on both sides to support the wall. The stone was transported on rails laid on a bridge that was constructed alongside the sheet-pile wall. Progress was much slower than anticipated and the erosion in front of the works consequently much more severe. The piling thus had to be done in highly turbulent water in a scour hole that preceded the sheet-pile construction and therefore more stone was needed for stabilization. On the inshore side, the railway was installed on made-ground, which often subsided, and derailments frequently occurred, thus escalating the problems. Later, the work method was adapted. The preceding scour was prevented by laying a 10 m wide stone protection on the bottom and the railway foundation was improved. Thus the problems were overcome.

Apart from the above-mentioned problems, a disadvantage of this type of steam driven equipment is that failure of the engine (damage) leads to major break-down of the complete works. The system is less flexible than one using manual labour.

Learning how to adapt the methods and use of the new equipment also stimulated the development of new methods. The engines could handle heavier units and reach higher production capacities. The advantages of this are:

Heavier units:
- can deal with higher flow velocities,
- give reduced material losses.

Higher production capacities:
- give a shorter critical phase,
- permit more progress in a still water period.
- lead to shorter execution time, thus greater production during the workable periods and reduce the risk of incidental bad weather.

Owing to these new techniques larger projects and projects with more critical conditions became feasible.

For instance, in 1932 a very large closure was realized in the Netherlands when the former Zuiderzee was cut off from the sea by the Enclosure Dike. The 32-km long dam crossed two main gully systems. During the execution of the works large deposits of boulder-clay, (a glacial till), were found. This material appeared to be very stable in the flow and could be handled by large cranes. A complete set of newly-designed floating cranes and transport barges was built and the closure was entirely constructed by these large floating units.

Another important change in the closure design was the development of mathematical modelling. Originally, designing had been a matter of experience and feeling, but calculations now started to replace the trial and error system (see Section 5.2.4). This reduced the risk of failure and was essential for the very large projects. For the damming of the enormous tidal basin, the Zuiderzee (now called IJsselmeer) in 1932, the differential equations for tide-propagation had to be solved. Professor Lorentz, a Nobel Prize winner in physics, achieved this. Three questions had to be answered before the job started:

- How would the tide change when the works were in progress, as this would affect the closing conditions?
- How would the tide change when the works were completed, as this would affect the design water level of the dike?

• What other design condition would affect the profile of the dike in this new equilibrium state of the sea (storm set-up and waves)?

Another challenge was presented in 1944, when for military reasons (World War II), the island of Walcheren was inundated by the bombing of the surrounding dike in four places.

This action dislodged the enemy troops and opened the fairway to Antwerp for the allied army fleet but at the same time it demolished the sea-defences and opened the low-lying island for tidal penetration. Restoring the sea-defences had to be done quickly in order that the island would not be permanently lost. Again, the mathematical basis for calculating tide-propagation improved. The four gaps, (three of these affecting one storage-basin), each with its own tidal amplitude and phase, and the propagation over inundated land with obstacles and ditches, and partial drying out at low tide, were a very complex system for a mathematical approach. And this was needed to establish the most favourable order of progress and also to ascertain risks that would arise if different path should occur in practice. Moreover, owing to the progressive erosion of gullies, the hydraulic resistance changed with time.

Immediately after the bombing the gaps in the dike were still relatively small. With the tide flowing in and out twice daily with ranges of 3.5 to 4 m, erosion deepened the gaps and a system of gullies was scoured out eating back into the inland area (Figure 2-14).

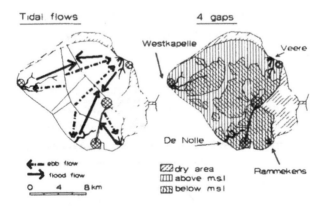

Figure 2-14 Walcheren - four gaps on one island

There was no material or equipment available (wartime) and the areas were covered with mines. In June 1945, when at last a start could be made, closure of the gaps was nearly an impossible task. The traditional methods of closure failed because they progressed too slowly, or because the equipment and materials could not cope with the circumstances. The four gaps had to be closed simultaneously within a period of four months (before winter storms) and these closures were inter-related.

The only available suitable means to achieve these closures were the caissons of the Mulberry Harbour, used temporarily a year before during the invasion of the Allied Army in Normandy (France). After providing scour-protection in the gaps, a variety of large units, like pontoons, caissons, concrete and steel vessels, and even large quantities of anti-torpedo-nets, were dropped or positioned in the gaps. The job was not finished before the winter and conditions worsened. Several times, initial success was followed by failure a few days later due to storm surges and piping. By the end of January 1946 however, the gaps were closed. A very good description of the difficulties encountered is given in the novel "Het verjaagde water" by A. den Doolaard.

Much experience was gained in the handling of caissons and vessels in closure gaps and ideas for the design of purpose-made caissons developed. The closure process could be improved by either creating a gap profile in accordance with the shape of the caisson or constructing a caisson to fit the requirements of the desired gap profile. In addition, the sinking could be controlled in a better way by regulating the water inlets by means of valves and separate chambers.

Different plans to improve the sea defences of the delta area of the Netherlands were drawn up and several closures were made. In 1950 the river mouth of the Brielse Maas was closed, using a purpose made caisson. In 1952 the Braakman, an estuary along the river Western Scheldt, was closed using two caissons, one of which was equipped with sluice gates. These temporary gates could be opened after the positioning of the caisson in the gap in order to reduce the water head in the basin after the closure and thus restrict the forces.

2.5.5 Period after 1952

A new flood disaster occurred in the southern North Sea on the 1st of February 1953 (according to Dutch tradition this flood should be named: St-Ignatiusvloed). A storm surge, together with spring tide-high water inundated 2000 km² of land in the Dutch Delta, creating 73 major dike-breaches and very many smaller ones. Again, all available technical experience and equipment and improvisation had to be used on many sites simultaneously, to close these gaps before the next winter season. Initially, not all the gaps had the same degree of difficulty or dimension. However, various gaps could not be dealt with immediately because of the disrupted infrastructure and as a result they scoured to tremendous dimensions. This is illustrated in Figure 2-15 for the Schelphoek breach on the Isle of Schouwen along the Eastern Scheldt.

This was one of the major dike breaches that occurred in the flood disaster on the 1st of February 1953. The scouring process continued during the actual closure works as well. The gap increased from an initial 40 m width to 525 m after 6 months, while the maximum depth increased from 10 m to over 35 m.

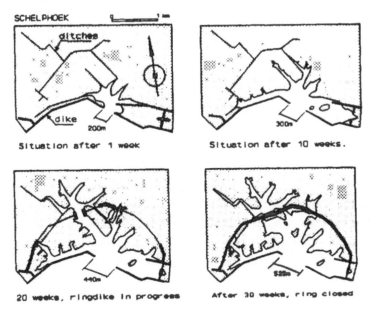

Figure 2-15 Development of erosion gullies

A typical example of successful quick improvisation is the closure of the gap at Ouderkerk on the IJssel. The storm surge at this spot reached a level of 3.75 m above MSL, overtopping the dike. The unprotected inner slope of the dike slid down over a length of about 40 m and the top layer of the dike slid and scoured away. However, the slope protection on the outer side remained intact up to the level of +1.70 m, as it rested on a centuries-old clay-core. Six hours later, at tidal-low water (still reaching a level of +2.00 m), two small vessels were positioned on the outer slope, which broke the force of the falling water, although piping underneath was severe. Jute-bags filled with sand were carried in by hand and a small embankment was created on top of the remains of the dike. At the next high water (+2.80 m), the emergency provision remained intact and could be strengthened.

The many very difficult circumstances led to various innovative actions, which resulted in repair within 10 months. Table 2-2 illustrates the enormous achievement:

Once more, the experience was used in later developments of the closing technology. This is shown by the following example. The principle of a temporary closure made in 1953 near Kruiningen (Waarde) was copied on a much larger scale, in 1985 to close a major estuary in Bangladesh (Feni River). In this case 1,000,000 bags filled with clay, totalling about 20,000 m^3 and stored in 12 stockpiles along the alignment, were carried by 12,000 Bangladeshi labourers into the 1000 m long gap to construct a dam in 5 hours.

Date	no. of gaps closed	remaining gaps	inundated area (km²)
2 February	3	70	2000
8 February	+ 8 = 11	62	2000
15 February	+ 6 = 17	56	2000
1 March	+20 = 37	36	1400
1 April	+17 = 54	19	800
1 may	+ 7 = 61	12	220
1 June	+ 4 = 65	8	150
1 July	+ 3 = 68	5	150
1 November	+ 4 = 72	1	100
December	+ 1 = 73	-	getting dry

Table 2-2 Closure scheme of gaps after the flood disaster of 1953

The disastrous flooding in 1953, with all its negative aspects (1835 people drowned), had an offspring in the decision making process for the reconstruction of the sea defence in the Netherlands. In order to avoid the need to strengthen all existing dikes it was decided to shorten the lengths of the defence works by closing the estuaries. This was accomplished during the succeeding 25 years. Although many closures were beyond the scope of the current experience, it was considered possible to develop the required methods during that period, working from the small to the large-scale projects. This period was therefore characterized by many experiments, a lot of research and the introduction of new materials and technology.

At a later stage it was decided to adapt the plan to the changed views regarding ecological importance and the largest estuary (Eastern Scheldt) was provided with a storm-surge barrier, which took another 8 years to construct. Since parts of the closure dam had already been made and the creation of the new design and its execution were parallel, many problems arose in this period. A lot of new ideas had to be generated and tested. The much-improved computer and measuring facilities played important roles. As a result of all these efforts, the present day designer has many rules, formulas, graphs and test-results at his disposal.

3 THE DESIGN PROCESS

In this chapter, reference is made to the design methodology used at the Faculty of Civil Engineering and Geosciences of Delft University of Technology. In the context of the subject "breakwaters and closure dams", some aspects of the design process have been omitted from this book. It is assumed that certain decisions have already been taken at a different level, be it only on a preliminary basis. For the breakwater, these decisions concern the question whether a new port should indeed be built and, if so, at which location, and for what kind of traffic. For the closure dams discussion of the pros and cons of a closure, such as the environmental, social and other consequences, the location and function of the final dam, is beyond the scope of this book. This does not mean that no strategic choices have to be made. However, the strategic choices no longer refer to the questions of whether and where the structure should be built, but rather to how it should be built.

3.1 General

In the design process both the functional as well as the structural design has to be looked to. This implies that one has to design a construction which fulfils the functional requirements, but also that the construction will not fail, collapse or be damaged seriously with a predefined probability. The objective of the design process is to find a concept that meets the requirement(s) and that can be realised, not only in terms of technical feasibility, but also in terms of cost - benefit ratio and social and legal acceptance. This implies that the solution of the design process must combine the following elements:

- Functionality
- Technology (what is feasible)
- Environment (what is allowed or accepted)
- Cost and benefit
- Paper work (drawing board)
- Matter (actual construction)

3.2 Abstraction level

In any design process various levels of abstraction can be discerned. In most cases it is sufficient to distinguish three levels:

- Macro level: the system
- Meso level: a component of the system
- Micro level: an element of one of the components

A few examples are presented in Table 3-1.

The indication of three levels does not mean that a very complex problem should always be divided into three levels. It is very useful to discern one level that is higher than that on which the actual work takes place and one level that is lower. This enables the designer to refer certain questions to a higher level in the hierarchy and it enables him to leave certain non-essential items to a later stage or to a lower level in the organisation.

General terms	Macro level System	Meso level Component	Micro level Element
Example 1[a]	Harbour in the global and regional transport chain	Harbour layout	Breakwater
Example 1[b]	Harbour layout	Breakwater	Crest block
Example 2[a]	Regional water management plan	Fresh water basin	Closure dam for fresh water basin
Example 2[b]	Fresh water basin	Closure dam (location, cross section)	Closing method
Example 2[c]	Decision to construct the Delta project	Dam in Brouwers-havense Gat	Closing method north-gap
Example 2[d]	Dam in Brouwers-havense Gat	Closing method north-gap	Design of caisson

Table 3-1 Examples of different scale levels

When considering the planning of a port, one may distinguish various levels of abstraction including:

- Design of a world or regional concept for the transport of certain commodities
- Design of regional or national economic plans
- Design of a national or provincial zoning policy
- Design of an overall port plan with intermodal facilities
- Design of the breakwater for such a port plan
- Design of a quarry to provide stone for the breakwater
- Design of the workshop for maintenance of the equipment of the quarry

Similar levels of abstraction can be distinguished for the design of a closure dam.

3.3 Phases

During the design process, one can also recognise certain phases that in some countries are related to the general conditions of contract between employer and consultant. Therefore the phases may vary from country to country. The contractual contents of each phase are subject to modifications in the same way. A logical set of phases is:

Initiative
Formulation of the ultimate goals of the design object as part of the system.
Feasibility
Review of the system with respect to technical, economic, social and environmental consequences and feasibility. Requirements are formulated on the component level.
Preliminary design
Giving shape to the system on broad lines, including determination of the exact functionality of the components and definition of requirements at the element level.
Final design
Composition of a set of drawings and specifications for the system in which the final shape of the components is fixed and the functionality of the elements is determined.
Detailed design
Composition of a set of drawings and specifications in which the final shape of the elements is fixed.

This concept can easily be schematised in a matrix in which each row represents one of the phases and shows which activities will take place at the various levels of abstraction. The columns show how the levels of abstraction in the project become more concrete throughout the phases. The matrix also shows that working on the elements does not start before one reaches the preliminary design phase and certain decisions have been taken about the purpose and function at the system level and about the purpose at the component level.

Following this line of thought helps to ensure that the proper approach is chosen at each stage so that neither too much nor too little detail is sought.

3.4 Cyclic design

Each activity in the design process, which is represented by a cell in Table 3-2, is a cyclic process in its own right, consisting of a number of steps:

Phases	Abstraction Level		
	System	Component	Element
Initial	Purpose		
Feasibility	Functionality	Purpose	
Preliminary Design	Shape	Functionality	Purpose
Final Design	Specifications	Shape	Functionality
Detailed Design		Specifications	Shape

Table 3-2 Schematisation of the design process

Analysis:

Assembling of available data and arrange for the provision of missing data;

Drawing up a set of criteria that the design must fulfil (List of Requirements), crosschecking all with respect to cost and functionality.

Synthesis:

Generation of conceptual ideas and alternatives that broadly meet the requirements.

Simulation:

Detailing of concepts and alternatives (by calculation, simulation, or modelling) up to a level that makes them mutually comparable. Again a crosscheck with respect to cost and functionality is required.

Evaluation:

Assessment of the concepts and alternatives, comparison on the basis of cost and benefit.

Decision:

Selection of the best option. If more than one option is acceptable, repeat the process in further detail, until a final decision can be taken. This may involve some toggling between the abstraction levels in a particular phase of the design process.

3.5 Consequences of systematic design

The effects of the systematic design procedure on the purpose of the present book are obvious. It makes no sense to draw a cross-section of a breakwater when neither the depth of the water in which it is to be built nor the acceptable wave action in the lee of the structure is known. One has to start by considering the purpose of the system, i.e. its national or regional socio-economic role in the global transport system. From there, one goes down a step to the port, still as part of the system:

- which cargo flows are foreseen
- which kind of vessels will carry the cargo
- what are the requirements for access from the seaward side and from the landward side
- what will be a proper size of the port
- what will be a suitable location

Only if these questions have been answered, can one start to think in more detail about the breakwaters, starting with a rough layout and an indication of the required functions. Only in the final stage of the design process, can the actual design of the cross-section be made, including decisions about crest level, slope, and choice of materials and construction method.

Similar considerations apply to the design of a closure dam. Starting from the decision that a watercourse or dike gap has to be closed, the most suitable location or alignment is still to be selected. One must have insight into the hydraulic system of the flow, the subsoil conditions in the area and probably the infrastructure of the region (road connections), before one can start considering where and how the final dam should be made. For the closing process it may be even more important to realise at which abstraction level one is working, since the closure dam often is a structure with a temporary function. As soon as the watercourse has been closed, a new situation has been created. The final design for the scheme may involve a different step. For instance, the definite sea defence dam could be made in the lee of the temporary closure structure, enabling the construction elements of the closure dam to be used elsewhere. Consideration may also be given to splitting the actual closing operation into two or three compartments to keep the construction process and the construction materials within a workable scale.

Considering these remarks, one can conclude that a study book on the design and construction of breakwaters and closure dams deals with the final stages of the design process for the structure itself. Notwithstanding, for a proper understanding of what one is doing, throughout the process the link has to be maintained with the higher abstraction levels. If one fails to do this, the risk emerges that one teaches students to apply prescriptive recipes, instead of designing creative solutions. For this reason, relatively much attention will be given to the link with the purpose and functionality of the system. At the same time, it will be clear that certain details of the design need not be worked out in the early stages. It makes no sense to plan a working harbour in detail before the closure method has been chosen.

3.6 Probabilities

No construction can be designed in such a way that the construction will never fail. However, the probability of failure has to be very small. The probability of failure of a structure is partly a financial problem (the extra cost of lowering the probability of failure has to be lower than the capitalised cost of failure), and partly depends on non-monetary values, like loss of lives, ecological damage, etc. In case probability of failure is mainly a financial problem, the optimum probability of failure can be computed; this will be explained later. In case many non-monetary values are at stake (e.g. a dike protecting an urbanised area or a natural reserve), an objective optimisation is not possible, and usually a political choice is made regarding the allowable probability of failure.

After the feasibility study and preliminary design, the details of the design have to be filled in. As discussed before, this will be done during the stage of the *detailed design* and sometimes already during the stage of the *final design*. Basically this means that each structural part should not fail or collapse with a probability, as follows from the boundaries as set in the feasibility study.

3.6.1 Basics of a probabilistic analysis and the use of safety coefficients

A structure fails when the load is larger than the strength, in other words if:

$$Z = R - S < 0,$$

where R is the strength and S is the load[1]. Usually R consists of a number of parameters (e.g. material properties) and S consists of a number of load values.

In a very simple design, this problem can be solved easily. For example if one needs to design the cable in a crane, the design force in the cable F is equal to the design mass, multiplied with the acceleration of gravity. The strength of the cable depends on the intrinsic strength (σ) of the cable material, multiplied with the cross sectional area A of the cable:

strength: $R = A \cdot \sigma$

load: $S = M \cdot g$

$$Z = R - S = A\sigma - Mg$$

For critical conditions (brink of failure) $Z = 0$. The critical cross sectional area (which in fact is the design parameter) is

$$A_{crit} = \frac{Mg}{\sigma}$$

M is the mass of the nominal load to be lifted (design load). This is a clear input parameter, it is defined by the client; σ is prescribed in the specifications and g is the gravitational acceleration. Because there are always uncertainties, in the traditional design process a safety coefficient γ is added:

---

[1] S as a symbol for load and not for strength does not seems logical, but it is according to international agreement. R and S are acronyms related to the French words *Résistance* and *Sollicitation* ("asking"). We will adhere to this agreement, despite the confusion at first glance.

$$A_{crit} = \gamma \frac{Mg}{\sigma}$$

The magnitude of γ is usually given in professional codes and standards; if not, it is usually based on experience (in case of breakwater design PIANC has issued values of γ to be used in the design; see chapter 7.5).

The safety coefficient γ covers the following uncertainties:
- the actual mass being different from the nominal mass;
- deviations in the value of g, the accelerationof gravity;
- the actual strength of the material σ being different from the specified strength;
- the actual cross section of the cable A being different from the specified cross section.

In more complicated cases, and specifically when there are no codes or when experience is lacking, a probabilistic approach should be implemented, which will be explained later (see Appendix 10).

3.6.2 Additional problem in coastal engineering

Unfortunately in the design of coastal structures there is a complicating factor. For example the stability of armour units depends on the wave height (H_s), the mass of rock or concrete, the slope of the structure, and many other parameters. In a stability calculation, the wave height is the load parameter, while the other parameters (mass of rock or concrete, slope, shape of the armour, etc) are strength parameters. Often, the strength parameters are Gaussian distributed with a relatively small standard deviation. So, at the strength side of the equation, the problem is very comparable to the cable example mentioned above.

But for the load parameter (H_s) an "average" value cannot be determined. It has to be a significant wave that does not occur too often. And related to the wave height there is also the wave period (which is usually also present in the more advanced design equations). It means that the definition of our "design wave" or "design storm" is a key problem in our design.

The choice of the probability of the "design storm" is usually the most important parameter decision in the design process. In choosing this probability two cases have to be distinguished:
1. It is a pure economic problem.
2. Also human lives and other non-monetary values are taken into account, like protection of a museum or a religious site.

In the first case one can calculate the optimal "design storm", in the second case this cannot be calculated but is subject to political decision making. For breakwaters it is usually a pure economic problem. In case of failure there will be damage: the cost of repairing the direct damage plus the loss of income during non-operation of the breakwater (consequential damage). The details of the economic optimization will be explained in Appendix 7 of this book.

Often such an economic optimization is not made. This is usually due to the fact that decisions on the investments for a breakwater project are not based on proper life-cycle analysis, but on the budget available or on the (short-term) rate or return on the initial investments. Therefore in practice often a (political) decision is made on the return period of the design storm, based on ad-hoc considerations.

3.6.3 Determination of a design storm

Usually the design storm is related tot the economic lifetime of the structure. For breakwaters, an economic lifetime in the order of 50 years is very common. As a result, decision makers often suggest using the once in 50 years storm as a design storm.

The first task for the design engineer is to explain to the decision maker that this does not mean that the design storm will occur after exactly 50 years, but that every year there is a probability of 1/50 (i.e. 2%) that the design storm will occur. And that may be next year.

The second task for the design engineer is to explain that the probability of serious damage during the lifetime of the construction is given by the Poisson distribution:

$$p = 1 - \exp\left(-f\, t_L\right)$$

in which:

p probability of occurrence of an event one or more than one times in period t_L
t_L considered period (e.g. the lifetime of the breakwater) in years
f average frequency of the event per year

So, the assumed lifetime of 50 year and a storm frequency of 1/50 per year gives

$$p = 1 - \exp\left(-\frac{1}{50} \cdot 50\right) = 1 - \exp(-1) = 0.632$$

This means that there is probability of 63% that the construction will fail during its lifetime. It is clear that this is unacceptable. More acceptable values would be between, say, 5% and 20%. The actual choice depends largely on the purpose of the structure and on the risk involved. In this book, some examples have been worked out based on the relatively high value of 20%[2]. This must not be interpreted as a recommendation, but just as an example!

It means that the storm frequency becomes:

$$f = -\frac{1}{t_L}\ln(1-p)$$

$$= -\frac{1}{50}\ln(1-0.2)$$

$$= 0.0044 = \frac{1}{225}$$

In case one accepts a probability of failure of 20% during a lifetime of 50 years, one should apply a 1/225 (= $4.4 \cdot 10^{-3}$) per year storm. So realize that in spite of the fact that we did allow (a rather high) 20% probability of failure during lifetime, still we use a design storm with a probability of $4.4 \cdot 10^{-3}$ per year in our calculations.

In the above text, it has been assumed implicitly that the probability of storms has some statistical distribution, but that all other parameters (notably the strength parameters) are fixed, deterministic values. Of course, this is not true. The combined effect of all these uncertainties will be discussed in Section 7.4. It will be shown that the effect of the uncertainty in strength parameters is much less than the uncertainty in the storm occurrence, but not negligible. Because determination of the parameters of the design storm is extremely important for the design, this will be discussed separately in Section 5.3.

---

[2] The value of 20% is selected because this value is also used in the examples in various PIANC publications; from economical analysis often will follow that values of p in the order of 5% are more economic.

4 CONSIDERATIONS AT SYSTEM LEVEL

In this chapter the actual design of breakwaters and closure dams is linked to considerations and decisions that in fact belong to a different abstraction level than does the design itself. From these links, it is often possible to derive considerations with respect to the functionality of the structure under consideration. Attention is paid to the side-effects of the construction works, which may lead to a reconsideration of decisions taken earlier. For students, this chapter is an indispensable tool to establish the quantified functional requirements for the design of a breakwater or closure dam. It is therefore essential to study this chapter in detail before any design exercise is attempted.

4.1 General

In Chapter 3, it was indicated that a design problem should be considered at various levels of abstraction, starting with the system. In this chapter we attempt to discuss some of the aspects at system level, when the system is either a port or a scheme to close a river or estuary. The breakwater or the closure dam is then an element of that system. By discussing the system, we attempt to approach our design problem from a slightly more abstract position. This refers to both the functions and requirements, and to the side effects of the project.

4.2 Functions of breakwaters and examples

Breakwaters can fulfil a variety of functions, the most important of which are:
- Protection against waves (Section 4.2.1). This can be subdivided into firstly, protection of ports and shipping and secondly, shore protection.
- Guiding of currents (Section 4.2.2)
- Protection against shoaling (Section 4.2.3)
- Provision of dock or quay facilities (Section 4.2.4)

4.2.1 Protection against waves

Ports and shipping

Vessels at berth

The function of protection against wave action must be split into sub-categories. The best-known protection function relates to navigation and over the years breakwaters have been used in port construction. However, the status of the vessels or installations that are to be protected makes a big difference to what is required. In other words, one must have an idea how vulnerable the area to be protected is before deciding what degree of protection must be provided.

In general, a vessel is at its most vulnerable when it is moored alongside a rigid structure such as a quay or a jetty or alongside another vessel. The acceptable wave height is related to the size of the vessel, on one hand, and the height, period and direction of the waves, on the other hand. Thoresen [1988] gives suggestions for ships at berth in head seas. These values are slightly modified in Table 4-1 according to the experience of the authors. The acceptability of the conditions refers to both damage to the vessel and damage to the structure.

Type of vessel	Maximum H_s in m
	At berth (head sea)
Pleasure craft	0.15 - 0.25
Fishing vessels	0.40
Dredges and dredge barges	0.80 - 1.00
General cargo (< 30,000 dwt)	1.00 - 1.25
Dry bulk cargo (< 30,000 dwt)	1.00 - 1.25
Dry bulk cargo (up to 100,000 dwt)	1.50
Oil tankers (< 30,000 dwt)	1.00 - 1.25
Oil tankers (100,000 to 200,000 dwt)	1.50 - 2.50
Oil tankers (200,000 to 300,000 dwt)	2.50 - 3.00
Passenger vessels	0.70

Table 4-1 Maximum wave heights for ships at berth

Loading and unloading operations may impose extra restrictions. It will be clear that loading and unloading liquid bulk cargo via a flexible hose allows larger ship movements than placing containers in a slot. Velsink and Thoresen approach this question from a different angle. Thoresen gives values for acceptable ship movements; Velsink [1987] gives limiting wave heights for different directions. The approach of Velsink relates more directly to the functional requirements of the breakwater. Therefore, his data are given in Table 4-2. A comprehensive review of the problem of ship movements is given in PIANC report II-24 [1995].

Type of vessel	Limiting wave height H_s in m	
	0° (head or stern)	45° – 90° (beam)
General cargo	1.0	0.8
Container, Ro/Ro ship	0.5	
Dry bulk (30,000-100,000); loading	1.5	1.0
Dry bulk (30,000-100,000); unloading	1.0	0.8 – 1.0
Tankers 30,000 dwt	1.5	
Tankers 30,000 – 200,000 dwt	1.5 – 2.5	1.0 – 1.2
Tankers >200,000 dwt	2.5 – 3.0	1.0 – 1.5

Table 4-2 Maximum wave heights for loading and unloading operations

How often the exceeding of these limits is accepted is not indicated in the above figures. In other words, they do not indicate for what percentage of time loading and unloading operations may be interrupted, or how often specific berths must be left by vessels needing to find a safer place to ride out a storm. This question must be answered on the basis of a thorough economic analysis, including the risk of negative publicity for the port. Such studies are beyond the scope of this book, but nevertheless the answer to the question must be known when the design of the actual breakwater is started. The point stressed here is that these considerations will lead to the definition of Service Limit States (SLS) that are usually different from the Ultimate Limit State (ULS), which concerns the survival of the structure under extreme conditions.

Figure 4-1 shows the layout of a harbour where the breakwater typically protects the harbour basin, including berths for loading and unloading.

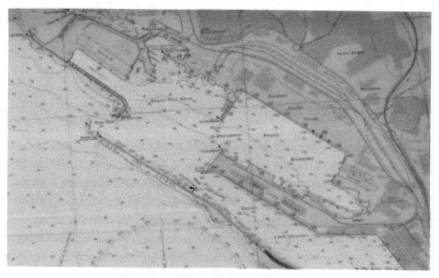

Figure 4-1 Harbour of Marseilles (France)

Sailing vessels

So far, we have considered the protection required by vessels at berth. Free sailing vessels are fortunately much less vulnerable.

National regulatory bodies, like the Netherlands Shipping Inspectorate, strictly control the operation and the design of ocean going vessels. The work of these national organizations is coordinated by the International Maritime Organization, IMO. In addition to the Government-related regulatory bodies, there are also private regulatory bodies that check the design of vessels, often on behalf of the insurers. Such private bodies include Bureau Veritas, Det Norske Veritas, and Lloyds. These bodies issue certificates of seaworthiness, with or without certain restrictions.

Ocean-going vessels with an unrestricted certificate are designed to cope with the highest waves. In severe conditions they may adapt their course and speed to the prevailing wind and wave direction, but in principle, modern vessels with an unrestricted certificate can survive the most severe conditions at sea. The situation changes when a free choice of course and speed becomes impossible, for instance because of the proximity of land, the need to sail in a specific (dredged) fairway, or the wish to come to a halt at a mooring or anchorage. The more confined the conditions, the stricter will be the limits with respect to wind, waves and currents.

What applies to vessels designed to sail the high seas without restriction does not apply to all categories of vessels. Some vessels have a certificate that limits their operation to certain areas (coastal waters, sheltered waters, and inland waters) or to certain periods in relation to certain areas (North Atlantic summer). Such restrictions refer not only to the structural aspects of the vessel, but also to skill and number of crew.

What does all this mean for the operation of a port, and for the functional requirements of its breakwater? Can a vessel enter the port under any circumstances? Obviously not, but we have already concluded that a sailing vessel is less vulnerable than a moored vessel. The functional requirements for a breakwater that protects only an entrance channel are thus much lower than those for a breakwater that protects a harbour basin. Still, the actual situation will change from place to place. If ships need the assistance of a tug during the stopping operation and the subsequent turning or mooring, the waves must be attenuated to a level that makes tugboat operation feasible. In general, one can assume that a significant wave height of 2 to 2.5 m is acceptable for tugs and their crews working on deck. If only tugs with an inland waters certificate are available, their operation may be restricted to significant wave heights of 1 to 1.5 m. If the limits imposed by the certificate are exceeded, often the insurers will not cover the cost of damage.

Figure 4-2 shows an example of a breakwater, which does not protect any berths.

Here again, decisions must be made as to how frequently interruption of the navigation due to closure of the port for weather conditions can be accepted. One must realize that pilotage also becomes a limiting factor under heavy sea and swell conditions. In general, delays and interruptions are accepted of one or two days per annum.

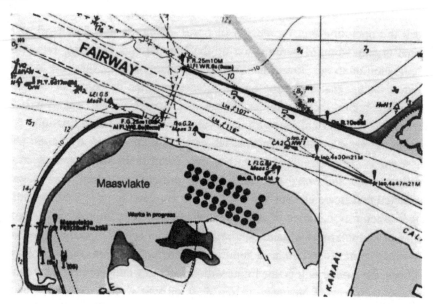

Figure 4-2 Breakwater at the Europoort entrance

Port facilities

A third condition that needs attention is the harbour basin itself, with the facilities that may suffer damage if the wave heights in the basin become too high. Quays and jetties and the equipment that is installed on them may be damaged, even in the absence of vessels. Here again, it must be decided whether any such damage is acceptable, and if so what chance of its occurrence is acceptable. It is evident that if the harbour installations are damaged, one is concerned not only about the direct cost of repair but also about the consequential damage due to non-availability of the cargo transfer systems. In this respect one may try and imagine what happens if the only power plant or refinery in a region must be closed because no fuel can be supplied.

Shore protection

From coastal engineering theory, we know that waves cause both, longshore transport and cross-shore transport. Both phenomena can cause unwanted erosion, especially on sandy shores.

As far as cross-shore transport is concerned, the erosion is often connected with changes in the equilibrium profile. A more gentle profile (after the erosion of dunes!) is associated with higher incoming waves, whereas a milder wave climate tends to restore the beach by landward sediment transport. Similarly, when erosion is due to a gradient in the longshore transport, the effect will be less when the wave heights are lower.

In general terms one can therefore conclude that the reduction of wave heights in the breaker zone will mitigate beach erosion. Such reduction of wave heights can be achieved by constructing offshore breakwaters parallel to the shore (Figure 4-3). However, from the literature it is known that one must be careful when using this solution. Due to wave set-up, the water level on the lee side of the breakwater rises, which causes a concentrated return current, (comparable with a rip current) between the breakwater sections (Bowder, Dean and Chen [1996]).

Figure 4-3 A system of detached breakwaters at Fiumicino, Italy

4.2.2 Guiding of currents

When approaching a harbour entrance, vessels are slowing down by reducing power. This is done because at high speed they require a rather long stopping distance and the vessels produce a high wave and a strong return current. A slower speed means that the vessel is more affected by a cross current (or a crosswind), since the actual direction of propagation is the vectorial sum of the vessels own speed and the current velocity. Thus, to sail a straight course into the port along the axis of the approach channel the vessel must move more or less crab-wise.

Closer to the shore, at the same time one must expect stronger tidal currents parallel to the shore. If the port entrance protrudes into the sea, there will possibly be a concentration of flow lines near the head of the breakwater.

The combination of the slower speed of the vessel with the potentially stronger crosscurrents at the harbour entrances poses manoeuvrability problems. In the lee of the breakwater tugs can assist the vessel, but it takes some time (about 15 minutes) before the tugs have made a connection with the vessel, and in the meantime the vessel continues to sail without external assistance. Assuming a speed of 4 knots, the vessel travels a distance of about 1 nautical mile (1850 m), before the tugs can control the course of the vessel. Only then can the remaining stopping procedure be completed. The vessel gives full power astern and it will stop within 1 to 1.5 times its own length.

This means that cross currents are critical over a considerable distance that extends from well outside the harbour entrance to the point where tugs assume control. It is not only the velocity of the crosscurrent that is important but also the gradient in the crosscurrent, since this forces the ship out of its course.

The entrance to the Port of Rotterdam is a good example of an entrance where the layout (plan) of the breakwater is designed to cope with the current pattern (Figure 4-4). In this case, the function of the breakwater is twofold: it guides the current and it damps the waves to a level at which the tugs can work.

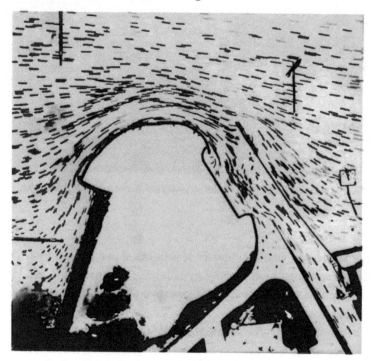

Figure 4-4 Flow pattern at the Europoort entrance

4.2.3 Protection against shoaling

Many ports are located at a river mouth or in an estuary. Coastal engineers are aware that the entrance channel has an equilibrium profile that is mainly determined by the tidal prism. (d'Angremond et al. Introduction to Coastal Engineering [1998]). If the natural depth in the entrance channel is insufficient for nautical purposes, one may decide to deepen the channel by dredging. Though this may be a very good solution, disturbance of the equilibrium means that dredging has to be continued throughout the life of the port. In a number of cases it has therefore been decided not to dredge, but rather to restrict the width of the natural channel and to force the channel to erode its bed. This may also be the functional purpose of a breakwater that is designed to guide currents. An example of the use of such a solution is the port of Abidjan (Figure 4-5 and Figure 4-6).

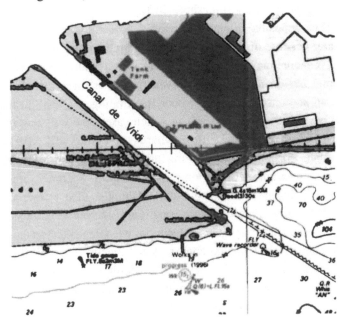

Figure 4-5 Entrance to the port of Abidjan

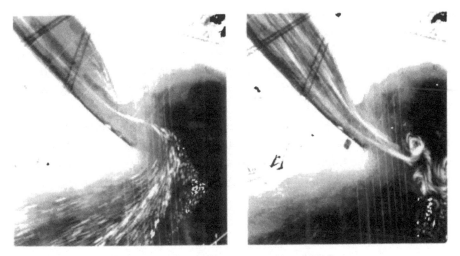

Figure 4-6 Flow pattern at the port of Abidjan

It is stressed here, that improvement of the efficiency of dredging and the lower cost of dredging operations have caused a shift away from building breakwaters towards accepting the annual cost of dredging.

Another challenge for those designing entrance channels into a port is the existence of the longshore current along sandy shores. Under the influence of oblique waves, a longshore current develops in the breaker zone. Due to the high turbulence level in the breaker zone, a large quantity of sand is brought into suspension and carried away by the longshore current (longshore drift).

The sand will be deposited at places where the velocity is less, i.e. where the water depth is greater because of the presence of the shipping channel. Thus a dredged or even a natural channel may be blocked after a storm of short duration and high waves or after a long period of moderate waves from one direction. To avoid this, a breakwater can be constructed. For proper functioning, the head of the breakwater must extend beyond the breaker zone, in which case, sand will be deposited on the "upstream" side of the breakwater, whereas erosion will take place at the downstream side. In coastal engineering this is the classical example of erosion problems due to interruption of the longshore transport. A good example is given in Figure 4-7, which shows the actual situation in IJmuiden (The Netherlands).

Even if the breakwater is present, sedimentation of the port's entrance channel may occur. This happens when so much sediment has been deposited on the upstream side of the breakwater that the accumulated material reaches the end of the breakwater and passes around it's head. Dredging is difficult in such cases because of the proximity of the breakwater. An example of a breakwater that is too short is the breakwater of Paradip (India), shown in Figure 4-8.

4.2.4 Provision of dock or quay facilities

When the breakwater is directly protecting a harbour basin (and therefore already quite high), it is especially attractive to use the crest of the breakwater for transport of cargo and passengers to and from moored vessels. Special facilities must be provided in this case to enable the vessels to berth alongside the breakwater. These facilities may consist of a vertical wall on the inside, or a piled or non-piled jetty connected to the breakwater.

In this case, it must be ascertained that the conditions on or directly behind the crest of the breakwater are safe. Again a distinction can be made between operational conditions (Service Limit State or SLS) and extreme conditions like survival of the installations (Ultimate Limit State or ULS). Further details of acceptable conditions relating to run-up and overtopping are given in Chapter 10.

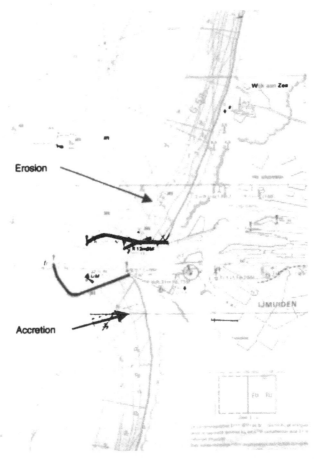

Figure 4-7 Port and breakwaters at IJmuiden

4.3 Side effects of breakwaters

4.3.1 Failure modes

From the above it is clear that failure to fulfil the functional requirements (at system level) may be due to inadequacies:

- In the layout of the breakwater (for example, location, length, orientation, width of the harbour entrance.). Such deficiencies may lead to undesirable disturbance in the harbour basin, unsafe nautical conditions or undesirable accretion or erosion.
- In the shape of the cross-section (crest level, permeability for sand and waves). This will lead to similar problems and also to unsafe conditions at the crest of the structure.

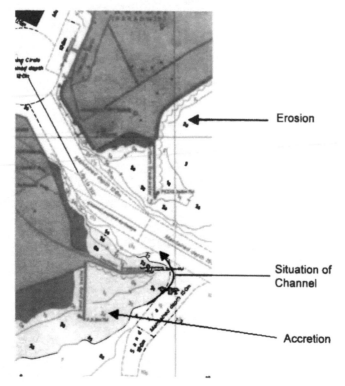

Figure 4-8 Siltation at entrance to port of Paradip

- In the structural design of the cross-section (stability under severe design conditions: ULS) or due to other unforeseen conditions that are listed in most textbooks on probabilistic design (see Chapter 15).

The present book will mainly discuss failure modes of the last two categories. It is stressed here that the choice of the crest level in relation to the functional requirements is one of the most important design decisions.

4.3.2 Nautical characteristics

Since breakwaters usually have a function connected with navigation, it is of the utmost importance to ensure that the layout of the breakwater(s) and channel creates safe nautical conditions. A first impression may be obtained by following the PIANC/IAPH guidelines (Anonymous [1997]).

In practice, a design prepared on the basis of guidelines must always be checked with the aid of navigational models. In this respect there is a choice between physical scale models, real time computer simulation and fast time computer simulation. A discussion of the merits of these methods is beyond the scope of this book.

In this respect, mention must be made of another side-effect of a breakwater that may influence the nautical environment: reflection of waves. Reflection of short waves may cause a choppy sea in the neighbourhood of the breakwater, which is a nuisance to smaller (often local and inland) vessels.

4.3.3 Morphology

Although one of the purposes of a breakwater may be to interrupt the longshore sediment transport in order to prevent the siltation of a port entrance, a coastal engineer cannot close his eyes to the consequences of this phenomenon in a larger space and time frame. Accretion and erosion of the coastal zone on either side of the breakwater will most likely pose a serious threat to the community in the region and possibly to the ecosystem as well. It goes without saying that such consequences have to be assessed and quantified, and that remedial measures have to be designed, planned and executed. In this respect, one may think of:

- an adequate sand-bypassing system
- replenishing the eroding beach with sand dredged during maintenance operations
- use of material dredged during port construction as a buffer against future erosion

4.4 Functions of closure dams and side effects

A number of purposes and side effects are listed below. Sometimes it is difficult to determine why a specific effect is termed a side effect and in historic cases it has turned out that what were initially side effects, became important aspects of the situation that was created.

Main purpose of closing a watercourse:
- land reclamation
- shortening the length of sea defence
- creating of fresh water reservoir
- creation of a tidal energy-basin
- creation of a fixed level harbour dock
- creating a construction dock
- providing a road or rail connection
- repair of a dike breach
- control of upland flow
- creating fish ponds
- cutting off river bends

Various possible side-effects (dependent on circumstances):
- change of tide (amplitude, flows) at the seaward side of the dam
- change in bar and gully topography, outside the dam
- disappearance of tides on the inner side of the dam
- change in groundwater level in adjoining areas

- alteration of drainage capacity for adjoining areas
- loss of fish and vegetation species
- loss of breeding and feeding areas for water birds
- rotting processes during change in vegetation and fauna
- stratification of water quality in stagnant reservoir
- accumulation of sediments in the reservoir
- impact on facilities for shipping
- impact on recreation and leisure pursuits
- change in professional occupation (fishery, navigation)
- social and cultural impacts

In the past, watercourses were mainly closed for the purposes of land reclamation and controlling the water levels on marshy land. In both cases this was linked to agricultural development. It is typical of these damming activities that the control of river and storm surge levels becomes essential. Follow-up action, like the repair of dike breaches and sometimes the cutting off of river bends has been necessary throughout the ages. The other purposes mentioned, like generation of tidal energy, harbour and construction docks, dams for road or rail connection and fish ponds are incidental works and have a smaller impact on the surroundings. Today, since the quality of life is becoming an important aspect for society, certainly in the industrially developed countries, damming activities are initiated to serve various other purposes. These include the creation of fresh water storage basins, the prevention of water pollution in designated areas, the provision of recreational facilities and the counteraction of salt intrusion or groundwater flow.

Depending on the circumstances, there will always be a number of side effects. These are sometimes temporary, but sometimes generate long-term developments that are difficult or impossible to predict with any degree of accuracy. The above list gives an indication of possible effects but does not pretend to be complete.

Below, a number of closures, some constructed centuries ago, are briefly described and comments are given on their purposes and side effects, in so far as these can be ascertained.

4.4.1 Closure of the rivers Rhine and Meuse

As mentioned in the historic review, the rivers Rhine and Meuse were dammed in the period 1200 to 1300. The Rhine debouched into the North Sea near Katwijk and, choked by sediment, regularly inundated the coastal area behind the dunes. First a closure dam was made on the borderline between the provinces of Holland and Utrecht, near Zwammerdam, resulting in inundations in Utrecht. Around the year 1200, after several years of conflict, the ruler of Utrecht dammed the river Rhine further upstream at Wijk bij Duurstede. The purpose was clearly to protect the inland zone behind the dunes near Katwijk from inundation. The flow was diverted via the Lek river-branch. Of course, this dam had side effects. It excluded the downstream area from further silting up and the outer delta at the river mouth at Katwijk lost its sediment feeder. In the centuries that followed the coastline in the locality retreated by several kilometres and the Roman fortress "Brittenburg" disappeared into the North Sea.

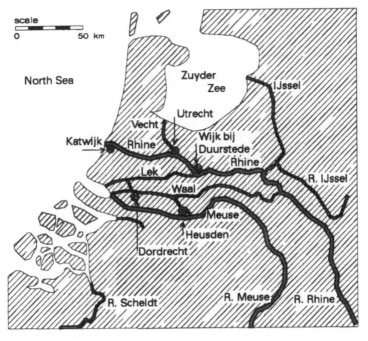

Figure 4-9 The Rhine Meuse-delta before the year 1000

The damming of the river Meuse (Maas) followed a different scheme. The town of Dordrecht had obtained staple rights (the right to store and sell certain goods) along the river Merwede. However, the payment of toll dues could be avoided by sailing along the River Meuse (see Figure 4-10, Oude Maas).

Probably because of this, the river Meuse was dammed in the year 1270. Although not problematic at first, in extreme conditions this distorted the discharge capacity of the delta and ultimately led to a major inundation after the dike breached in 1421 (St. Elizabeth's flood). This resulted in the permanent loss of the most developed agricultural area of Holland (the Grote Waard polder) by erosion of the topsoil. The region changed into a unique large tidal freshwater basin.

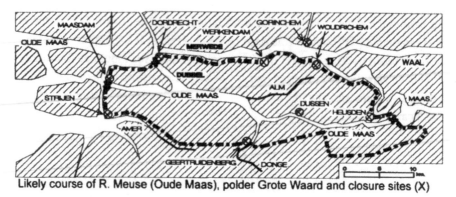

Likely course of R. Meuse (Oude Maas), polder Grote Waard and closure sites (X)

Figure 4-10 Situation after damming the River Meuse

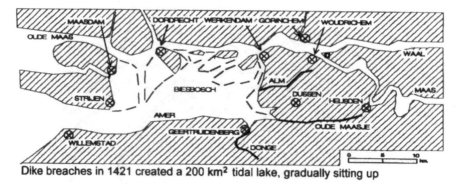

Dike breaches in 1421 created a 200 km² tidal lake, gradually sitting up

Figure 4-11 Situation after the St. Elizabeth's Flood

In the period 1000 to 1400, very many areas were surrounded by embankments and drainage of these areas by rivers ceased. Whether or not the results of all these activities should be considered positive or negative is debatable. For nearly a thousand years all sediments carried down by the rivers were evacuated to the sea instead of regularly settling on the marshy land. The drainage lowered the water-table and this caused the peaty soil to settle. It changed the morphology of the landscape and its flora and fauna. Started as a simple water-level control system, it turned out to be a threat to the country. Gradually, the sea took large areas of the sinking ground. The side-effects, certainly when considered over very long periods, were tremendous. The people of today inherited a vast area below sea level that is continuously threatened by water and entirely dependent on its pumping capability for the evacuation of the water.

In some cases however, nature got an opportunity to show what could have happened otherwise. This natural restorative process is well demonstrated in the example of the lost "Grote Waard". The enormous lake created by the 1421-flooding, that is called the Biesbosch, formed a settlement basin and after 550 years this lake was nearly completely silted up again and restored as a marshland. In order to prevent recurrence of the flooding, two main artificial rivers were dredged, the Nieuwe Merwede and the Bergse Maas, the latter restoring the historic discharge route of the river Meuse. Apparently the old scheme (at system level) could not be maintained.

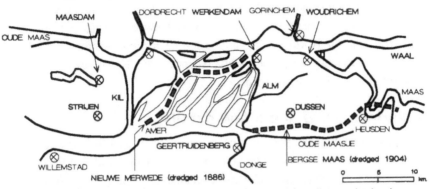

500 years sedimentation refilled the basin and drainage-"rivers" were dredged

Figure 4-12 The Biesbosch area

4.4.2 Side effects of the Enclosure Dike (Afsluitdijk)

As mentioned in Chapter 2, closing the Zuiderzee by the Enclosure Dike completely changed the tidal conditions on the seaward side, in the shallow sea called Waddenzee. The amplitude of the tide gradually increased to more than twice the former tide with the progress of the closure. This was studied before the works started. However there was a more difficult question to be answered:

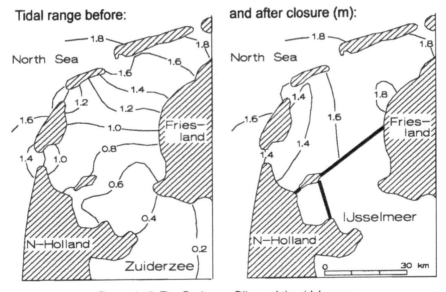

Figure 4-13 The Enclosure Dike and the tidal range

How will the sea outside the dike adapt to the new conditions in the long run and change its topography and morphology?

By now, 60 years later, we know that this coastal water with its tidal flats and gully system is closely dependent on the exchange of water and sediment with the North Sea. Every change in the tidal volume passing between the islands separating the Waddenzee from the North Sea has a long-term effect on the balance of shoals and channels. Consequently, even the coastal balance in the North Sea on the outside of the islands must have been distorted.

4.5 Various dams and a few details

In this book various examples of closure works have been referred to. These are listed below with their name and/or the location, together with the year of closure. The list is not a complete list of closures executed in the past but is given because of its relevance to this book. Many of the closures are situated along the continental coastline of the North Sea. These locations are numbered 1 to 25, in Table 4-3. Only one example outside Europe is discussed here, no. 26.

No.	name or location	country or area	year	method or means
1	Hindenburgdam	Sylt-Schleswig (Germany)	1925	Sheet-pile wall
2	Dagebuell	German Bight (Schleswig)	1633	sunken vessel
3	Meldorf, various gaps	Sylt-Schleswig (Germany)	1978	sand closure, sunken barges
4	Closure-dike Lauwerszee	Waddenzee (Netherlands)	1969	concrete caissons
5	Closuredike Zuiderzee	IJsselmeer (Netherlands)	1932	boulder clay (crane pontoons)
6	4 Dike breaches Walcheren	Walcheren (Netherlands)	1945	vessels and caissons
7	Veerse-Gat dam	Walcheren - Noord Beveland Deltaworks	1961	caissons with gates
8	Storm-surge-barrier	Eastern Scheldt Deltaworks	1986	gates between monoliths
9	Schelphoek, var. gaps	Schouwen (Netherlands)	1953	caissons with gates
10	Brouwersdam, 2 gaps	Schouwen-Goeree, Deltaworks	1972	caissons, blocks (cableway)
11	Haringvliet-Sluices	Goeree-Voorne, Deltaworks	1971	concrete blocks (cableway)
12	Prim. dam Brielse Gat	Brielse Maas (Netherlands)	1950	caisson
13	Braakman	Zeeuws-Vlaanderen (Netherlands)	1952	sluice caisson
14	Sloedam	Walcheren-Zd. Beverland (Netherlands)	1871	sinking willow mattresses
15	Ouwerkerk	Duiveland (Netherlands)	1953	caissons
16	Grevelingendam, 2 gaps	Flakkee-Duiveland, Deltaworks	1964	small caissons, quarry stone
17	Oudenhoorn	Voorne-Putten	1953	caisson with side trapdoors
18	Kruiningen, var. gaps	Zd.-Beveland (Netherlands)	1953	caissons; sandbags
19	Krammer closure	St. Philipsland	1987	sand closure
20	Bath	Zd. Beveland (Netherlands)	1953	ship
21	Markiezaatskade	Bergen op zoom (Netherlands)	1983	quarry stone, vertically
22	Volkerakdam	Flakkee-N.-Brabant, Deltaworks	1969	caissons with gates
23	Nieuwerkerk/IJssel	Hollandse IJssel (Netherlands)	1953	small ship
24	Ouderkerk/IJssel	Hollandse IJssel (Netherlands)	1953	sand bags and two vessels
25	Papendrecht	Alblasserwaard (Netherlands)	1953	sand bags, quarry stone, clay
In other areas several major closure projects have been realized also, as for instance:				
26	Feni	Bangladesh	1985	bags filled with clay

Table 4-3 Various dams

5 USE OF THEORY

This chapter repeats some of the theoretical knowledge on fluid mechanics, tides, waves and sediment transport. The text is intended mainly to refresh the reader's knowledge of results and formulae. It also attempts to present a direct link between the more theoretical considerations and practical applications. Derivations have largely been omitted. The results of very specific model investigations that present empirical relations (such as the stability of rubble mounds) are not treated here, but rather in other dedicated chapters. If the content of this chapter is not familiar, the reader is referred to specialized textbooks or lecture notes on these subjects. Most of the relevant lecture notes are indicated in the special section of the list of references.

5.1 General

It is impossible to discuss the design of breakwaters or closure dams without referring to certain subjects from the theory of Fluid Mechanics and/or Geo-technology. Therefore, some important elements from both subjects that should be readily available to the reader are briefly treated in this chapter.

5.2 Hydraulics of flow

5.2.1 Upland discharges

Damming a watercourse requires thorough knowledge of the hydraulic behaviour of the total water-regime of which the course is a part. A quick overview of the conditions that may occur as a result of tides and upland flows is given in this section. Usually the impact of the closure on this regime is such that during closure the hydraulic boundary conditions near the closure site will change. How great the change will be has to be determined before the works start. Calculation of this is the first part of the technical design process. Depending on circumstances, the calculation required may vary between a simple calculation and very complex physical and/or mathematical modelling.

Upland rivers are dammed for different purposes. In the case of reservoir construction the river is completely blocked in order to store the upland discharge. Only the excess water is evacuated via a spillway. Construction of the dam will have to be done in a low discharge period, but some discharge will always take place. Since in the final stage great differences in the water level will occur, the foundation of the dam has to be very stable, especially against the effects of piping and sliding. The flow is therefore usually diverted temporarily to enable removal of the sedimentary subsoil and improvement of the riverbed. Analyses of discharge data and of the probabilities of their extremes are needed for each design of the various construction phases.

Lowland rivers usually have a meandering course or may possibly be comprised of a system of channels and bars (braided river). Cutting off a meander or concentrating the flow in one channel requires the closing-off of the superfluous channel(s). During the progress of the closure, the discharge will gradually take its new course. Likewise, the fall in water-head over the closure gap will gradually increase. This reaches its maximum when the full discharge follows the new course and is equal to the fall in head over the entire length of the new channel. Of course, this fall depends on the quantity discharged by the river. The probability of occurrence of fluctuations in this discharge determines the design condition for the closing method. A sudden peak-flow in the upper river (e.g. due to a rainstorm) will raise the water levels and change the head loss and thus the flow velocity. Both, level and flow affect the conditions governing the closure. Rating curves will provide a timely warning of changes.

The velocity in the channel sections will be different from the gap velocity. Different formulae are used to determine the velocities in channel and gap.

In calculations for the channel sections, upstream and downstream of the closure, Chezy's formula for open channel flow applies:

$$u = C\sqrt{R\,i} \qquad\qquad\qquad (5.1)$$

In this calculation, the fall in head is balanced by friction along the channel section. For a narrow gap in which the fall in head concentrates, the flow is calculated by converting potential energy into kinetic energy. For such a gap the weir formulae detailed in Section 5.2.3 have to be used. An elaborated example is given in Section 16.1.

5.2.2 Hydraulics of tides

In coastal areas, the tide may be the governing factor, which of course is not the case in upland rivers. Since the tidal conditions vary enormously over the earth, a general description of tidal characteristics follows.

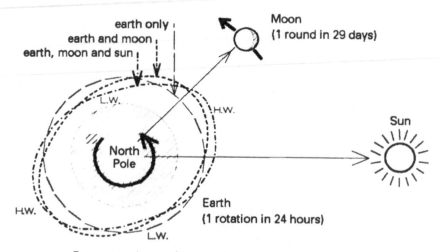

Exaggerated water layer on the earth globe for a given position of the moon

Figure 5-1 Initial tidal wave by the moon and the sun

The tidal wave is initiated by the earth's own rotation and the revolution of the moon about the earth. Gravitational forces between earth and the celestial bodies result in an uplift of the sea level at the side of the body and also the diametrically opposite side. This results in a drawing down of the level perpendicular to this uplift. Relative to a fixed earth, the sun circles around in 24 hours and the moon in 24 hours + 50 minutes and so do the respective uplifts. Since the only line of latitude on earth with a continuous sea-surface is situated at about 65 degrees South (passing south of S. America, S. Africa and Australia), the initial double-wave circles around the southern oceans.

The circumference of this parallel is about 16,000 km long, so the wavelength is about 8000 km. The wave of the moon (called M2) thus passes every 12 hrs 25 min., the wave of the sun (S2) every 12 hrs. Although the mass of the moon is much smaller than that of the sun, because its distance from the earth is much shorter, the effect of the moon is much greater. The two effects coincide when earth, sun and moon are in line, which is during new moon and during full moon (M2+S2 = spring tide). These principles are illustrated in Figure 5-2.

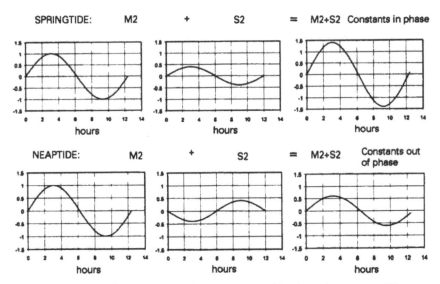

SPRINGTIDE: M2   +   S2   =   M2+S2 Constants in phase

hours   hours   hours

NEAPTIDE: M2   +   S2   =   M2+S2   Constants out of phase

hours   hours   hours

Figure 5-2 Adding semi-diurnal constants resulting in spring or neap tide

The symbols M2 and S2 stand for tidal constituents. All tidal components are written in the mathematical form: $= A \cos (\omega t - \alpha)$, in which

A stands for the amplitude (half the wave-height),

ω determines the frequency,

α determines the phase of the wave at $T = 0$.

Due to the 25 minutes difference between the periods of M2 and S2, the two go out of phase and seven days later the effects are opposite, which is during the first and the last quarter (A {M2 + S2} = A {M2} – A {S2} = neap tide).

The waves on opposite sides of the earth are not equal, which means that there is a difference in height between the consecutive daily high and also between the daily low waters. If a water level record is analyzed by harmonic analysis, this effect appears as a few diurnal components with a period of about a day. The most important of these are called O1 and K1. Combination of these diurnal components intensifies or weakens the resulting daily inequality periodically (Figure 5-3).

The resulting wave travels along the Southern Ocean from east to west. From there, it enters the Pacific, Indian and Atlantic Oceans going north. As a result of the earth's rotation, the flow-vector pattern in a large basin will tend to rotate (Coriolis acceleration), clockwise in the Northern Hemisphere and anti-clockwise in the Southern Hemisphere.

For the tidal wave, this may result in a circular propagation along the circumference of the basin, while in the centre the tide is negligible (the amphidromic point). This rotation (path of the wave-crest) is the opposite of the effect on the flow, anti-clockwise in the northern and clockwise in the Southern Hemisphere. The time of travel required to reach any point on earth determines the time lag between HW (and LW) and the phase of the moon's position. Consequently the same time lag exists for the spring-neap variation of the moon.

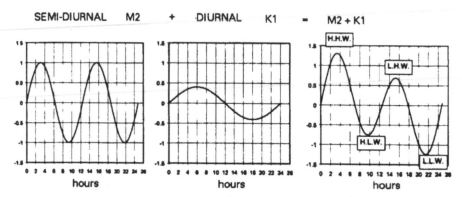

Figure 5-3 Adding a diurnal to a semi-diurnal constant

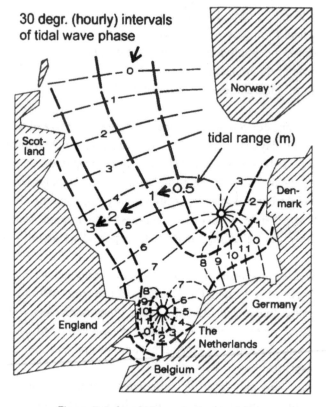

Figure 5-4 Amphidromy in the North Sea

The travelling speed can be approximated by the celerity formula[1]:

$$c = \sqrt{g\,d} \tag{5.2}$$

which for the oceans is about 200 m/s. Travelling up the Atlantic to the North Sea, for instance, takes 24 hrs. On the continental shelf the depth of the water diminishes to 200 m. In the Southern North Sea the depth is about 25 m, so the wave speed is reduced to 15 m/s. The tidal wave enters the North Sea around Scotland, moves south along the east coast of Great Britain, then east towards Denmark and north to the Norwegian coast. In the North Sea, there are two major amphidromic rotations. In the Pacific Ocean, there are six of these rotations, which complicates the picture of tidal propagation.

The astronomical effects such as gravitational forces and rotations are the basis for the tidal wave, but other effects influence its shape. While travelling, the wave changes in consequence of variations in depth and owing to resonance and reflection. Some areas are influenced by tidal waves approaching from various sides, as for instance north of Australia where the waves travelling via the Pacific Ocean and via the Indian Ocean meet. Finally, the tidal wave may be influenced by meteorological phenomena. Half-yearly constant winds like the monsoon-winds change the tide with the same frequency as the declination of the sun, which causes the winter-summer variation. The resulting tide at any seashore therefore has its own typical characteristics, related to the astronomy, but very much adapted during its travelling across the oceans and seas and definitely shaped by the geometry of the area. Generally, the M_2 and the S_2 are the governing components. Then the tide is called "semi-diurnal". At various locations, the diurnal components intensify and the tide is then called a "mixed tide". In very rare cases, the diurnal effects overrule the semi-diurnal components. In such cases the tide is called a "diurnal tide", showing only one HW and one LW per day. The spring-neap variation, resulting from the semi-diurnal components, is absent in this case. Diurnal components show a similar effect. To illustrate the differences, a mathematical example is given in which four harmonic constants are added up: (M2+S2+K1+O1). The four graphs differ only in the amplitudes of the constants. Identical angles and phases are taken, while the sum of the four amplitudes is about the same. The resulting graphs show a typical semi-diurnal tide, two mixed tides and a typical diurnal tide.

The semi-diurnal tides are found along the Atlantic coasts generally, the mixed tides along the Indian and large parts of the Pacific Oceans, while the diurnal tides occur more incidentally, e.g. in Vietnam, parts of Indonesia and Thailand (see Figure 5-5).

---

[1] In spite of the large depth in the ocean, the formula for the shallow water conditions is used because the wave length of approximately 8000 km is much larger than the water depth.

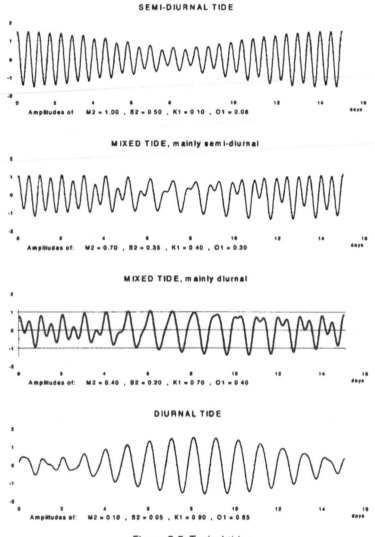

SEMI-DIURNAL TIDE

Amplitudes of   M2 = 1.00 , S2 = 0 50 , K1 = 0 10 , O1 = 0.08

MIXED TIDE, mainly semi-diurnal

Amplitudes of: M2 = 0.70 , S2 = 0.35 , K1 = 0 40 , O1 = 0.30

MIXED TIDE, mainly diurnal

Amplitudes of: M2 = 0.40 , S2 = 0.20 , K1 = 0 70 , O1 = 0 40

DIURNAL TIDE

Amplitudes of: M2 = 0 10 , S2 = 0 05 , K1 = 0 90 , O1 = 0 65

Figure 5-5 Typical tides

Finally, the wave enters estuaries and river-mouths. Shallows, funnel-shapes and upland discharges have impacts on the penetrating wave. Generally, these result in increased tidal amplitude. The top of the wave (high water) may then propagate faster than the trough (low water) and the wave front gets steeper. In the harmonic analyses, this shows as quarter-daily constants, like M_4 and M_6. In extremes, the top may overrun the trough and the tide enters as a tidal bore.

The relation between the water level variation and the flow velocities is an important characteristic of the tide. In relatively small basins (length shorter than 0.05 times the tidal wave length), the two variables will be 90 degrees out of phase. At the moment of high water, the basin is full and the inflow stops. This situation is reversed at low water. At the moment of mean level, the flows have their maximum. This is not true for propagating waves. The slack water/still water after ebb or flood may lag behind for some hours. If so, the maximum flood flow occurs during higher water levels on average than the ebb flow. The mass of water entering the estuary during the flood period, the flood volume, has to flow out during the ebb period with lower levels. Ebb velocities are therefore generally the largest and follow the deep gullies.

In some shallow river-mouths a typical phenomenon occurs. On the average, the flood volume entering during high water discharges during the low water period. During spring tides this tidal volume is larger than during neaps, while the difference in water depth between high and low water is greater. Then, during spring tides the flood may bring more water than the ebb can return. This is particularly true in shallow rivers with high tidal ranges. Water stays behind and the mean level starts rising until spring has reached its maximum range. Thereafter, going to neaps, the difference in depth reduces and the system balances out by lowering the mean level. In the harmonic analysis this effect appears as a component with a fortnightly period, called MSf, which progresses in line with the spring-neap variation. This fortnightly constant is pictured in Figure 5-6.

It will be clear that the tidal characteristics at any point on earth have to be analyzed very thoroughly before a closure of a tidal basin or within in a tidal basin can be designed. Since much of the final appearance of the tide is determined by the geometry of the area near the closure site, the closure itself will have an influence on the shape of the tide. Selection of the boundaries of the area to be studied has to be carefully done in order to make sure that they extend beyond the area influenced by the works.

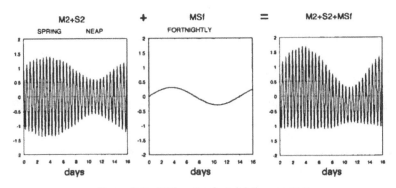

Figure 5-6 Adding the fortnightly constant

5.2.3 Flow through gaps

During the progress of closing a watercourse a distinct constriction in the flow develops. This constriction can be vertical by the creation of a sill, or horizontal by the construction of dam-heads, or by a combination of both. Chezy's law for flow in open channels no longer applies. Depending on the dimensions of the gap and the water depth, different flow patterns may occur. Consequently, different formulae for the calculation of the magnitude of the flow and the discharge capacity of the gap apply.

In the case of a horizontal constriction, the flow velocity can be approximated by the formula:

$$u = \sqrt{2g(H-h)} \tag{5.3}$$

in which g is gravity and $H - h$ stands for the difference in head over the gap. All potential energy is transferred into velocity and friction is ignored (conservation energy). When the water flows through the gap the flow lines contract and strong eddies develop behind the gap.

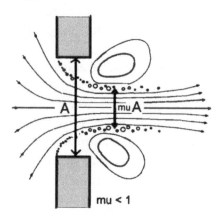

Figure 5-7 Flow pattern in a gap

Where the flow is narrowest, the velocity reaches the above value, which is its maximum. Generally, this will be downstream of the gap. In the gap the average velocity is lower by a factor μ. This μ is the ratio between the cross-sectional flow areas of the gap (A) and the flow gorge.

The discharge capacity of the gap is:

$$Q = \mu A u = \mu A \sqrt{2g(H-h)} \tag{5.4}$$

If the dam heads have a rounded shape of considerable dimension, the flow lines follow the dam head contours and energy losses due to friction are rather small: μ approaches the value 1. An example of $\mu = 1$ is found for sand closures where the shape of the dam head is adapted by the flowing water itself. For closures made by tipping quarry-stone a steep, narrow dam head is formed. Then, the flow in the gap separates from the dam head. The profile of the gap does not contribute fully to the effective discharge, so the value of μ becomes less than 1. Moreover, a lane of vortices develops in the gap in the shear layer between flowing and stationary water, resulting in high-energy losses. Instead of a "μ", a discharge coefficient m is taken, which takes account of both effects. The value of m may be as low as 0.8 for these closures (see Figure 5-7 and Figure 5-9).

Figure 5-8 Flow through a closure gap

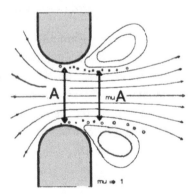

Figure 5-9 Flow in a gap with round dam heads

For the closure operation, these aspects mean that the highest flow velocities develop downstream of the gap, above the protected bottom. The dam is attacked at the point where the flow separates and the shear layer starts developing. The most severe attack on the bottom occurs in the vortex lane along the wake. Since the highest velocity occurs downstream of the actual closing gap, in most cases it will be necessary to protect the original bottom against scouring. In extremely shallow gaps, it is possible that the bottom protection in the gap is sufficient to cause the flow pattern to change into super-critical flow when the head difference during closure increases. In this case, the calculation is identical to that for a vertical closure (see below).

A vertical constriction gives a completely different flow pattern. Assuming a gap of infinite length, the flow pattern can be denoted two-dimensionally in a vertical section. The flow may be sub-critical flow or super-critical flow, depending on the relative levels of the water and the sill crest.

For sub-critical flow the discharge formula reads:

$$Q = mBh\sqrt{2g(H-h)} \qquad\qquad (5.5)$$

in which H is the energy level of the upstream water and h is the water level at the downstream side of the gap, both measured with reference to the crest level of the sill and B is the flow-width (= gap length measured along the dam axis).

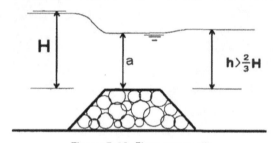

Figure 5-10 Flow over a sill

In fact, the formula is derived by applying Bernouilli's law on the inflow-side, using the water depth a instead of h. However, a is unknown. Calculating with h decreases the head-loss $(H - a)$ and increases the cross-profile $(B * a)$. The resulting error that is introduced in this way is included in the discharge-coefficient, as determined by the shape and the roughness of the sill. Therefore it is called m instead of μ, and may reach values higher than 1. The magnitude of m goes from 1.3 for wide smooth sills with gentle slopes to 0.9 for sharp crested sills.

The mean flow velocity above the sill is:

$$u = \frac{Q}{Ba} = m\frac{h}{a}\sqrt{2g(H-h)}. \qquad\qquad (5.6)$$

which is generally slightly more than calculated by $m\,(2\,g\,z)^{1/2}$.

If a closure gap with sub-critical flow is further constricted by progressively raising the sill, a situation will develop in which the flow becomes critical. This is the case when the vertical distance between sill level and downstream water level is 2/3 of that at the upstream side. The discharge capacity is fully dependent on the upstream energy level in relation to the level of the sill. Then, $z = 1/3\ H$.

$$Q = m\,B\,a\,\sqrt{2g\frac{1}{3}H} \qquad (5.7)$$

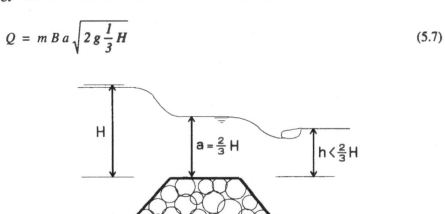

Figure 5-11 Critical flow over a sill

If the downstream water level decreases further, the discharge capacity is no longer influenced and a remains constant viz. 2/3 H. So,

$$Q = m\,B(\frac{2}{3}H)\sqrt{2g(\frac{1}{3}H)} \quad \text{and} \quad u = m\sqrt{2g(\frac{1}{3}H)} \qquad (5.8)$$

The value of m depends on the inflow condition and the roughness of the sill, as the flow is independent of the downstream condition. Thus, since a is known and used in this case, the maximum value of m is 1. In the case of sharp crested rough sills, this value may go down to 0.8. The water falls down along the sill's downstream slope and changes into sub-critical flow via a hydraulic jump. Depending on conditions (roughness of the sill and Froude number) this may show different flow patterns.

For the closure operations, these downstream conditions may be more critical than the conditions on the crest of the sill. Suction forces will attack the downstream slope and endanger the stability of the sill. The down-pouring jet flow may reach the bottom in the case of a closure in relatively shallow water and endanger the bottom stability.

Finally, a combination of vertical and horizontal closure frequently occurs. Then, a situation may exist in which the gap has a vertical constriction caused by a sill and dam heads on both ends, which are progressively built out. This three-dimensional situation is more complex than the separate ones and difficult to describe mathematically. The general approach is to add the two effects as if they were occurring independently. However, the peak velocity downstream of the gap, usual for horizontal closure, may be compensated by the increase in water depth after the flow has passed the sill.

5.2.4 Modelling

To assess the hydraulic situation in the various stages of the closing process, extensive modelling has to be carried out. The modelling techniques that are available vary from physical scale models to electric analogous models and mathematical models. For many years, physical scale models were the best option. All closures in the Delta Project in the Netherlands have been accompanied by extensive physical model testing. Such models were built on different scales for different purposes. With the rapidly increasing capacity of digital computing techniques, mathematical models have replaced the physical scale models to a large extent. Sometimes, additional physical model tests are recommended to find the proper input values in a mathematical model (discharge coefficients, scour, etc.). Whatever model technique is used, one must be aware of the typical model effects when interpreting the results.

Any model investigation starts with an inventory of the boundary conditions and the estimation of the boundaries of the area that will not be affected by the closure. In tidal basins the latter is sometimes rather difficult to calculate, but when the effect is less than a few centimetres the result is sufficiently accurate. The resulting error will be less than the influence generated by the natural variation in tidal conditions.

External influences may be:
• Upland discharge(s) coming down the river(s)
• Tidal waves passing along or entering the basin to be closed
• Drainage water from surrounding areas
• Direct rainfall in the basin
• Wind set-up and draw down in the area outside the basin
• Wind set-up and draw down within the basin
• The earth's rotation (Coriolis acceleration)

The conditions, which determine the hydraulic behaviour within the basin, are:
• The network of gullies and shallows (if applicable)
• The flow profiles (depth/width relation) of the gullies
• The storage (level/surface area relation) of the shallows
• The hydraulic resistance of the gullies
• The existence of density (mainly salt/fresh water) currents
• The occurrence of tidal bores

The mathematics available to describe the hydraulic situation includes the basic formulae for conservation of mass and of moment.

$$\frac{\delta Q}{\delta x} + B_x \frac{\delta H}{\delta t} = 0 \tag{5.9}$$

$$\frac{\delta Q}{\delta t} + \frac{\delta(\alpha\, Q\, u)}{\delta x} + g\, A\, \frac{\delta H}{\delta x} + \frac{g\, |Q|Q}{C^2\, A\, R} - W_x = 0 \tag{5.10}$$

These can best be solved by a numerical computation of unsteady flow in a network of watercourses. The extent of the network, the number of storage areas, the schematization of the flow profiles, the terms in the equations ignored (if any), and the correctness of the estimated hydraulic resistance determine the accuracy of the calculation. On the other hand, if the accuracy is not affected too much the mathematical system (and thus the quantity of deskwork) can be simplified. Besides, these simplifications are indispensable for a first orientation, and for the determination of an order of magnitude.

An extreme means of simplification is to ignore the conservation of momentum in the basin. What remains is the mass balance. For the gap the weir-formulae apply, while the water level in the basin is assumed to be horizontal over the full surface area at any time. To represent drying shallows, a variable area can be taken for different heights. The water level changes over time and this change, multiplied by the surface area of the water, must balance the quantity of water flowing through the gap during equal time lapses.

The tidal wave propagates at infinite speed within the basin and it neither reflects nor dampens. In addition, the simplification assumes that throughout the basin the water moves towards and away from the gap basin without driving forces and without friction.

The deviation is acceptable if the basin is relatively short in relation to the length of the tidal wave ($\leq 0.05\ L$), and of course friction should be reasonably low, which in practical terms means that the area is well criss-crossed by gullies.

The quantity of water flowing through the gap per unit of time is determined by multiplying the flow velocity by the cross-sectional profile of the gap. The velocity changes as result of the changing head loss over the gap, while the cross-sectional profile changes as result of the changing water level in the gap. If a vertical closure is considered, it is necessary to ascertain whether the flow condition is critical or non-critical for every time step, as different formulae to calculate the flow velocity apply. More gaps than one can be dealt with by a simple summation of the water quantities, as long as it may be assumed that there is no head difference between any of the gaps within the basin. In this case, there is no propagating wave from one gap to the other.

One gap with variable depths can be schematized by the summation of a number of gaps, each having different dimensions (length and depth).

In vertical closures, various construction stages can be defined, each having a specified sill level. However, during the actual construction there will be intermediate stages in which only part of a new layer is present. Moreover, a failure may create a local depression in the sill level. In these depressions higher flow velocities will exist. Therefore, the calculation should always be executed with a gap of variable depth. **The plain horizontal sill is a theoretical case and does not represent a determining practical situation.**

When a dike breach is being closed, the storage basin may be an inundated area in which it is very difficult to define a system of gullies and shallows. Moreover, owing to erosion of the unstable ground, the system may quickly change. The Chezy value for this overland flow may be as low as 30. The area will fill during high water periods but will not drain fully during low water. In these calamitous situations, it is of the utmost importance to prevent the development of scouring gullies and to maintain the high resistance of the terrain (see Section 2.5.5, the gap called Schelphoek).

As an example, in Chapter 16 a calculation is made to illustrate the change in the tidal characteristics during the closure of a tidal basin.

5.2.5 Forces on floating objects

Various closing methods make use of floating equipment and submerged or floating means of closure. Some of these have substantial dimensions and when operating in the closure gap they obstruct the flow. In consequence, the gap dimension is smaller than anticipated and higher current velocities than expected occur. In calculating the forces acting on these structures and vessels this factor should be incorporated. Various factors determine the force due to the flow.

- the shape of the object
- whether it is floating or submerged
- the keel clearance
- the projected obstructed cross profile of the flow
- the dimension in the flow direction
- the roughness of the object
- the Reynolds number

The force can be approximated by the formula:

$$F = \tfrac{1}{2} \rho u^2 C_D A \qquad\qquad (5.11)$$

in which all items except 4 are covered in the C_D. The estimation of this value may be rather difficult and it has a considerable impact on the resulting force. Moreover, the value is not a constant for a vessel, but may increase when the keel clearance decreases.

The C_D-value can vary from 0.3 for a submerged torpedo-shaped body, up to 4 for a large rectangular vessel in shallow water. Keel clearance in closure gaps is often very limited. If keel clearance is limited the C_D-value is related to the μ-value for flow through narrow gaps. A vertical component of the force will also develop and the vessel will be drawn down and may eventually touch the bottom.

The use of large vessels and floating cranes in stationary positions is determined by their ability to withstand the flow forces with the anchoring configuration in use (number of mooring lines, wires and winches, type of anchors and soil conditions). In addition, motions and forces are also induced by wave action. The flow is generally the determining factor in closure works, as are the waves in breakwater construction.

When caissons are transported by towing or pushing, these external forces for flow and waves must be likewise distinguished since they have a bearing on the required pushing power, the navigability and the internal strength of the structure.

5.2.6 Stability of floating objects

Any object of arbitrary shape submerged in an ideal fluid is exposed to pressures that act in a direction normal to the surface. If the total surface is divided into a large number of small surface elements, the forces acting on each element can be found by multiplying the area of the element with the external pressure. Integration of these elementary forces over the total surface of the body yields a resultant force. The force components in the horizontal plane nullify each other as can be proved mathematically. In the vertical direction, the resultant force proves equal to the weight of the fluid displaced by the submerged body (Archimedes). It is called the buoyancy and is directed upwards.

An object is floating when the buoyancy force is equal to the weight of the object. The body will position itself in such a way that the vector representing the weight of the body is equal but opposite to the vector that represents the buoyancy and works along the same line. The point of application of the weight is the centre of gravity G, the point of application of the buoyancy is the centre of buoyancy C. Within the condition "floating" two options can be distinguished:

- the object is completely submerged (example: submarine) and
- the object is partially submerged (example: surface vessel).

The position of the floating body can either be described as a stable equilibrium or as an unstable equilibrium. We refer to the term stable equilibrium if a small deviation from the equilibrium position will result in a return to that original position. We use the term unstable equilibrium if a small disturbance will not result in a return to the original position.

It is evident that a stable position exists when C is located vertically above G. Any small rotation of the object will result in a correcting turning moment (Figure 5-12). This condition can only be achieved when the mass density of the object is not distributed uniformly over its volume. One of the best examples is a submarine. Such vessel is only stable indeed if there is ballast at the keel or bottom.

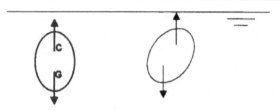

Figure 5-12 Stability of a submerged object

A stable position can also exist, however, if C is located below G as is often the case when we consider objects with a uniform distribution of the mass density such as for instance a rectangular piece of wood (Figure 5-13). The centre of gravity G and the center of buoyancy C can easily be found by construing the geometric centres of the complete body and the submerged part respectively.

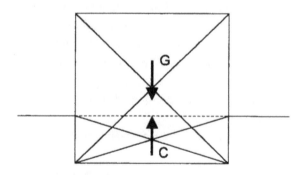

Figure 5-13 Stability of a floating object

When we follow the definition of stable equilibrium, we should tilt the body slightly and see whether the turning moment by the shift of the gravity versus buoyancy vector is a stabilizing or a de-stabilizing factor. For this purpose, we first investigate the condition of a light weight box (Figure 5-14).

In the original (horizontal) position C is located in the centre line of the box, halfway the draft, and G is also located in the centre line, halfway the total height of the box (Figure 5-14, left). If the box is tilted slightly, rotation takes place around a point (or rather a line perpendicular to the page) at the intersection of the water line and the centre line of the box. This causes G to move slightly to the right. One would expect C to move slightly to the left, but this is not the case because the upward pressure against the bottom of the box is not longer uniformly distributed. Due to the trapezoidal shape of the presseure diagram, C moves to the right, and in the example of a light weight box so much that the joint action of the buoyancy vector and the gravity vector yields a stabilizing moment. If the disturbance from the eqilibrium position becomes too large, the box will loose stability and turn to its next stable equilibrium. This may be the case when a large external horizontal force is applied, for instance wind on the superstructure, a wave or flow force on the submerged part or pull by the wire of a tugboat.

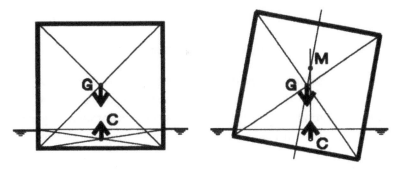

Figure 5-14 Stability of a light weight box

The horizontal shift of the point of application of the buoyancy causes the working line of the buoyant force to intersect the centre line of the box in a point M, which is located above C. The distance between M and C (MC) can be calculated by rotating the body over an angle φ, and by subsequently calculating the magnitude and position of the force vectors.

It appears that $MC = \dfrac{I}{V}$,   (5.12)

in which I is the moment of inertia of the water plane about the axis of the body, and V is the displacement of the body.

This theory is often used to calculate the rolling stability of vessels. In that case, I and V can be calculated easily (see also Figure 5-15):

$$I = \int_{-\frac{1}{2}b}^{+\frac{1}{2}b} y x^2 dx, \text{ and } V = \frac{G}{\rho g}$$ (5.13)

For our box with width B and length L, I turns out to be $\dfrac{1}{12}LB^3$.

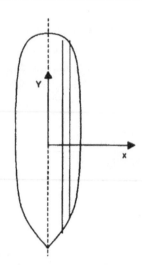

Figure 5-15 Moment of inertia

It can now easily be understood that stability of rotation depends on the distance between M and G. This distance is called the metacentric height m_c. If M is located above G, the equilibrium is stable, if M is below G, it is unstable. In general, a value of 0.5 m for m_c gives sufficient stabilising moment for caissons used in hydraulic engineering.

If the mass density of the light weight box of Figure 5-14 increases, the following changes take place:

- G moves downward and C moves upward, which causes an increase in stability
- m_c decreases, (because I remains the same and V increases), which decreases stability.

This results initially in a smaller stability that will cause the box to float diagonally if the mass density is about half the density of water. A further increase of the weight of the box will turn it stable again, though the draft has increased considerably.

5.3 Waves

5.3.1 Linear wave theory

When waves are treated in a textbook, it usually starts with a description of linear wave theory. Although this theory is sometimes quite far from reality, it is useful for a proper understanding of the basics. On the assumption that the wave height is low in comparison to wave length and water depth, it is possible to describe the pressure field and flow field and to analyze changes in the behaviour of waves when they travel from deep water into shallower water. It should also be noted that such theory is referring to monochromatic, regular waves. It is therefore possible to distinguish the following basic elements (Figure 5-16):

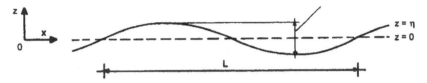

Figure 5-16 Definitions of a regular wave

H: wave height in m
T: wave period in s
L: wave length in m

The application of the linear wave theory leads to a well-known set of equations, of which the most important ones are given below:
Surface elevation:

$$\eta = \frac{H}{2} \cos\left(\frac{2\pi x}{L} - \frac{2\pi t}{T}\right) \qquad (5.14)$$

Or if $2\pi/T$ is substituted by ω, and $2\pi/L$ by k,

$$\eta = \frac{H}{2}\cos(kx - \omega t) = \frac{H}{2}\cos\theta \qquad (5.15)$$

Wave celerity:

$$c = \frac{L}{T} \qquad (5.16)$$

and:

$$c = \sqrt{\frac{gL}{2\pi}\tanh\left(\frac{2\pi h}{L}\right)} \qquad (5.17)$$

Wave length:

$$L = \frac{gT^2}{2\pi}\tanh\left(\frac{2\pi h}{L}\right) \qquad (5.18)$$

Owing to the specific properties of the hyperbolic functions, it is possible to simplify the above formulas into approximations that are valid for deep ($h/L > 1/2$) and shallow ($h/L < 1/25$) water.

For deep water (subscript $_0$), the following applies:

$$L_0 = \frac{gT^2}{2\pi} = 1.56T^2 \qquad\qquad (5.19)$$

and

$$c_0 = \frac{gT}{2\pi} = 1.56T \qquad\qquad (5.20)$$

For shallow water, the wave celerity becomes independent of the wave period:

$$c = \sqrt{gh} \qquad\qquad (5.21)$$

Still, within the linear small amplitude wave theory it is possible to derive the periodic changes of the position, the velocity and the acceleration of water particles. In deep water the particles follow a circular pattern; in shallower water the circular motion is transformed into an elliptical pattern. This orbital motion can be described by splitting position, velocity and acceleration into horizontal and vertical components (ξ and ζ, u and w, and a_x and a_z respectively).

$$\xi = -H/_2\frac{\cosh[2\pi(z+h)/L]}{\sinh(2\pi h/L)}\sin\theta \qquad \zeta = H/_2\frac{\sinh[2\pi(z+h)/L]}{\sinh(2\pi h/L)}\cos\theta \qquad (5.22)$$

$$u = \omega H/_2\frac{\cosh[2\pi(z+h)/L]}{\sinh(2\pi h/L)}\cos\theta \qquad w = \omega H/_2\frac{\sinh[2\pi(z+h)/L]}{\sinh(2\pi h/L)}\sin\theta \qquad (5.23)$$

$$a_x = \omega^2 H/_2\frac{\cosh[2\pi(z+h)/L]}{\sinh(2\pi h/L)}\sin\theta \qquad a_z = -\omega^2 H/_2\frac{\sinh[2\pi(z+h)/L]}{\sinh(2\pi h/L)}\cos\theta \qquad (5.24)$$

Note that at the water surface ($z = 0$), the expressions for ζ and η become identical, and that at the bottom, for $z = -h$, all vertical movements become zero due the influence of sinh(0).

Finally, there is a similar expression for the pressure at a level z below the water surface:

$$p = -\rho g z + \rho g \eta \frac{\cosh[2\pi(z+h)/L]}{\cosh(2\pi h/L)} \qquad\qquad (5.25)$$

In this expression $\rho g z$ represents the hydrostatic pressure, and $\rho g \eta$ cosh (..)/cosh (..) the harmonic component of the pressure, because $\eta = \frac{1}{2} H \cos \theta$.

When the conditions for the linear (small amplitude) wave theory are no longer fulfilled deviations will occur. The wave becomes asymmetrical, with a sharp and high crest and a long and shallow trough. This is solved mathematically by adding a higher harmonic component. This means that other (higher order) theories have to be applied, taking into account this deformation of waves. Miche [1944] presents a comprehensive mathematical review. A review of the validity of various wave theories is given by Le Mehaute [1969]. See also Figure 5-17.
However, it is remarkable how well the linear wave theory works, even beyond the strict limits of its validity!
The theory fails, however, when waves approach the stage of breaking, either due to a high steepness (H/L) or due to shallow water depth (H/h). Theoretical limits are $H/L < 0.14$ and $H/h < 0.78$. These limits occur when the particle velocity in the crest exceeds the wave celerity, in other words when the water particles tend to leave the wave profile. This initiates a process of energy dissipation that considerably reduces the wave height.

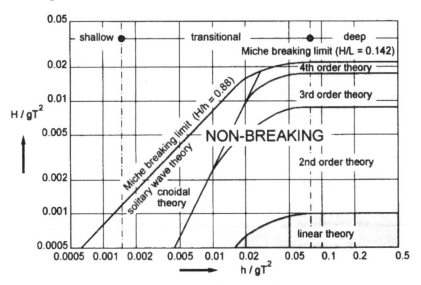

Figure 5-17 Validity for waves theories

5.3.2 Refraction, diffraction, shoaling, breaking and reflection

Refraction

When waves travel from deep water into shallower water, some significant changes occur. From equations (5.18) and (5.22), it can clearly be seen that the wave celerity decreases with depth. When a wave approaches underwater contours at an angle, it is evident that the sections of the crest in the deeper parts travel faster than those in the shallower sectors. This makes the wave crest turn towards the depth contour. This bending effect is called refraction, in analogy with similar phenomena in physics (light, sound). The effect is shown in Figure 5-18.

In the refraction theory it is assumed that no wave energy moves laterally along the wave crest. The energy remains constant between orthogonals normal to the wave crest. The direction of the orthogonals changes proportionally to the wave celerity according to the law of Snellius:

Figure 5-18 Example of wave refraction

$$\sin \alpha_2 = \left(\frac{c_2}{c_1}\right)\sin \alpha_1 \tag{5.26}$$

By applying this equation, it is possible to construct a field of orthogonals over a given bottom configuration for a given wave direction and wave period. During that process, the distance b between the orthogonals may vary. Considerations of energy conservation show than that:

$$\frac{H_2}{H_1} = \sqrt{\frac{b_1}{b_2}} \tag{5.27}$$

The factor $\sqrt{\frac{b_1}{b_2}}$ is also called the refraction factor k_r, which is used to calculate the change in wave height when a wave approaches the shore. Designing a field of orthogonals is a cumbersome job, especially when it has to be repeated for various wave directions and various wave periods. The outcome is not always satisfactory; certainly not when owing to the bottom topography some of the orthogonals intersect. This leads to infinitely high wave heights!

Diffraction

When we discussed refraction we assumed that no lateral transfer of wave energy would take place along the wave crest. This assumption is correct as long as the lateral gradient is not too great. However, this assumption is no longer valid when an infinitely high gradient occurs. For instance, when at one location the wave energy is allowed to pass, while next to it the wave propagation is prevented by an obstruction (island or breakwater). In such cases, there is some lateral transfer of wave energy. The phenomenon can clearly be distinguished in Figure 5-19.

The theory of wave diffraction is solved mathematically by the application of the "Cornu-spiral". Practical guidelines for considerations about diffraction are given in the Shore Protection Manual [1984].

Figure 5-19 Diffraction behind a detached breakwater

Shoaling

If we assume that waves approach the coast perpendicularly, refraction does not occur. In this case, we can consider the energy flux towards the shore, and assuming that no energy is dissipated, this leads to the following expression:

$$\frac{H}{H_0} = \sqrt{\frac{1}{\tanh(2\pi h / L)} + \frac{1}{1 + \frac{(4\pi h / L)}{\sinh(4\pi h / L)}}} = k_{sh} \qquad (5.28)$$

Like the refraction coefficient, the shoaling coefficient can be used to calculate the wave height H at a given location with a given depth h when the deep water wave height H_0 and the wavelength are known. The result is shown in Figure 5-20, which clearly indicates that when a wave enters shoaling water (right side of graph), wave heights are first slightly reduced, before they increase considerably when the waves come nearer to the shore. Similarly, one can notice how wave length and wave celerity decrease. The process sketched here is reversible, in the sense that waves proceeding from shallow water into deep water regain their original properties.

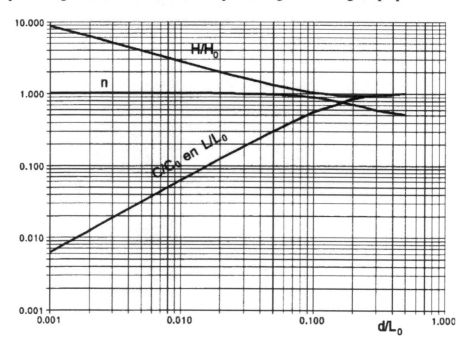

Figure 5-20 The effect of shoaling on wave parameters

This procedure is only applicable when no energy is dissipated, or in other words when no breaking takes place. It is evident that closer to the shore breaking must take place since wave heights increase considerably and the water depth reduces. So at a certain point the theoretical limit of $H/h = 0.78$ would certainly be exceeded.

Breaking

In Section 5.3.1 it was mentioned that traditional wave theories that are based on conservation of energy fail when energy is dissipated, or in other words when waves start breaking. When waves approach the shore, the wave celerity reduces. In addition, we have seen that the wave height increases owing to shoaling and the orbital motion within the waves is also changing. If the orbit in deep water was a circle, in shallow water it becomes an ellipse with a horizontal axis larger than the vertical axis. The vertical axis of the orbit at the water surface is equal to the wave height. Since the wave height increases owing to shoaling, the vertical motion of the water particles at the surface must also increase. Moreover, the horizontal movements grow in relation to the vertical movements, which means that there must be a significant increase in the particle velocity near the surface. When this particle velocity exceeds the wave celerity, the crest of the wave is no longer stable, so that breaking must occur. Since most wave theories are based on energy conservation, their validity ends when energy dissipation takes place. Breaking of an individual wave may be initiated by the wave being too steep ($H/L > 0.14$) or by the water being too shallow ($H/h > 0.78$).

Reflection

Surface waves follow not only the general physical laws on refraction, but also those on reflection. This means that a structure reflects the incident wave in such a way that the angle of incidence is equal to the angle of reflection. A curved convex structure may concentrate wave energy in a limited area just as a curved mirror or a lens focuses light.

The reflection can be influenced by the nature of the surface. A vertical solid wall (a caisson or a quay wall) produces 100% reflection, whereas a sloping wall or a rough permeable/porous structure considerably reduces the amount of reflection. A sloping beach is considered to have 0-reflection.

This is a point that must be taken into account when considering wave penetration into a harbour. The unrest in a harbour basin will be greatly influenced by the reflection from the boundaries.

In theoretical terms, a normally incident wave can be described by (see Section 5.3.1):

$$\eta_i = {H_i}/{2} \cos\left(\frac{2\pi x}{L} - \frac{2\pi t}{T}\right)$$ (5.29)

with the subscript i indicating the incident wave.

In a similar way, the reflected wave (travelling in the opposite direction) can be described by:

$$\eta_r = {H_r}/{2} \cos\left(\frac{2\pi x}{L} + \frac{2\pi t}{T}\right)$$ (5.30)

Assuming that a fraction r of the incident wave is reflected, the total disturbance of the water surface can be described as:

$$\eta_{tot} = \eta_i + \eta_r =$$

$$= (1+r)\frac{H_i}{2}\cos\left(2\pi x/L\right)*\cos\left(2\pi t/T\right) + (1-r)\frac{H_i}{2}\sin\left(2\pi x/L\right)*\sin\left(2\pi t/T\right) \qquad (5.31)$$

For the nodes ($x = \frac{1}{4}L \pm n\,\frac{1}{2}\,L$), this becomes:

$$\eta_{tot} = (1-r)\frac{H_i}{2}\cos\left(2\pi t/T\right) \qquad (5.32)$$

and for the antinodes ($x = 0 \pm n\,\frac{1}{2}\,L$):

$$\eta_{tot} = (1+r)\frac{H_i}{2}\cos\left(2\pi t/T\right) \qquad (5.33)$$

For 100% reflection (or r=1) this leads to the well-known standing wave or clapotis. Another interesting substitution is $t = \frac{1}{4}\,T \pm n\,\frac{1}{2}\,T$!

Application of the same goniometric exercise to the formulae for the orbital velocities, the pressures and the accelerations leads to similar results, although the position of nodes and antinodes alternates.

5.3.3 Irregular waves in deep water

Contrary to the wave theory discussed in the previous sections, waves in nature are not small in amplitude and do not show the regular character with respect of height H and period T.

Irregularity takes place on at least two distinctly different time scales that are characterized by the short term and the long-term variations respectively. The easiest way to distinguish these two phenomena is to assume for the time being that during a particular storm, the wave pattern is stationary. In other words, we neglect the gradual growth and decay of the wave field, and we consider the storm more or less as a block function. Even then, the wave motion is irregular as is demonstrated by the wave record of Figure 5-21.

Short-term statistics

Individual waves can be differentiated according to international standards by considering the water surface elevation between two subsequent upward or downward crossings of the Still Water Level (SWL)[2]. The time span between these crossings is the wave period; the range between the highest and the lowest level is the wave height. In this way, a height and a period can be defined for each individual wave I ($0 < i < n$) from the wave train. Since all heights and periods of individual waves are different, it is logical to apply statistical methods to characterize the set of data. The easiest way is to determine the statistical properties of the wave heights only.

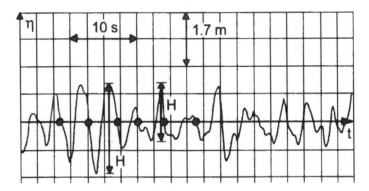

Figure 5-21 Irregular wave

It appears that in deep water, the probability of exceeding the wave heights follows a Rayleigh distribution:

$$P(\underline{H} > H) = e^{\left[-2\left(\frac{H}{H_s}\right)^2\right]}$$
(5.34)

in which H_s, the significant wave height is equal to the average of the 1/3 highest waves. It can also be defined as the wave height that is exceeded by 13.5% of the waves. A third definition and method to determine H_s is given later.

A graphical presentation of the exceedance curve is often made on so-called Rayleigh graph paper (Figure 5-22). On such paper, a straight line through the origin represents a data set that follows the Rayleigh distribution. In view of the definitions, the value of H_s can be read at the 13.5% exceeding value. In this way, the strength of the storm considered is apparently determined by just one value: H_s. A stronger storm would lead to a steeper distribution curve, which is again defined by a specific value of the significant wave height.

---

[2] Still Water Level is the water level in the absence of the waves, it is taken equal to the mean water level during the recording period

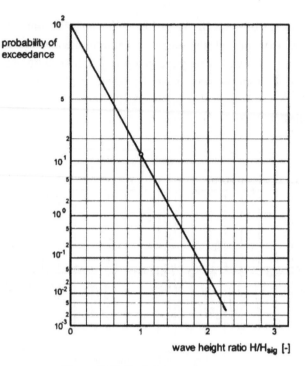

Figure 5-22 Rayleigh graph paper

Wave periods are generally treated in a slightly different way. It is possible to consider the irregular surface level $\eta(t)$ to be the sum of a large number of periodic waves:

$$\eta(t) = \sum a_i \cos(2\pi f_i t + \varphi_i)$$ (5.35)

in which
a_i = amplitude of component i
$f_i = 1/T_i$ = frequency of component i
φ_i = phase angle of component i

The spectral energy density $S(\omega)$ can then be expressed as:

$$S(\omega) = \tfrac{1}{2}\sum^{\Delta\omega} a_i^2 / \Delta\omega$$ (5.36)

The energy contained in the whole frequency range is proportional to $\overline{\eta^2}$, and denoted by m_0. Consequently:

$$m_0 = \overline{\eta^2} = \frac{1}{2}\sum_{i=1}^{n} a_i^2 \qquad (5.37)$$

The determination of $S(\omega)$ in practice is based on a more mathematical concept via the auto-correlation function $R(\tau)$ and its Fourier transform. In this process, some mathematical hiccups may occur. It is therefore recommended to check whether a direct analysis of the wave height distribution yields the same significant wave height as the spectral approach. In other words, check whether

$$H_s = 4\sqrt{m_0} = H_{13.5\%} \qquad (5.38)$$

When the wave energy spectrum has been established in this way, in most cases it is possible to distinguish a frequency $f = 1/T$ or period T where the maximum energy is concentrated. This value is called the peak period T_p. Of course, one can also count the total number of waves (n) during the recording period and thus define an average period T_m. An example of a typical wave spectrum is given in Figure 5-23.

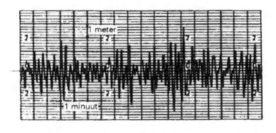

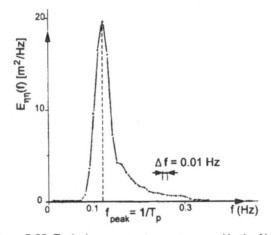

Figure 5-23 Typical wave spectrum, measured in the North Sea

It is stressed that the spectral analysis and the Rayleigh distribution are only valid to analyze a stationary process. One must therefore be careful to choose a measuring period that is not so long that one is almost sure that the wave climate will change during the observation period. On the other hand, one must choose a long enough observation period to ensure that the sample leads to statistically reliable results. It has become common practice to measure waves during a period of 20 to 30 minutes at an interval of 3 or 6 hours.

Summarizing, one can state that the short-term distribution of wave heights, i.e. the wave heights in a stationary sea state shows some very characteristic relations:

Standard deviation free surface	$\sigma_\eta = \sqrt{m_0}$	1	0.250
RMS height	H_{rms}	$2\sqrt{2}$	0.706
Mean Height	$\overline{H} = H_1$	$2\sqrt{\ln 2}$	0.588
Significant Height	$H_s = H_{1/3}$	4.005	1
Average of 1/10 highest waves	$H_{1/10}$	5.091	1.271
Average of 1/100 highest waves	$H_{1/100}$	6.672	1.666
Wave height exceeded by 2%	$H_{2\%}$		1.4

Table 5-1 Characteristic wave heights

Note: these relations are valid only for deep water, i.e. in the absence of breaking waves.

In a similar way, periods can be related:

Name	Notation	Relation to spectral moment	T/T_p
Peak period	T_p	$1/f_p$	1
Mean period	T_m	$\sqrt{(m_0/m_2)}$	0.75 to 0.85
Significant period	T_s		0.9 to 0.95

Table 5-2 Characteristic wave periods

Long-term statistics

It has been pointed out above that it makes no sense to determine the significant wave height and a spectrum if the wave train is not part of a stationary process. Therefore, waves are measured during a relatively short period (15 to 30 min.) at regular intervals of 3 to 6 hours. Each record is considered to be representative for the whole interval. In this way, a new time series is developed, consisting of the significant wave height (and period) per interval. This time series may cover a period from several months to several years.

The result of series of wave observations covering a longer period will therefore again become a set of random data that represent the long-term wave climate of the location. In this set, one may distinguish different patterns.

It is possible that there is a real stationary condition throughout a year (passaat or trade winds), or that there may be distinct seasons (summer/winter, monsoon periods), while superimposed on these there may be incidents like the passage of a storm, a hurricane or a cyclone (**Figure 5-24**).

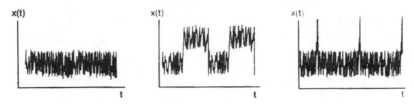

Figure 5-24 Typical types of wave statistics patterns

Depending on the purpose of the analysis, one must decide what to do with a long series of wave observations. In a number of cases, when one is interested in workability or in the accessibility of a port, one can simply analyze all data and determine the so-called Serviceability Limit State or SLS. For such considerations, the incidents need not be important, provided that a timely warning is received and one can take some measures. The case is different when one is interested in the maximum loads exerted on a structure during its lifetime. Exceeding of the design load may cause serious or even irreparable damage to the structure. In that case, extreme value theories must be applied to determine the Ultimate Limit State (ULS). For such analyses one uses only the highest observations from the set, as observed once per month or once per year. For some types of structure (rubble mounds) it is sufficient to characterize the load conditions by the intensity of the storm, i.e. by H_s. This simplification is justified because the damage of a rubble mound breakwater progresses slowly. When a caisson-type breakwater or another structure with a brittle failure behaviour is considered, failure may be caused by just one extremely high wave. This means that one has to establish an extreme value distribution for individual waves. This can be done by combining the long term and the short term wave statistics.

Whatever the case, it will be difficult to collect sufficient data from actual wave observations, simply because the period of wave observations is too short to establish a reliable prediction of extreme events. Thus it is necessary to use long-term wind records or visual observations of wave heights made on board ships to try to establish a long-term distribution of wave conditions. Actual measurements can then be used to calibrate the model that is used to determine the wave conditions indirectly. For details about the extreme value theory, one is referred to E.J. Gumbel [1958] or to J.K. Vrijling [1996].

Observation periods and storm duration

As mentioned above, for determination of the ULS, one has to apply extreme value theories. One method to do so is to take the highest observation value from a set of given duration (e.g. a month or a year). The disadvantage of taking the highest value in a month is that in such an analysis an unwanted seasonal effect may be introduced. Taking the highest value in a year has the disadvantage that the amount of data available becomes very restricted. Therefore, an analysis is often made of all storms in the record. A storm is defined as a period with a more-or-less constant and relatively high wave height. In the PoT-analysis (Peak over Threshold) a storm is explicitly defined as a period of time during which the wave height is higher than a given threshold value; the magnitude is defined by the highest wave observation within the storm. This method is worked out as an example in Appendix 10.

Even when in this way a number of storms is identified, an extrapolation must be made to arrive at an assessment of the (rather rare) design storm. For this extrapolation, one may apply the Exponential, the Gumbel or the Weibull distribution. For details, one is referred again to Appendix 10.

The final result depends on the threshold value adopted and on the statistical distribution used in the extrapolation. In general, the most reliable results are obtained with relatively high threshold values. Even then, the result depends on the statistical method used. Table 5-3 shows the results of the calculation from Appendix 10, with a threshold value of 4.0m, which results in an average of 5.3 storms per year.

Statistical Method	$H_{s\,100}$
Exponential	8.31 m
Gumbel	7.81 m
Weibull	7.77 m

Table 5-3 Example of predicted design wave heights

The difference between the methods is in the order of half a meter. N_s is the number of storms per year in the database. Lowering the threshold level will increase the number of storms in the analysis, but at the same time make the analysis less reliable.

Relation between wave height and wave period

When analyzing series of wave data it is always good to make a number of checks. One of those checks is to study the relation between H_s and T_p, as calculated for each wave record. Plotting the data in a H/T diagram may lead to a result such as that shown in Figure 5-25.

One must realize that substituting T_p in the deep-water wavelength formula leads to a deep-water wavelength L_{op}. The fictitious wave steepness can then be expressed as $s_{op} = H_s/1.56\,T_p^2$. Values of s_{op} are seldom higher than 5 or 5.5%. These high values are representative for waves that are generated by one typical nearby wind field. Low values of s_{op} indicate swell from remote wind fields. The low values are associated with low wave heights and very long periods. Values below 1% are rarely reported, maybe partly because many measuring instruments do not measure this type of wave accurately.

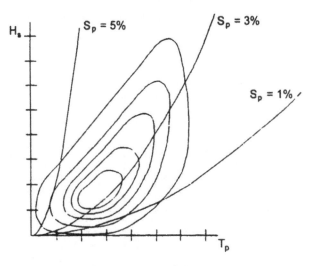

Figure 5-25 *H/T* diagram

Transformation of irregular waves in shallow water

In Section 5.3.2, attention was paid to refraction, diffraction, shoaling and breaking, which are processes that take place when a wave enters shallower water. In the small amplitude approach, it was possible to distinguish individual waves with a single height and a single period, and to study changes in their direction and height on the basis of the ratio between water-depth and wavelength.

For irregular wave fields, the analysis of changes in the wave pattern is much more complicated when the waves enter into shallower water. Owing to the variability of the wave period, there are no fixed orthogonals that can be calculated. Because of the variability in wave heights, one must expect that some waves will break owing to their extreme height, whereas other (lower) waves remain unaffected. This means that the Rayleigh distribution cannot be valid in shallow water. Recently, this has been demonstrated by H.W. Groenendijk [1998] in his Master's thesis and by M.R.A. van Gent [1998].

Battjes and Janssen [1978] laid the basis for a numerical approach to the problem. Holthuijsen and Booij have continued this approach. Their work has resulted in a set of mathematical models. The most recent of them, SWAN is not only used in the Netherlands, but also elsewhere in the world. The method is based on considerations of energy flux through the boundaries of elements that cover the area of interest. By adding criteria for wave breaking, the models replace separate calculations for shoaling, refraction and breaking. Diffraction and reflection are not yet included, however.

Delft Hydraulics (WL) has performed calculations with a one-dimensional version of this program (ENDEC) to study the behaviour of irregular waves approaching an idealized straight coast in a perpendicular direction. Some of the results are given in Figure 5-26, where H_s/h is plotted against h/L_0 for four values of s_{op}, and various values of the slope of the foreshore.

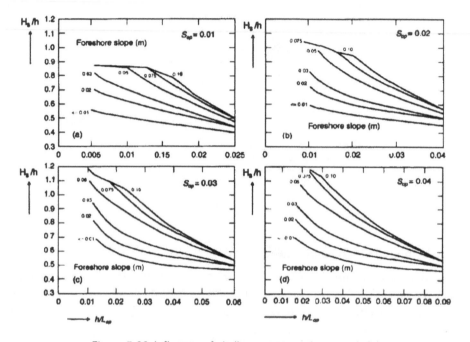

Figure 5-26 Influence of shallow water on the wave height

It is remarkable that relatively high values of H_s/h are found close to the shore (even > 1), although the theoretical limit for regular waves was 0.78. This indicates that the results must be used with great care. These high values of H_s/h occur for small values of h/L_{op} and steep foreshores. Moreover, it is stressed that the calibration of these results in prototype conditions has not yet been carried out and should have a high priority.

In an attempt to make these results applicable to a practical breakwater design problem along a coast comparable to the Dutch North Sea coast, one may use the following boundary conditions:

Peak period	$T_p = 10$ s
Deep water wave length	$L_{op} = 156$ m
Typical water depth	$h = 6$ to 10 m
Typical foreshore slope	$m = 0.01$ to 0.02
Typical deep water wave steepness	$s_{op} = 0.04$

This leads to values of H_s/h between 0.5 and 0.6

This agrees fairly well with the rule-of-thumb approach of many Dutch coastal engineers, who state that the maximum value of H_s is approximately $0.55h$. This means that the limited water depth very effectively protects structures in shallow water. It is emphasized, however, that steeper slopes of the foreshore will certainly reduce this protection. It is also emphasized that the construction of a breakwater may lead to erosion, and consequently greater water depths and more severe wave attack. Finally, attention is drawn to the fact that in many locations the occurrence of a storm or hurricane causes not only high waves, but also a storm set-up. Since these two phenomena are not independent stochastic events their joint occurrence must be taken into account. Owing to the higher water levels resulting from the storm surge, water depths will also be greater and higher waves can therefore penetrate to the location of the structure.

A short remark must also be made on changes that occur to the shape of the wave spectrum after breaking. Where the spectrum in deep water will show one or two distinct peaks, these peaks disappear largely in shallow water. This may have serious consequences for the behaviour of coastal structures. The reader is advised to consult the latest literature in this respect.

5.4 Geotechnics

5.4.1 Geotechnical data

The construction of dams and breakwaters takes place along shores and river mouths with a great variety of subsoil conditions. Damming river branches and tidal basins generally takes place in alluvial deposits of sand, gravel, silt and clay, stratified in various compositions and of variable characteristics. Bedrock underlies these layers at various depths. Sometimes local deposits as boulder clay or cobbles may be encountered. In several deltaic regions layers of peat may also be present. These soil layers form the foundation bed for the structure and need to withstand groundwater flow, waves loads and differential water pressures. During the construction of dams and breakwaters, they are the sub-base for the vehicles and equipment driving across the site, while when submerged they will be exposed to eroding currents and waves. Therefore, a thorough knowledge of the soil types present, including their characteristics and the stratification is a prerequisite for the design and execution of breakwater construction and closure operation.

Geotechnical data can be obtained from sample-borings and penetrometer-tests executed in the field and analyzed in laboratories. The number of borings is usually limited in relation to the area considered and to obtain a reasonable picture of the sub-bottom, an overall examination, together with historical and geological information, is required. Particularly in deltaic areas and tidal inlets, former gullies filled up with different types of soil may give sudden changes in the sub-bottom profiles of soil layers. Very localized deviations from the general picture as well as obstacles (fossil trees, large boulders) seldom appear in the results of a field survey. The most important parameters of the sub-bottom are stratification, soil type and phreatic levels. Laboratory tests on samples of every layer of clayey soil will give the values for cohesion, angle of internal friction, Atterberg limits and water content. Granular soil types are characterized by the grading of the grain sizes, the sharpness (roundness) of the grains and the pore volume (relative density).

Cohesion and friction-angle are the most important parameters for soft soils. The field tests are done by using Standard Penetration Tests (SPT), giving the number of blows needed to hammer a pin down into the ground or by using the Dutch Cone, that gives the force required to push a cone down into the soil. These values are sometimes translated via relation-tables into values for cohesion and friction-angle. Far more accurate is triaxial testing of soil samples. Triaxial tests can be drained or undrained and consolidated or unconsolidated, which leads to different values for the same soil. Which of these is the most appropriate depends on the purpose.

A triaxial test, on undrained and unconsolidated (UU), corresponds to a field situation in which the soil is surcharged without possibility of compression. Pore water cannot escape. This is true for saturated soil of very low permeability that is quickly charged. It this case, the resulting friction-angle is zero. There is cohesion only. Deformation takes place without volume change. Line UU in the τ-σ diagram (Figure 5-27).

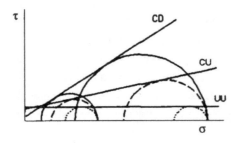

Figure 5-27 Stress relations determined by soil testing

If the test is done on a consolidated soil the sample is first charged by lateral support (water stress) and open drainage. Compression leads to increase of strength. Then the drain is closed for undrained testing (CU) and surcharge is applied to determine the moment of failure. Next, the drain is re-opened, the water stress is increased to a higher level and further consolidation is waited for. (For soils with low permeability this procedure takes a very long time and the method is adapted, but the result is the same.) Then the drain is closed again and testing by another surcharge gives a new failure level. Without the intermediate drainage the same would happen as in the UU situation: friction-angle zero, but on a higher cohesion level (the sample is stronger due to the initial consolidation). The meaning of these test-results (CU) in practice is that the same soil, when taken from a deeper level, is adapted to the higher water-stress existing at that level and is initially stronger. However, if charged above that strength, it also has no friction angle if water cannot escape. The line CU is a succession of points, each true only for a sample depth corresponding to the actual water pressure in the bottom. Usually, data for cohesion and friction-angle given in soil-test reports are these consolidated, undrained tested values: c and φ.

If all testing is done with open drainage (CD), water can escape and thus over-stresses in the pore-water will disappear if loading is slower than the permeability allows. The stresses in the grains remain. This happens mainly in sandy soil and cohesion will be close to zero. The values for cohesion and friction-angle give the ultimate strength of the soil under long lasting surcharge. Usually, these data are presented in test reports by an index: c' and φ'. In the construction phases of dams and breakwaters, the surcharge is applied quickly and soft soil layers have little or no time to adapt themselves because of the low permeability. Therefore, usually the values c and φ, for undrained and consolidated tests have to be used for dam construction in deltaic areas.

For granular material (sand), the relative density and the permeability are the most important parameters. The grain structure itself is strong enough to withstand considerable surcharges. Problems may arise if some grains want to rearrange to form slightly more dense packing and the pore water cannot escape. The latter is the case if the permeability is low. This applies to sand with a high content of fines. Generally, the 10% finest part of the sieve curve determines the permeability.

Apart from the subsoil, the materials used for the dam construction form part of the total soil mass to be considered. Sand-fill, rock masses and clay cores are a surcharge on the one hand and on the other hand are subject to instability. Stability criteria have to be determined not only for the final design but also for the various construction stages. As described above, surcharging compressible soils with low permeability will result in an increase in the water-stress which fades away slowly and thus the rate of loading may be a determining factor. Therefore, the planning of the construction is important and to ensure stability during construction, considerable waiting periods may be required between subsequent construction phases. If the waiting times are too long, it may become necessary to apply costly remedial techniques to avoid mishaps.

Several mathematical models are available for calculating the soil-mechanic behaviour. Of course, the accuracy of the results and predictions depends largely on the schematization of the soil layers in the model, the uniformity of the soil parameters, the correct input of the execution phases and timing, and the validity of the basic assumptions implemented in the model used (such as sliding planes being circular). A good soil investigation should be quite extensive. Obviously permeability and pore volume are very important parameters, but these are the most difficult to determine because usually the sampling instrument will have disturbed the sample. A soil investigation programme, which includes both field and laboratory work is a rather expensive part of the design process, which is office work. Savings on the number of borings and the extent of the programme, however, may lead to very costly mishaps in the construction phase of the project.

5.4.2 Geotechnical stability

The most important property of the soil is its bearing capacity, which frequently is a determining condition for the design and a limiting factor for the operations. The soil mechanics problems, related to bearing capacity, are:
• Sliding
• Squeeze
• Liquefaction

Sliding:

In dam and breakwater constructions, sliding is the instability of an earthen, or quarry stone embankment along the side slope or at the construction front. The driving force of the slip is the surcharge on the subsoil by the own-weight of the embankment. The steepness of the slope is an important parameter. In any dam construction two aspects have to be considered in particular. One is the water level variation at the dam site, the other is the erosion of the soil in front of the dam.

Water level variations influence the amount of surcharge. A body made of quarry run will have a specific weight above water of about 17 kN/m^3, however, when submerged its weight will be only 10 kN/m^3. As a part of the dam may be submerged during high water, the weight gradually increases during falling water levels owing to loss of buoyancy. Such a dam, built out by tipping with dump trucks gets very steep slopes naturally, (gradients up to 1:3). Although extended without problems during the high water period, the dam may suddenly fail during low water.

A dam, made of sand and constructed by hydraulic filling, results in far more gentle slopes. Below the waterline, this will be in the order of 1 in 5 to 1 in 15 (depending on the grain size of the sand), and even flatter above water. The fill provides fully saturated material, which has a weight of about 20 kN/m^3, and 10 kN/m^3 for the submerged part. In spite of the greater unit weight, the flatter slopes tend to provide a bit of extra security against sliding.

The possible erosive action alongside and in front of the embankment resulting from the high flow velocities of the currents around the dam head are more difficult to predict. Erosion pits may develop rather quickly and take away part of the soil that is assumed to provide the counterweight that is necessary to keep the dam stable. These erosion pits seldom appear on the design drawings. Moreover, it is difficult to predict their shape, size and depth. However, a designer has to include these conditions in his considerations that govern a safe operational procedure. Preventive measures, such as placing a protective layer of quarry run ahead of the progressing works, will avoid the erosion. For a permanent structure like a breakwater, more attention must be paid to the design of the protective layer, probably by including a granular filter or geotextile.

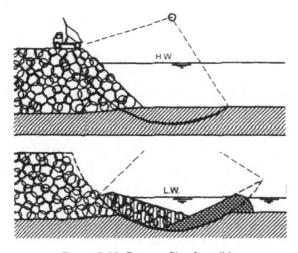

Figure 5-28 Dam profile after slide

The method of Bishop is generally used to calculate the risk of sliding. In this mathematical model the subsoil and the surcharge layers are schematized in vertical slices. The forces keeping the individual slices in place are detailed by attaching values for cohesion and angle of internal friction to the different types of subsoil and the construction layers. The instability is assumed to take place in a circular plane intersecting the horizontal layers. Each layer contributes to the resistance by a different value that depends on the length of the circle segment in the layer and the friction force it can yield. When sliding occurs, all the vertical slices within the circle turn together until a new equilibrium is reached. Basic assumptions for the calculation are the correctness of the soil mechanics values of the soil layers and the dam, the uniformity of the composition of every layer and the circular shape of the plane that first reaches over-stressing. The method determines ratio between the shear resistance along the full plane and the driving force generated by the total weight. When the old method is followed, a safety coefficient of 1.0 to 1.1 is used for temporary constructions and 1.2 to 1.3 for permanent constructions. Nowadays it is common to use partial factors for materials and forces. The ground parameters are corrected by NEN 6740 and φ has a multiplier of 1.15 for permanent work and 1.0 for temporary work. Some risk remains during the construction phase.

In practice, sliding should be prevented or used intentionally. Preventive measures must cover all circumstances during the construction of the initial closure dam profile. During a later stage of construction, the dam profile will be enlarged by further heightening and widening and finishing of the slopes. In many cases, with the knowledge that the soil will gain strength after some time, the construction of this final profile may be scheduled in such a way that stability is not critical. The calculations for the initial and final profiles require different cohesion values.

Sliding is used intentionally if a soft layer of limited thickness is situated on top of a good bearing subsoil. There are two design choices. Either to remove the soft layer by dredging (and backfill the created trench with good sand) or to construct on top of the soft soil and press the embankment down into it by using its own weight, which involves sliding. The first choice is the safest and should always be used for critical parts and operations of the closure dam construction. Nevertheless, the second method is used sometimes, although there is a risk that the soft soil will be only partly pressed away and that in later stages deformation will continue. For such a procedure the safety factor should be much lower than 1, as the embankment has to fail during construction in favourable conditions. Controlled failure is far more difficult to achieve than ascertained stability.

The subsoil encountered during damming activities (period 1970 to 1976) in the rear of the tidal basin called the Eastern Scheldt, consisted of large areas of peat, criss-crossed by former stream gullies of the River Scheldt that had been filled by clay and sand. Part of the Markiezaatsdam was hydraulically filled with sand up to a specified profile and several slips leading to the formation of heaps of peat along the side of the dam occurred. Another part was built of stone tipped from dump-trucks. This dam dropped almost instantaneously down into the soft material. It was then decided to replace the subsoil from the site of the final gap. Although the closure had been designed as a vertical closure, consolidation time between layers would have been too short. In a later design for the nearby Oesterdam in comparable conditions, hydraulic filling was again prescribed, but this time specifying that the profile should be given gentle slopes and a limited height in first stage. The filling required some extra equipment and the rehandling of sand, but no failures occurred. This proved that the consolidation method was a viable option, but that to ensure success it needed thorough planning and the definition of intermediate construction profiles.

Squeeze:

It is clear that in a situation in which very weak subsoil is surcharged by very stable dam material, instability may be restricted to the subsoil. Instead of sliding along a plane, deformation of the soil layer occurs that is comparable to that in ice-cream squeezing out between two wafers. Over-stressing will than start at one location and locally lead to deformation without changing volume. The stress transfers to the surrounding area in various directions so the deformation expands. The dam body on top sinks into the subsoil and the same volume of weakened soil escapes on the edge of the dam. This type of instability is called squeeze. As this failure starts by over-stressing in one point and then progressively expands via deformation, the mathematical approach is different from that used for the slip along a plane in basically undeformed bodies. For the calculation of instability by squeeze a different mathematical model, (e.g. Plaxis) has to be used.

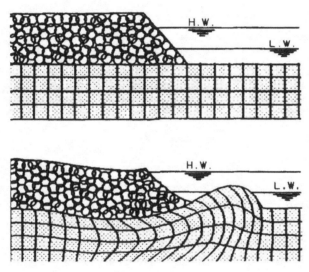

Figure 5-29 Squeeze

The figures show that slip and squeeze are completely different. This is more difficult to establish by observation in the field. The profile after instability is seldom as ideal as the one sketched and in both cases a heap of subsoil material rises up in front (or at the side) of the dam or breakwater. Taking samples at the toe of the slope will show the facts. In practice, the difference between slip and squeeze has few implications. A choice has to be made in the design between soil improvement and deliberately induced subsidence with associated waiting times.

Liquefaction:

Liquefaction may occur in loosely packed sands. This is aggravated by low permeability of the soil. The strength of a sand mass is determined by the transfer of the forces from grain to grain. In saturated sand, the voids between the grains are filled with water.

The water pressure is hydrostatic and does not bear or support the grains. If, for some reason, a shear force leads to movement of the grains, they will try to obtain a denser packing, which is possible only when some of the water is driven out of the pores. At that moment, the water is taking part of the load from the grains. If the permeability of the sand is low, for instance, because of a high content of fines, the water cannot escape and the grains lose their stabilizing contact forces. A fluid mass of water and sand grains, with a density of 1800 kN/m^3 results and it behaves like a heavy liquid (quicksand).

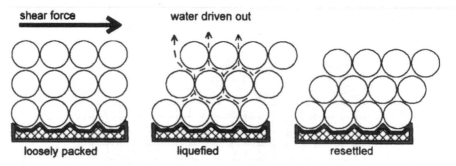

Figure 5-30 Liquefied sand

From the above, it is clear that liquefaction may occur in loosely packed sand of poor permeability only after an initial event has triggered a disturbance of the grain structure. In nature, such deposits of sand occur in areas where the conditions for the settling of the sand particles was ideal, there having been no turbulence, no waves and not too much current flow. Artificially made bodies of sand, placed under water, for instance by hydraulic filling, always have a loosely packed grain structure. Then, the quantity of fines determines sensitivity to liquefaction. The initiating event may vary in nature. If the grains are in a stress situation, for instance because of a developing scour hole, a little vibration or shock may start the flow. This applies also to wave induced varying loads transferred into the soil by a rigid structure like a pier or a caisson. Unlike slip-circle and squeeze, liquefaction will result in a flow slide with very gentle slopes at the toe.

In practice, many stability failures have occurred in closure operations and during breakwater construction. Usually they resulted from variations in soil characteristics, changes in the works program, limited soil information and unforeseen conditions. Apparently, soil conditions on the sites where these works are executed, give little margin for error. Of course, it is important to know what should or should not be done in such cases. After a failure, the soil is disturbed, the soil structure is distorted, over-stressed water is still present and the new situation will have a very vulnerable equilibrium state. Shifted material will be in the wrong place, for instance at the toe of the slope, and the top layer of the dam will be too low. However, every corrective action to take away the wrong toe material or extend the dam by recharging the top will lead to continuation of the failure. The only possible measure to improve the situation is to let the soil mass consolidate and let the stresses in the water diminish. This takes time and can be only slightly speeded up. If it can possibly be installed, artificial drainage by vertical drains may be helpful. Otherwise, it is better to bypass the area and thereby provide time for the water pressure to reduce to normal. Next, preferably working from the toe side uphill and very gently layer by layer, the profile of the dam can be restored. It must again emphasized that it is of the utmost importance to avoid failure of the foundation by adequate design and by adequate analysis of all construction phases.

5.4.3 Settlement

Closure dams are constructed from various materials. Sand is used for dam bodies that are built out into the flow when the current velocities are modest. Furthermore sand is used to enlarge (widen and/or heighten) the profile of the dam after an initial profile has been made by using other materials. Because of their resistance to erosion, lumps of clay and boulder clay are suitable for closures with medium current flow velocities. Otherwise, quarry stone or concrete blocks are used for both closure dams and breakwaters. Apart from the clay, these materials will show little settlement of the created profile. However, the clay may exhibit considerable settlement and, when used for the initial profile of the closure, will generally be strengthened afterwards by placing a sand profile alongside it. The main problem with settlement therefore is the subsoil.

The magnitude of the settlement can be estimated by using the formula of Terzaghi:

$$\varepsilon = \frac{1}{C} \ln\left(\frac{p + \Delta p}{p}\right) \qquad (5.39)$$

in which p is the grain stress, Δp the surcharge and C is a constant depending on the soil properties. For peat C may be as low as 3 to 10, while for clay 10 to 20 is used. Compressible soils consisting of loam, clay or peat may show considerable settlement after some time. The development over time can be calculated on the basis of the permeability of the soil. Water has to be driven out of the soil and escape to the nearest draining soil layer. As in many cases the permeability is very low and the distance for the water to travel may be long (thick layer), the settlement may take several years to complete. The problem of settlement therefore is mainly a matter of the determination of the final design profile rather than a matter for the various construction phases. If the long period of settlement after completion of the works is unacceptable for operational reasons, one may decide to enhance consolidation by applying artificial drains combined with a temporary surcharge.

In practice, the expected settlement is often added to the design height of the structure. Consequently, the surcharge onto the subsoil increases initially with the extra height of dam material and may put the profile at risk of failure. The over-dimension is not necessary in the first instance and in a critical situation, may be postponed until the subsoil has gained strength by some degree of consolidation. Then, after some time, the top layer of the dam has to be completely reconstructed. Alternatively, the settlement-sensitive soil layer can be dredged away and be replaced by better soil (sand).

An example of such a case was the closure of two small tidal creeks near Odafado in Nigeria as part of a canalization project. The subsoil consisted of a thick layer of rather soft silty clay and considerable settlement was expected. The extra height added to the dam to compensate for the settlement increased the load and thus the settlement. In the end the design engineer decided to select the alternative and have the soft soil removed and replaced by sand. Unfortunately, this increased the total construction cost disproportionately. At the suggestion of the contractor, and after consultation with the local authorities, it was decided to execute the closure process without soil exchange, by stepwise building of the final dam. For the initial closure profile the lowest possible dam height was selected (for instance by making a very gentle outer slope to reduce on wave run-up). In addition, the closure was based on the old Dutch methods that used trees and branches ballasted with soil, which was the least heavy dam material that could be used. Of course, regular maintenance was required afterwards to cope with the settlement.

5.4.4 Groundwater

Damming watercourses will influence the groundwater pressures in the vicinity of the dam and in the dam itself. These changes may be short-term variations, e.g. in relation to the daily tide levels, or long term adaptations, such as lining up with the rise in mean level of a basin. Differential pressures lead to flow of the groundwater. As long as the Reynolds-number is smaller than about 5, the flow velocity according to Darcy's law is:

$$u = k\,i \tag{5.40}$$

in which k denotes the permeability of the soil and i the pressure head loss per unit length. The flow velocity u has a complex relation to the actual velocity of the water in the pores of the soil. The quantity of water passing through the soil is given by the u multiplied by the full cross sectional area A of the soil, as if there were no soil particles.

$$Q = k \, d \, b \, \frac{\Delta H}{L} = u \, d \, b = u \, A \qquad\qquad\qquad\qquad (5.41)$$

In granular soil like sand, for instance, the pore volume will be in the order of 40% only. This means that if the soil is considered as a solid mass provided with 40% volume of straight narrow tubes, the real flow velocity will be 2.5 times u. The actual flow pattern is very complex because the water has to force itself through the grain structure by acceleration and slackening. The actual flow velocities act on the grains and may transport loose fines that do not form part of the structure. Differential water levels consequently may lead to:

- instability of the total soil mass if the force exerted by the pressure head is more than the weight of the soil layer (macro stability).
- instability of individual grains at the surface of the soil layer due to the out-flowing water (slope stability under seepage conditions)
- instability of grains within the soil layer (micro stability), leading to transport of the finest soil particles.

6 DATA COLLECTION

Chapter 6 summarizes data that must be collected before a large hydraulic project can be designed or constructed. It gives readily available sources of data and it discusses some methods that can be used for the collection of the relevant data. This chapter is meant to illustrate to students the preparatory work that has to be done in the early stages of project preparation.

6.1 General

In Chapter 5, we recalled some of the theory that we need for designing structures like breakwaters and closure dams. In practice, however, it is not only important to master the theory, it is also equally important to assess the physical presence of certain phenomena that are known from theory. For each subject, this chapter pays attention to the availability and, if necessary, the collection of relevant data on the prevailing hydraulic, geotechnical and other conditions.

For design purposes, we are sometimes interested in extreme events, specifically, when we try to establish the loading conditions for the Ultimate Limit State. In many cases, it is impossible to use direct observations since our records are too short to make a sensible assessment of extreme events. In such cases, we have to rely upon an indirect approach, in which we use data that have been recorded over a sufficiently long period. In this respect, specific reference should be made to meteorological data (wind direction, wind speed, barometric pressure, temperature, rainfall, visibility, humidity) that have been collected (even in remote areas) over a much longer time-span than those of wave heights, tidal currents and river discharges. Meteorological stations, airports, hospitals and even missionary outposts may come up with a surprising wealth of data. Calculation and calibration can then transform such data into the required data.

6.2   Meteorological data

Although meteorological phenomena generally do not play a direct role in the design of hydraulic structures, they are indirectly important. In the previous sections, we have discussed the importance of barometric pressure and wind data as generators of surges and waves. In a similar way, precipitation plays a major role in the generation of river discharges. Wind plays a direct role, when one considers forces on ships and structures or the effects of spray from breaking waves.

Other factors may be important as well, though their role is less obvious. In this context we must mention visibility, which is important because any marine operation is seriously hampered by fog. Temperature and humidity are important to equipment (cooling, corrosion), but also for the hardening of concrete, and the formation of cracks during the hardening process. Freezing conditions and the presence of ice must be taken into account where applicable.

Data are generally available from national meteorological offices; measuring instruments are easily available in the market.

6.3   Hydrographic data

6.3.1   Bathymetry

Before starting a design job, it is essential to have proper maps of the seabed or riverbed. In most cases, hydrographic charts are useful but their scale is such that the information is not detailed enough for engineering purposes. The same applies to standard river navigation charts. Another problem is that the datum of those charts is often related to tidal levels (MLWS or similar) and may not be the same for the entire chart. This can pose serious problems when quantities of dredged or fill materials have to be determined on the basis of such charts. The advantage, however, is that for most regions in the world such reliable charts have been available for at least 200 years. The hydrographic departments in many countries preserve these old charts. They can provide very valuable information about long-term morphological developments in the region.

In practice, this means that most projects require that specific maps should be made. Nowadays a hydrographic survey poses few problems. One needs a survey launch equipped with an echosounder, (D) GPS positioning, a reliable radio tide gauge and a data logger. When the sea is rough and the survey launch makes considerable movements due to waves, one may add a heave compensator. With the aid of modern software, without much human effort a plotter input can draw the maps. Because the data are available in a digital form, it is also possible to use them for many kinds of arithmetical exercises, such as calculating volumes to be dredged or filled. They can also be used to assess erosion or accretion between subsequent surveys.

It is wise to pay attention to the type of echosounder used. When a high frequency (210 kHz) is used for the measuring beam, the depth indicated is the top of a soft layer. When lower frequencies (30 kHz) are used, the beam penetrates into the soft mud layers. By using a dual frequency instrument, one can obtain an impression of the thickness of such layers of soft mud.

6.3.2 Tides

(See also Section 5.2.2.)

Vertical tides

The tidal constants for most important harbours in the world are known. They are published annually either by a national hydrographic office or by the British Admiralty. Via the Internet many tidal predictions can be obtained. An international operating site is http://tbone.biol.sc.edu/tide/. Also via the French Hydrographic Service (http://www.shom.fr/) on line data are available. For the Netherlands one is referred to http://www.getij.nl/.

For minor ports, one has to rely on national or local authorities, and the reliability of data provided may not always be as good as required. Setting up a local observation point and performing hourly observations of the water level during a period of one month can yield a provisional insight. The application of harmonic analysis techniques easily leads to a reasonable estimate of the most important tidal constants. Only when one is interested in the long periodic components is a longer observation period required.

Horizontal tides

Tidal currents are sometimes indicated on hydrographic maps, the standard of accuracy of which is usually insufficient for design or planning purposes. Some hydrographic departments issue flow atlases with more comprehensive information (DIENST DER HYDROGRAFIE [1994], BUNDESAMT FUR SEESCHIFFFART UND HYDROGRAPHIE [1990], U.S. NAVAL OCEANOGRAPHIC OFFICE [1970]). Usually it is necessary to make dedicated flow measurements, which are time consuming since they have to be continued for at least 13 hours. It is therefore advisable to analyse flow phenomena by using a mathematical model and to use field measurements mainly to calibrate the model.

The above mentioned website http://tbone.biol.sc.edu/tide/ provides also some data on horizontal tides.

6.3.3 Storm surges

Tidal water level variations can be predicted accurately on the basis of astronomical facts. In addition to the tidal variations, there may be meteorological effects that influence the water levels. Since meteorological effects cannot be predicted long in advance, they must be taken into account on a statistical basis. If no direct observations are available, one may use records of wind velocities, barometric pressures and hurricane or cyclone paths to estimate the probability of extreme water levels.

6.3.4 Waves

There are few places in the world where long series of wave observations are available. This is simply because reliable instruments for wave recording did not exist until recently (say 1970). The oldest observations of waves were carried out on board of ships that had a voluntary agreement with a meteorological office to carry out certain observations. Although the observations of waves were visual observations, their accuracy is acceptable since the officers were well trained and could compare the observed state of the sea with standard pictures provided by the met-office. These observations were collected and sorted according to locations spread over the oceans. In this way, the first wave atlases were edited (DORRESTEIN [1967], HOGBEN AND LUMB [1967], BOUWS [1978]). More recently, a large collection of similar data has been assembled and edited (YOUNG AND HOLLAND [1978], HOGBEN, DACUNHA AND OLIVER [1986]). A disadvantage of these data sets is that the oceans have been divided into relatively large areas, so that detailed information close to the shore is still not readily available. More detailed information can be obtained via commercial wave data banks, like Argoss (http://www.waveclimate.com).

Direct measurement of wave heights in the preparation phase of a project will never provide the required long-term data. However, it is still useful to have such direct observations, if only to calibrate the calculation methods used to transform indirect observations into local wave data.

Modern methods for wave measurement are

- electric (resistance or capacitor type) wave gauges, mounted on a platform or a pile
- acceleration type gauges, mounted in a floating buoy
- pressure gauges, mounted on the seabed
- inverted echosounder, measuring the distance from sea bed to water surface
- remote sensing techniques (from satellite)

One problem is that one wants to measure wave heights in relatively deep water, so that shoaling or breaking does not yet affect the measured heights. This makes all pile or platform mounted gauges relatively expensive, unless use can be made of an existing facility. The use of pressure gauges is not recommended because the actual wave pressure at bottom level is a function not only of the wave height, but also of the ratio between wave length and water depth. One of the most popular instruments is the Dutch Waverider Buoy, fabricated by Datawell, Haarlem. This device measures the vertical acceleration of the water surface, and transforms this by double integration into a vertical motion. This makes the observation of very long swell $(T > 20$ s$)$ difficult and not fully reliable. Remote sensing techniques are still in an experimental stage, but their prospects are certainly good, because observation of remote wave fields can result in timely warning of bad weather.

An example of the transformation of measured wave data or of wave data from a general database into exceedance curves that can be used to determine design wave heights is given in Appendix 10.

6.4   Geotechnical data

The geotechnical data required for a breakwater or closure dam certainly includes all data required to assess the bearing capacity of the subsoil, both during construction and later. Stability of the works must be ascertained during all phases. Furthermore, one wants to predict settlement as a function of time in order to ensure that the required crest level is available permanently. In many cases, the works will be accompanied by substantial dredging, so that soil properties for this purpose must also be known. If erosion or scour is expected to occur, it is necessary to establish the resistance of the existing seabed to this threat. Table 6-1, after the CUR/RWS Manual (CUR/RWS [1995]) gives a good impression of the geotechnical failure mechanisms and their relation with basic geotechnical data.

Macro instability			Macro failure	Macro instability	Geotechnical information	
Slip failure	Lique- faction	Dynamic failure	Settle- ments	Filter erosion	Name	Symbol
A	A	A	A	A	Soil profile	-
A	A	A	A	A	Classification/grain size	D
A	A	A	B	A	Piezometric pressure	p
B	B	B	A	A	Permeability	k
A	B	B	A	B	Dry/wet density	ρ_d, ρ_{sat}, ρ_{sub}, $\gamma = \rho g$
-	A	B	-	-	Relative density, porosity	n, n_{cr}
A	B	B	-	C	Drained shear strength	c, ϕ
A	-	-	-	C	Undrained shear strength	s_u
B	-	-	A	-	Compressibility	C_c, C_s
A	-	-	A	-	Consolidation coefficient	c_v
B	B	A	A	-	Moduli of elasticity	G, E
B	A	A	A	-	In situ stress	σ
-	A	B	A	-	Stress history	OCR
B	A	A	B	-	Stress/strain cirve	G, E

Table 6-1 Required soil data for the evaluation of the geotechnical limit states (after CUR)

This means that for a major project a general geological analysis must always be carried out to determine the geophysical and hydro-geological conditions. The most important aspects that require attention are:
- Geological stratification and history
- Groundwater regime
- Risk of seismic activities

Basic data can be obtained from the national geological services and, in more general terms, from scientific libraries and universities. Most of the available information will refer to land and little to estuaries and sea. Such basic geological data will provide general insight into what can be expected in the area of interest. Usually this is insufficient for engineering purposes, so soil investigations are always necessary. These investigations may include following methods:
- Penetration tests (CPT or SPT) to establish in-situ soil properties
- Borings to take samples from various depths for further analysis in the laboratory
- Geophysical observations

These specific investigations are expensive and difficult because they must be done at sea, under the direct influence of tides, waves and currents. This makes any penetration test or boring a time consuming and risky affair. Therefore there is a tendency to limit the number of such local tests. This imposes the risk that discontinuities that occur in between the measuring locations will not be recognized. This is the reason why such local observations should be combined with a geophysical survey. The geophysical survey uses electro-resistivity and electro-magnetic and seismic techniques to obtain a continuous image of the soil conditions in the tracks sailed by a survey vessel. The disadvantage is that there is usually no direct link between measured data and the geotechnical soil properties. Combination of geophysical survey and point measurements eliminates the disadvantages of both. The geophysical data ensure that no discontinuities are overlooked, while the borehole and penetration tests provide the link with the actual engineering properties of the soil.

Table 6-2 gives a complete view of the available in-situ test methods and their applicability.

	Geophysical methods (Section C.1)				Penetration methods (Section C.1.2)					Borings (Section C.1.3)			
	Seismic	Electr. resist.	Electro-mag-netic	nuclear	Cone penetr. test (CPT)	Piezo conc. test (CPTU)	Stand. penetr. test (SPT)	Field vane test (VST)	Press. meter test (PMT)	Dilato meter test (DMT)	Dist. samples	Undist. samples + Lab. tests	Moni-ring wells
Soil profile	C	C	C	-	A	A	A	B	B	A	A	A	-
Classification	-	-	-	-	B	B	B	B	B	B	A	A	-
Piezometric pressure	-	-	-	-	-	A	-	-	B	-	-	-	A
Permeability	-	-	-	-	-	B	-	-	B	-	C	A	C
Dry/wet density	-	-	-	A	C	C	C	-	-	-	C'	A	-
Relative/density	-	-	-	-	B	B	B	-	C	C	-	A	-
Friction angle	-	-	-	-	B	B	B	C	C	C	-	A	-
Undr. shear strength	-	-	-	-	B	B	C	A	B	B	-	A	-
Compressibility	-	-	-	-	C	C	-	-	-	C	C	A	-
Rate of consolidation	-	-	-	-	-	A	-	-	A	-	-	A	C
Moduli of elasticity	A	-	-	-	B	B	B	B	B	B	-	A	-
In situ stress	-	-	-	-	C	C	-	C	B	B	-	A	-
Stress history OCR	-	-	-	-	C	C	C	B	B	B	-	A	-
Stress/strain curve	-	-	-	-	-	C	-	B	B	C	-	A	-
Ground conditions													
Hard rock	A	-	A	A	-	-	-	-	A	-	A	A	C
Soft rock-till, etc.	A	-	A	A	C	C	C	-	A	C	A	A	A
Gravel	A	B	A	A	C	C	B	-	B	-	A	C	A
Sand	A	A	A	A	A	A	A	-	B	A	A	C	A
Silt	A	A	A	A	A	A	B	B	B	A	A	A	A
Clay	A	A	A	A	A	A	C	A	A	A	A	A	A
Peal-organics	C	A	A	A	A	A	C	B	B	A	A	A	A

A: High applicability B: Moderate applicability C: Limited applicability

Table 6-2 In situ test methods and their perceived applicability (after CUR)

6.5 Construction materials, equipment, labour

6.5.1 Construction materials

The most important construction materials for closure dams and breakwaters are quarry stone and concrete.

Quarry stone

Quarry stone is natural rock, obtained from quarries. There are three or four basic types of quarry:
- Commercial quarries for the production of ornamental stone (e.g. marble tiles, etc.)
- Commercial quarries for the production of fine aggregates for concrete, road construction, etc.
- Commercial quarries for the production of large sized rock for hydraulic engineering
- Dedicated quarries for the same purpose

It is evident that the quarries for ornamental stone are not relevant to our purpose. The quarries for the production of aggregates are, in general, not equipped to supply the size of stone required for large hydraulic engineering projects, although the fine material that they produce can be used as filter material.

In some parts of the world, where a regular demand for larger size stone exists, some quarries have specialized in this field. They offer stone in standard weight categories, usually according to the national standards of their customers. Such quarries exist for instance in Belgium, Germany, Norway, Sweden and Scotland.

The relevant properties of the stone are widely known and listed in catalogues.

The situation is different if a large project is to be executed in an area where no such quarries exist. In that case, a rock formation has to be found that can be used to open a quarry that is specifically dedicated to the project. The following data should always be obtained:

- Specific weight and density of the material
- Durability in air and in water (fresh and saline)
- Resistance to abrasion
- Strength (tensile and compressive)
- Maximum size that can be obtained and distribution (yield) curve

In general, these data are so important that it is worthwhile to employ geological specialists to find a suitable location for a quarry. It is even recommended that one or more test blasts should be carried out before a final decision is taken to open a quarry. Apart from the technical data on the rock, it is necessary to be sure that the quarrying operation is acceptable from social, environmental and legal points of view. Since quarry stone and quarries are quite essential for any major hydraulic engineering project, more details are provided in Appendix 2.

Concrete

When a large project is to be executed in a remote area, it is also essential to be sure of the quality and availability of other construction materials. For closure dams and breakwaters, it is almost impossible to avoid the use of concrete. It is therefore recommended that data on the availability and quality of cement, aggregates, water and reinforcing steel should be collected. It is also essential to study the climatological conditions to see if special measures are required for curing the fresh concrete.

In this respect, it is also important to be aware of any local codes and standards, and whether the obligatory sections thereof will interfere with procedures planned by the designer or contractor.

6.5.2 Equipment

There is a high degree of interdependence between the design of a breakwater or closure dam and the construction method. In the same way, the equipment to be used depends largely on the construction method and vice versa.

Similarly, questions relating to the maintenance and repair of the final structure must be discussed during the design stage. Do we rely upon regular inspection and maintenance, or do we opt for a more or less maintenance-free structure? For construction stages (and a closure dam is usually a construction stage), one need not worry too much about maintenance and repair. Even if a minor part of the works should be lost, the contractor is still there with his equipment to take care of repairs. Nevertheless it makes sense to analyze the construction risk, since this may constitute a considerable part of the construction cost.

All the above leads to the main question of whether locally available equipment will be used or that the required heavy equipment is obtained from elsewhere. If local equipment is preferred, it is necessary to obtain a detailed assessment of the quality, capacity and cost of such equipment. If it is decided to mobilize the equipment, the questions that arise are how to get the equipment to the location, and whether there are any restrictions on temporary import. Local conditions like temperature (cooling of engines), dust (capacity of air filters), quality of fuel and lubricants, availability of spare parts also play a role.

6.5.3 Labour

When planning a large project, it is also essential to know whether there is a skilled local labour force and whether the employment of skilled and partly skilled expatriate labour is permitted. In many cases, it is necessary to provide special facilities for the accommodation of personnel. Such facilities must be available from the start of the actual construction. Poor working and living conditions will have a strong negative influence on the quality of the work.

7 STABILITY OF RANDOMLY PLACED ROCK MOUNDS

In Chapter 7, attention will be paid to the stability of individual stones on a sloping surface under wave attack. Due to the complex water movement of waves breaking on a slope, it is not yet possible to derive a satisfactory theoretical expression for the forces on and the stability of such stones. This means that this chapter contains a multitude of empirical formulae, all based on the results of small-scale experiments. Although it would not be wise to learn all these formulae by heart, it is necessary to understand them and their significance to the designer of a breakwater.

7.1 Introduction

Although the stability of individual stones on a slope under wave attack is certainly not the only criterion for the proper functioning of a rubble mound breakwater, it deserves great attention. This is because many breakwaters have failed due to a defective design in this respect.

7.2 Historic review

7.2.1 General

As indicated in Chapter 2, for many years breakwater design was a question of trial and error. It was shortly before World War II that, in an attempt to understand the influence of rock density, Iribarren developed a theoretical model for the stability of stone on a slope under wave attack. Iribarren continued his efforts throughout the years until his final publication on the subject at the PIANC Conference of 1965 in Stockholm.

In the meantime, in the USA, the US Army Corps of Engineers had developed a keen interest in the stability of breakwaters, and long series of experiments were carried out by Hudson at the Waterways Experiment Station in Vicksburg.

Where Iribarren focussed on a theoretical approach, assisted by some experiments, Hudson concentrated on collecting a large data set from hydraulic model experiments to derive conclusions from an analysis of those data. In both cases, experiments were carried out using the then standard techniques, i.e. by subjecting the models to regular, monochromatic waves.

The experiments comprised the construction of an infinitely high slope, covered with stones of a particular weight, shape and density. The slope was then exposed to a wave train with waves of a particular height and period, starting with low waves and increasing their height in steps, until loss of stability of the stones was observed. It must be kept in mind that loss of stability is not a clearly defined phenomenon. Some subjectivity is involved, in particular because the loss of the first stones cannot always be entirely attributed to wave action since it may at least in part, be due to the random position of the stone after construction. In the following sections, the work of Iribarren and Hudson will be explained in more detail.

7.2.2 Iribarren

Iribarren [1938, 1950, 1953, 1954, and 1965] considered the equilibrium of forces acting on a block placed on a slope. Since the considerations of Iribarren referred to forces, the weight of the block W is introduced as a force, and thus expressed in Newton. It is important to realize that in literature, one finds the block size indicated either by weight or by mass. Although this is confusing, it is the result of a less strict application of the ISO standard (mks system) in the past. When using the formulae it is wise to check whether g is introduced in the formula. In that case, the weight is calculated in N. If g is not present in the formula, the result is the mass of the block in kg.

The forces acting on a unit positioned on a slope at an angle α are (see Figure 7-1):
- Weight of the unit (vertical downward)
- Buoyancy of the unit (vertical upward)
- Wave force (parallel tot the slope, either upward or downward)
- Frictional resistance (parallel tot the slope, either upward or downward, but contrary to the direction of the wave force)

Iribarren resolved these forces into vectors normal and parallel to the slope. Loss of stability occurs if the friction is insufficient to neutralize the other forces parallel to the slope.

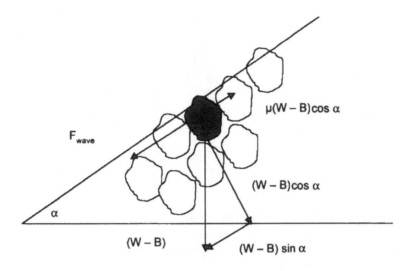

Figure 7-1 Equilibrium after Iribarren (downrush)

The parameters are:

W	= weight of block	[N]
B	= buoyancy of block	[N]
$W-B$	= submerged weight of block	[N]
V	= volume of block	[m³]
α	= angle of slope	[-]
μ	= friction coefficient	[-]
ρ_r	= density of block (rock)	[kg/m³]
ρ_w	= density of (sea) water	[kg/m³]
Δ	= $(\rho_r - \rho_w) / \rho_w$	[-]
H	= wave height	[m]
D_n	= characteristic size of stone = $V^{1/3}$	[m]
F_{wave}	= wave force	[N]
g	= acceleration of gravity	[m/s²]

Iribarren assumed a set of simple relations between F_{wave}, D_n, H, ρ and g as follows:

$$F_{wave} = \rho_w g\, D_n^2 H \tag{7.1}$$

$$W - B = \left(\rho_r - \rho_w\right) g\, D_n^{\,3} \tag{7.2}$$

and

$$W = \rho_r\, g\, D_n^{\,3} \tag{7.3}$$

It must be mentioned here that these relations may be criticized, since they are too simple. It must be expected that the shape of the block and the period of the wave play a role. Furthermore, the relation between the wave force and the wave height and stone size indicate the dominance of drag forces, whereas acceleration forces are neglected.

Nevertheless, considering the equilibrium for downrush along the slope, this leads to a requirement for the block weight:

$$W \geq \frac{N\rho_r g H^3}{\Delta^3(\mu\cos\alpha - \sin\alpha)^3} \tag{7.4}$$

For uprush, the formula changes into:

$$W \geq \frac{N\rho_r g H^3}{\Delta^3(\mu\cos\alpha + \sin\alpha)^3} \tag{7.5}$$

N is a coefficient that depends, amongst other factors, on the shape of the block, and its value must be derived from model experiments. The friction factor μ can be measured by tilting a container filled with blocks and determining the angle of internal friction.

In Iribarren [1965], recommendations are given for values of N and μ. The most important values are given in Table 7-1. The values of N refer to zero damage.

type of block	downward stability $(\mu\cos\alpha - \sin\alpha)^3$		upward stability $(\mu\cos\alpha + \sin\alpha)^3$		transition slope between upward and downward stability
	μ	N	μ	N	$\cot\alpha$
rough angular quarry stone	2.38	0.430	2.38	0.849	3.64
cubes	2.84	0.430	2.84	0.918	2.80
tetrapods	3.47	0.656	3.47	1.743	1.77

Table 7-1 Coefficients for Iribarren formula

It must be kept in mind that the coefficient N represents many different influences. At first, it is a function of the damage level defined as "loss of stability". It also includes the effect of the shape of the blocks, but not the internal friction, because this is accounted for in the separate friction coefficient. Finally, it covers all other influences not accounted for in the formula.

The friction coefficient μ seems to be on the high side, but is clearly related to the test procedure that Iribarren used. He found a large difference in friction, depending on the number of units in the slope. For details, one is referred to his original publication.

7.2.3 Hudson

Since 1942, systematic investigations into the stability of rubble slopes have been performed at the Waterways Experiment Station in Vicksburg, USA. On the basis of these experiments, Hudson [1953, 1959, and 1961] proposed the following expression as the best fit for the complete set of experiments:

$$W \geq \frac{\rho_r g H^3}{\Delta^3 K_D \cot \alpha} \tag{7.6}$$

The formula is applicable for slopes not steeper than 1:1 and not gentler than 1:4. The coefficient K_D represents many different influences, just like the coefficient N in the formula of Iribarren. At first, it is a function of the damage level defined as "loss of stability". It also includes the effect of the shape of the blocks and the internal friction. Finally, it covers all other influences not accounted for in the formula.

Recommended values for K_D have frequently been published and updated by the Corps of Engineers in the Shore Protection Manual. Subsequent editions of this manual thus reflect the changing insight over the years. In the 1977 edition (Anon. [1977]), the wave height H is defined as the significant wave height H_s, and the values for the most common types of blocks are given in Table 7-2.

type of block	number of layers (N)	structure trunk K_D		structure head K_D	
		breaking wave	non breaking wave	breaking wave	non breaking wave
rough angular quarry stone	1	**	2.9	**	2.3
rough angular quarry stone	2	3.5	4.0	2.5*	2.8*
rough angular quarry stone	3	3.9	4.5	3.7*	4.2*
tetrapod	2	7.2	8.3	5.5*	6.1*
dolos	2	22.0	25.0	15.0	16.5*
cube	2	6.8	7.8		5.0

* There is a slight variation of recommended K_D value for different slopes

** Use of single layer is not recommended under breaking waves

Table 7-2 K_D Values recommended given in SPM 1977

In the 1984 edition (Anon. [1984]), following a number of dramatic failures of rubble mound breakwaters, the use of H_{10}, the average of the highest 10 % of all waves is recommended. This is equal to 1.27 H_s.

type of block	number of layers (N)	structure trunk K_D		structure head K_D	
		breaking wave	non breaking wave	breaking wave	non breaking wave
rough angular quarry stone	1	**	2.9	**	2.2
rough angular quarry stone	2	2.0	4.0	1.6*	2.8*
rough angular quarry stone	3	2.2	4.5	2.1*	4.2*
tetrapod	2	7.0	8.0	4.5*	5.5*
dolos	2	15.8	31.8	8.0	16.0*
cube	2	6.5	7.5		5.0
akmon	2	8	9	n.a.	n.a.
Accropode * (1:1.33)		12	15		

* There is a slight variation of recommended K_D value for different slopes

** Use of single layer is not recommended under breaking waves

Table 7-3 K_D Values recommended given in SPM 1984

Be careful: probably too conservative!

A comparison between Table 7-2, and Table 7-3, shows a much more conservative design recommendation in 1984. Not only have the values of K_D been changed, but also the replacement of H_s by H_{10} is quite a dramatic change, certainly if one realizes that the wave height appears with a third power in the Hudson formula. In the opinion of many designers this results in too conservative an approach!

Hudson defines the K_D value for initial damage: 0-5% of the blocks in the armour layer. He counts the number of blocks from the centre of the crest down the outer slope to a level equal to the "no-damage wave height", $H_{D=0}$, below still water level. It is important, however, to know what happens when the wave height is greater than the zero damage wave height, in other words, when the structure is overloaded. The Shore Protection Manual gives data for various types of armour units and various levels of over-loading. These data are summarized in Table 7-4. From this table, it can easily be seen that traditional rubble mounds have an inherent safety coefficient because of the fact that complete failure occurs only at 50% overloading. This safety margin is considerably smaller when concrete armour units are used instead of quarry stone.

Unit		Damage (D) in percent						
		0-5	5-10	10-15	15-20	20-30	30-40	40-50
quarry stone smooth	$H/H_{D=0}$	1.00	1.08	1.14	1.20	1.29	1.41	1.54
quarry stone rough	$H/H_{D=0}$	1.00	1.08	1.19	1.27	1.37	1.47	1.56
tetrapod	$H/H_{D=0}$	1.00	1.09	1.17	1.24	1.32	1.41	1.50
dolos	$H/H_{D=0}$	1.00	1.10	1.14	1.17	1.20	1.24	1.27

Table 7-4 Damage due to over-loading

Be careful, damage due to breaking of units is not included
Damage percentage ≥ 30-40 often means failure

7.2.4 Comparison of Hudson and Iribarren formulae

When comparing the formulae of Iribarren and Hudson, the difference appears to be greater than it in fact is. The influences of wave-height, rock density and relative density are equal. The coefficients are different, but can easily be compared. The main difference occurs in the influence of the slope. A comparison of the two expressions within the validity area of the Hudson formula ($1.5 < \cot \alpha < 4$) reveals that the correct choice of coefficients leads to a minor difference between the two formulae only. It is evident that for very steep slopes (close to the angle of natural repose) Hudson cannot give a reliable result. It is also likely that for very gentle slopes waves will tend to transport material up the slope, a factor that was not considered by Hudson at all.

This becomes clearer when one takes the third root from both formulae. The stability expression then changes to:

Hudson: $\dfrac{H}{\Delta D} = \sqrt[3]{K_D \cot \alpha}$

Iribarren: $\dfrac{H}{\Delta D} = (\mu \cos \alpha \pm \sin \alpha) N^{-\frac{1}{3}}$ (7.7)

The coefficients in the formula are sorts of waste bins for all kind of unknown variables and unaccounted irregularities in the model investigations. Variables brought together in the coefficients K_D and N are:

- Shape of the blocks
- Layer thickness of the outer ("armour") layer
- Manner of placing the blocks
- Roughness and interlocking of the blocks
- Type of wave attack

- Head or trunk section of the breakwater
- Angle of incidence of wave attack
- Size and porosity of the underlying material
- Crest level (overtopping)
- Crest type
- Wave period
- Shape of the foreshore
- Accuracy of wave height measurement (reflection!)
- Scale effects, if any

In view of this, one cannot expect a good consistency in reported values of K_D. In fact, there is a tremendous scatter in the results, and this is no surprise. For the designer it means that he must be extremely careful when applying the formulae. When using the formulae, one must realize what influence uncertainties have on the final result. This applies to the selection of the coefficients, and to the choice of wave height and relative density. Small changes have a big influence on the required block weight.

Since there is no basic difference between the two formulae (as long as one applies the Hudson formula within the limits for the slope), one can work with either formula. Many designers prefer the Hudson formula because it is a little simpler to use and because there are far more experimental data on the coefficient K_D than on the Iribarren coefficients.

7.2.5 Application of Hudson formula

The Hudson formula gives the designer a clear picture of available means to improve the stability of the armour layer, and their effectiveness.

Increase of ρ_r

This can be done by selecting rock from a different quarry, or by producing concrete with heavy aggregates. Because of the cubed influence of Δ, it is very effective. At this point a warning must also be given that the density of water ρ_w has a similar, but reverse influence. Therefore, the actual density of the seawater must be used.

Increase W

When natural rock is used, there is generally a maximum block size that can be obtained from a quarry. When concrete units are used, one must take into account that very large units pose problems of structural strength. In this respect, reference should be made to Appendix 3: Concrete Armour Units.

Decrease slope

Be careful, the volume of material increases rapidly. Very gentle slopes may lead to upward loss of stability that is not covered by the Hudson formula. The method is only used for armour layers consisting of quarry stone. By reducing the slope, it may be possible to use cheaper local material, so that the extra volume of material is compensated. When concrete armour units are used, it can easily be demonstrated that it is always more economical to use the steepest possible slope (1:1.5).

Grout smaller blocks together with asphalt

This method has been used for the breakwater in IJmuiden, and at several other locations in the Netherlands. Special care is required to avoid the uplifting of grouted layers due to pressures building up under the armour layer (d' Angremond et al. [1970]).

Increase K_D by using specially shaped interlocking blocks

The most effective block shapes are the very slender blocks like Dolos. Because of breakage, their use is limited to smaller sizes. This reduces their applicability. It must be realized that special block shapes are costly because the higher cost of the moulds, the labour intensive use of the moulds, and the difficulties of handling and stacking the blocks.

The use of various kinds of specially shaped concrete armour units has become quite popular. There was a period during which no self-respecting laboratory or consultant could do without his own armour unit. This started with the development of Tetrapods by SOGREAH; it was followed by the Akmon, Dolos, and many others. The merit of the Tetrapod was that it demonstrated that by interlocking, a K_D value could be obtained that was about twice as high as the values for quarry stone, thus leading to half the weight. The disadvantage was the complicated shape, requiring an expensive mould. Since a patent protected the shape royalties levied by the inventor consumed part of the potential saving. This led to a Dutch initiative to develop a similar unit, and to levy no royalties. This resulted in the Akmon unit, yielding a K_D factor about equal to that of the Tetrapod. However, the complexity of the mould was similar to that of the Tetrapod.

Following the development of the Akmon, Merrifield and Zwamborn in S. Africa attempted to maintain the basic shape of the Akmon, but to increase the permeability/porosity by making the legs more slender. Initially, this was very promising, yielding K_D values of 20 and higher. At the same time, the sizes of vessels were increasing, requiring longer breakwaters, extending into deeper water with higher waves. This resulted in the design of some breakwaters (Sines, Portugal is the most striking example) with very large unreinforced units. In a very short period thereafter, a large number of breakwaters failed. It appeared that the mechanical strength of the concrete was inadequate to resist the forces, especially during rocking of the units against each other. This was never investigated in a model, and if tests had been carried out the results would have been of no use, because of scale effects.

In the meantime, it has been recommended that a more conservative approach should be adopted and that the rocking of slender units larger than 20 to 25 tons should be prevented. Another development was to avoid the use of slender units and to rely upon simple cubes. Although the required weight is greater than for the more sophisticated shapes, considerable savings are achieved on the cost of moulds, the cost of casting, and the storage and handling costs. More recently, SOGREAH developed a massive block, the Accropode®, which can be used in a single layer, provided it is carefully placed in a specific pattern. SOGREAH makes this unit available without the payment of royalties, on condition that both the design and method of placing are checked and approved by SOGREAH.

For details on concrete armour units, refer to Appendix 3.

7.3 Irregular waves, approach of Van der Meer

7.3.1 General

Between 1965 and 1970, the first wave generators that could generate irregular waves according to a certain predefined spectrum were developed. Model tests in the first years were aimed at ad-hoc investigations. Several researchers attempted to overcome the shortcomings of the Hudson approach by introducing more variables. Initially, their results diverged. In his PhD thesis at Delft University, Van der Meer [1988] succeeded in presenting an approach based on irregular waves that has gradually been accepted throughout the engineering community.

In the first place, he used a clear and measurable definition of damage. Initially, this was expressed by the parameter:

$$S = \frac{A}{D^2_{n50}}$$ (7.8)

in which:

A = the erosion area in a cross section in [m²]

$$D_{n50} = \left(\frac{W_{50}}{g\rho_r} \right)^{\frac{1}{3}} = \left(\frac{M_{50}}{\rho_r} \right)^{\frac{1}{3}} \quad \text{[m]}$$

W_{50} = mean weight of armour stones   [N]

ρ_r = density of armour stone   [kg/m³]

For a definition sketch, refer to Figure 7-2. The area A is often measured by using a rod with a half sphere of a specific size attached to it.

The erosion in the area A is partly caused by settlement of the rock profile and partly by removal of stones that have lost stability. Since the erosion area is divided by the area of the armour stone, the damage S represents the number of stones removed from the cross-section, at least when permeability/porosity and shape are not taken into account. In practice, the actual number of stones removed from a D_{n50} wide strip is between 0.7 and 1.0 times S.

If the armour layer consists of two layers of armour units, one can define limits for acceptable damage and failure. These limits are more liberal for gentler slopes, since in that case, the damage is distributed over a larger area. Critical values for S are given in Table 7-5.

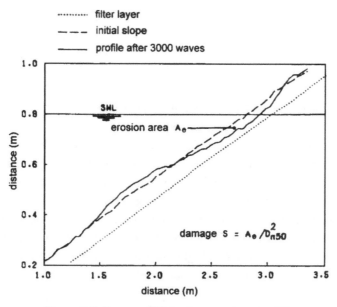

Figure 7-2 Damage(S) based on erosion area (A)

Slope	Initial Damage (needs no repair)	Intermediate Damage (needs repair)	Failure (core exposed)
1:1.5	2	3 – 5	8
1:2	2	4 – 6	8
1:3	2	6 – 9	12
1:4	3	8 – 12	17
1:6	3	8 – 12	17

Table 7-5 Classification of damage levels S for quarry stone

At a later stage, the definition of damage was slightly adapted. A value N is defined, which is the number of units displaced from one strip of the breakwater with a width of D_{n50}. The relation with S is established via the permeability/porosity. When the number of displaced units is counted, the settlement of the mound is omitted from the considerations of damage. The number N is often used when studying the stability of armour layers consisting of concrete units.

Van der Meer chose to express the stability in terms of $H_s/\Delta D_{n50}$, and then investigated the influence of several parameters that he considered relevant. These parameters are briefly discussed below.

Wave period

Van der Meer assumed the effect of the wave period to be connected with the shape and intensity of breaking waves. He therefore used the Iribarren parameter:

$$\xi = \frac{\tan \alpha}{\sqrt{s}} \tag{7.9}$$

in which:

$$s = 2 \pi H / gT^2 \tag{7.10}$$

Using the characteristic values for irregular waves in deep water H_s and T_p or T_m, this leads to the use of ξ_{s0p} and ξ_{s0m} respectively.

It must be noted that the value of H_s in the expression $H_s/\Delta D$ is measured at the location of the toe of the structure after elimination of any wave reflection.

Contrary to Hudson and Iribarren, Van der Meer found a clear influence of the storm duration. The longer the storm, the more damage. This can easily be explained by the model technique. Hudson and Iribarren used regular waves. A longer duration of the test series did not change the wave attack on the structure. In an irregular wave field, a longer storm duration leads to a higher probability of the occurrence of extremely high waves. Apparently, these extremely high waves are responsible for ongoing damage.

Van der Meer also finds a certain influence resulting from the permeability or porosity of the breakwater structure as a whole. He expresses this influence 'notional permeability' as a factor P, for which he indicates values based on a global impression of the stone size in subsequent layers (Figure 7-3). It is emphasized that, in fact, P is not a permeability parameter, although it is referred to as being the permeability parameter. It merely indicates the composition of the breakwater in terms of the mutual relations of the grain sizes in subsequent layers.

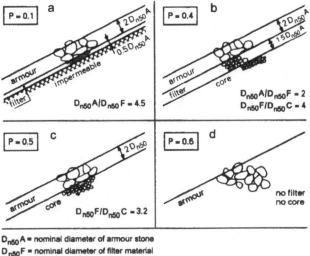

$D_{n50}A$ = nominal diameter of armour stone
$D_{n50}F$ = nominal diameter of filter material
$D_{n50}C$ = nominal diameter of core

Figure 7-3 Permeability coefficients for various structures

7.3.2 Quarry stone

After extensive curve fitting, Van der Meer concludes that for quarry stone, the stability is ruled by:

For plunging waves:

$$\frac{H_s}{\Delta D_{n50}} = 6.2\, P^{0.18} \left(S/\sqrt{N} \right)^{0.2} \, 1/\sqrt{\xi_m} \tag{7.11}$$

For surging waves:

$$\frac{H_s}{\Delta D_{n50}} = 1.0\, P^{-0.13} \left(S/\sqrt{N} \right)^{0.2} \sqrt{\cot \alpha} \; \xi_m^P \tag{7.12}$$

where:
H_s = significant wave height
Δ = relative mass density $(\rho_a - \rho)/\rho$ where ρ_a = mass density of stone and ρ = mass density of water
D_{n50} = nominal diameter, 50% value of sieve curve
N = number of waves
S = damage level $A/(D_{n50})^2$, where A = erosion area in a cross-section
α = angle of the seaward slope of a structure
ξ = surf similarity parameter $\tan(\alpha)/\sqrt{(H/L)}$
P = permeability coefficient

The transition between plunging and surging waves can be derived by intersecting the two stability curves, which yields:

$$\xi_{mcrit} = \left[6.2\ P^{0.31} \sqrt{\tan\alpha} \right]^{\frac{1}{P+0.5}}$$

(7.13)

Depending on slope and permeability, the transition lies between $\xi_{s0m} = 2.5$ and 4. The reliability of the formula can be expressed by giving the relative standard deviation σ/μ (in percent) for the coefficients 6.2 and 1.0. These relative standard deviations are respectively 6.5% and 8%, as compared to a reliability of the Hudson formula of 18%.

7.3.3 Concrete blocks

When testing armour layers of artificial material like concrete, it makes no sense to vary the slope of the breakwater. Since the block weight is not so strictly limited as it is for quarry stone (the quarry has a clear maximum block size), it is much more effective to increase the concrete block weight than to reduce the slope. This makes using the Iribarren number ξ in a formula less realistic, since this expresses the influence of both, wavelength or period and slope. All formulae for concrete units, except the Accropode® are based on a slope of 1:1.

Since the mechanical strength of the concrete blocks may play a role, it is useful to distinguish damage due to actually displaced units (their number is indicated by N_{od}, and damage due to blocks that might break because they are rocking against each other (their number is indicated by N_{or}).

The total number of moving units is equal to the number of displaced blocks plus the number of rocking blocks i.e. $N_{omov} = N_{od} + N_{or}$.

The value of N_{od} is compatible with the value of S, (compare equation (7.8). S is about double the value of N_{od}.

Van der Meer [1988] gives the stability for various frequently used blocks. He makes a distinction between displaced blocks and moving blocks. The difference appears to be a reduction of the stability number by 0.5. The scatter of data for cubes and Tetrapods is normally distributed and has a relative standard deviation $\sigma/\mu = 0.1$.

Note that D_n is the nominal diameter of the unit, or the cubic root of the volume. For various blocks this leads to:

Cubes $\quad D_n$ = equal to the side of the cube
Tetrapods $\quad D_n = 0.65\ D$ if D is the height of the unit
Dolos $\quad D_n = 0.54\ D$ if D is the height of the unit (waist ratio 0.32)
Accropode® $D_n = 0.7\ D$ if D is the height of the unit
Like the damage levels for quarry stone, damage levels can also be classified for concrete units as in Table 7-6.

Block Type	Slope	Relevant N-value	Start of Damage	Initial Damage (needs no repair)	Intermediate Damage (needs repair)	Failure (core exposed)
Cube	1:1.5	N_{od}	0	0 – 0.5	0.5 – 1.5	>2
Tetrapod < 25 ton	1:1.5	N_{od}	0	0 – 0.5	0.5 – 1.5	>2
Tetrapod >25 ton	1:1.5	N_{omov}	0	0 – 0.5	0.5 – 1.5	>2
Dolos < 20 ton	1:1.5	N_{od}	0	0 – 0.5	0.5 – 1.5	>2
Dolos > 20 ton	1:1.5	N_{omov}	0	0 – 0.5	0.5 – 1.5	>2
Accropode®	1:1.33		0			>0.5

Table 7-6 Classification of damage levels N_{od} and N_{omov} for quarry stone

Cubes

$$\frac{H_s}{\Delta D_n} = \left(6.7\frac{N_{od}^{0.4}}{N^{0.3}} + 1.0\right)s_{om}^{-0.1} \qquad \frac{H_s}{\Delta D_n} = \left(6.7\frac{N_{omov}^{0.4}}{N^{0.3}} + 1.0\right)s_{om}^{-0.1} - 0.5 \qquad (7.14)$$

Tetrapods

$$\frac{H_s}{\Delta D_n} = \left(3.75\frac{N_{od}^{0.5}}{N^{0.25}} + 0.85\right)s_{om}^{-0.2} \qquad \frac{H_s}{\Delta D_n} = \left(3.75\frac{N_{omov}^{0.5}}{N^{0.25}} + 0.85\right)s_{om}^{-0.2} - 0.5 \quad (7.15)$$

Dolos
Holtzhausen and Zwamborn [1992] investigated the stability of Dolos with the following result:

$$N_{od} = 6250\left[\frac{H_s}{\Delta^{0.74}D_n}\right]^{5.26} s_{op}^3 w_r^{20s_{op}^{0.45}} + E \qquad (7.16)$$

in which:
w_r = Waist ratio of the Dolos
E = Error term

The waist ratio has been made a variable in the Dolos design to enable the choice of a less slender shape with less chance of breaking. Waist ratios are between 0.33 and 0.4. The error term E represents the reliability of the formula. It is normally distributed and has a mean value equal to zero, and a standard deviation $\sigma(E)$:

$$\sigma(E) = 0.01936\left[\frac{H_s}{\Delta^{0.74}D_n}\right]^{3.32} \qquad (7.17)$$

Patented elements

in recent years a number of special units have been developed for use on steep slopes in single layer. These elements can be placed on a slope of 1:1.33. Examples of these elements are the Accropode[1]® developed by Sogreah, the Core-loc® developed by the US corps of engineers [Melby and Turk, 1997] and the Xbloc® developed by Delta Marine Consultants [Reedijk et al, 2003]. These elements can only be applied under special license from the developers. For the Core-Loc permission is even needed for model tests.

Placing of the elements has to be done in accordance with detailed specifications provided by the developer. For example, the placing method for Accropodes is given in Appendix 3. It seems that there is no influence of storm duration; and because these units are only applied on a slope of 1:1.333, the slope is not part of the stability formula. This formula thus becomes quite simple; $H_s/\Delta D_n$ equals a given number that is developed on the basis of model experiments (Table 7-7).

	Accropode	Core-loc	Xbloc[*]
Start of damage ($N_{od} = 0$)	3.7[a]		3.5[b]
Failure ($N_{od} > 0.5$)	4.1[a]		4.0[b]
Design value	2.5[a,b]	2.8[b]	2.8[b]

a) Based on 2D-tests by Van der Meer. b) Based on 2D-tests by developers.

Table 7-7 Stability numbers ($H_s/\Delta D_n$) for patented units

The values for "start of damage" and "failure" as given in table can be considered as stochastic variables with a standard deviation of 0.2. Because wave heights causing failure are only slightly higher than the wave height associated with "start of damage", a higher safety coefficient has to be applied than for normal rock structures. Van der Meer recommends for Accropode a design value of 2.5. The values for Core-loc and Xbloc are based on information from the supplier. They still require verification by independent research.

Before using any patented unit (and paying royalties) it is wise to see whether the patent is valid in the country of application and whetehr it has not expired in general.

7.4 Stability calculation

Basically, the computation for the stability of armour units can be made by substituting the significant wave height of the design storm in the stability equations like (7.11) or (7.14). In the computation, one can use a deterministic approach, an approach based on partial safety coefficients, or one can use a full probabilistic

---

[1] The Accropode patent has expired.

approach. In case of a design storm one determines the H_{ss} with the method as described in Chapter 5, using a probability of exceedance developed along the lines as described in Chapter 3.

In case of using the deterministic approach with partial safety coefficients one uses the H_{ss} with a probability of exceedance of once per lifetime (e.g. 1/50), and includes in the design equation two partial safety coefficients; one for the strength part, and one for the load part. The values to be used follow from international guidelines, like PIANC [1992]. It is also possible to use a self-developed design wave height (Chapter 3) and to use the safety coefficient for the strength parameters only.

In a probabilistic approach one re-writes the design equation as a Z-function (see Section 3.6), and determines the probability of $Z < 0$, given the distributions of all parameters in the equation. This is usually done in a probabilistic computer program, using either the FORM-technique (fist order reliability method) or the Monte Carlo technique. In this last case one has to know especially the distribution of H_{ss}.

These methods are worked out in the example in Appendix 10.5.

7.5 Special subjects

7.5.1 General

The basic elements relating to the stability of stone on a slope under wave attack have been discussed in the previous sections. However, there are some subjects that need further attention in order to fine-tune a structural design. These subjects will be treated in the subsequent sections.

7.5.2 Shallow water conditions

In Chapter 5, it was indicated that in shallow water when waves are breaking, the Rayleigh distribution is no longer valid. In deep water the wave height exceeded by 2% of the waves is about 40% higher than the significant wave height: $H_{2\%} = 1.4\ H_s$. In shallow water, this ratio may fall to a value between 1.2 and 1.3.

When considering the effect of shallow water depth on the stability of armour units on a slope, one can assume that the highest waves yield the most important contribution to the damage. One could therefore re-write the Van der Meer formulae for quarry stone by using $H_{2\%}$ instead of H_s. The formulae then becomes:

For plunging waves:

$$\frac{H_{2\%}}{\Delta D_{n50}} = 8.7 P^{0.18} \left(\frac{S}{\sqrt{N}} \right)^{0.2} \frac{1}{\sqrt{\xi_m}} \qquad (7.18)$$

For surging waves:

$$\frac{H_{2\%}}{\Delta D_{n50}} = 1.4 P^{-0.13} \left(\frac{S}{\sqrt{N}} \right)^{0.2} \sqrt{\cot \alpha} \cdot \xi_m^P \tag{7.19}$$

Note that ξ does not change, since the wave steepness is always measured in deep water!

If one is sure that $H_{2\%}$ is reduced owing to breaking, it is possible to account for this fact by using the re-written formulae (see Equations (7.18) and (7.19) and substituting the actual value of $H_{2\%}$. It is stressed, however that even if extensive breaking has been observed at the location of the breakwater, the conditions may be different during the design storm as a result of higher water levels (storm surge) or a lower bed level (erosion and scour).

7.5.3 Shape of quarry stone

Hudson had already indicated by varying values of K_D, that the angularity of quarry stone has an influence on stability. Latham et al. [1988] investigated the influence of the shape of individual stones on their stability. They used designations like "fresh", "equant", "semi-round", "very round", and "tabular". As compared with "standard" quarry stone, the coefficient in the Van der Meer formula changes slightly as shown in Table 7-7.

Rock shape	Plunging waves	Surging waves
Elongate/Tabular	6.59	1.28
Irregular	6.38	1.16
Equant	6.24	1.08
Standard v.d. Meer	*6.2*	*1.0*
Semi-round	6.10	1.00
Very round	5.75	0.80

Table 7-7 Effect of stone shape on stability

For a visual impression of block shapes, one is referred to Figure 7-4.

Similar investigations were carried out by G. Burger [1995]. In his master's thesis he indicates a relation between stability and l/d ratio of the quarry material.

7.5.4 Grading of quarry stone

When quarry stone is purchased from a commercial block-stone quarry, gradation is usually according to national standards. For Western Europe, one is referred to data in the CUR/CIRIA Manual (Anon. [1991]).

Converting mass into diameter is done by using the well-known D_n (nominal diameter) method:

$$D_n = \sqrt[3]{\frac{M}{\rho}} \tag{7.20}$$

In the Netherlands, such standardised grades are presented in Table 7-8.

Mass	D_n
10-60 kg	0.16-0.30 m
10-200 kg	0.16-0.43 m
60-300 kg	0.30-0.49 m
300-1000 kg	0.49-0.72 m
1000-3000 kg	0.72-1.04 m
3000-6000 kg	1.04-1.31 m
6000-10 000 kg	1.31-1.55 m

Table 7-8 Dutch standard grading

Elongate/Tabular (ET)
$p_r>0.015$

Irregular (IR)
$p_r=0.013-0.015$

Equant (EQ)
$p_r=0.011-0.013$

Semi-Round (SR)
$p_r=0.009-0.011$

Very Round (VR)
$p_r<0.009$

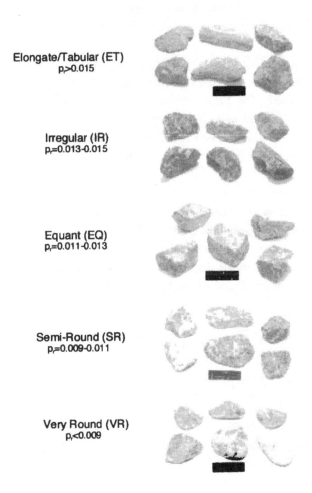

Figure 7-4 Visual comparison of block shapes (from CUR/CIRIA Manual)

If the stone is classified according to sieve diameter, one can determine D_s. Although sieving is not a practical method for the larger stones, one can establish a general relation:

$$D_n = 0.8 \, D_s \qquad\qquad (7.21)$$

The grading of a stone class is often defined as D_{85}/D_{15}. Common values are:

Type of grading	D_{85}/D_{15}
Narrow	< 1.5
Wide	$1.5 - 2.5$
Very wide (quarry run or riprap)	$2.5 - 5$ and more

Stability is usually investigated for grades classified as wide grades. The use of very wide grades will result in slightly more damage than narrow and wide grades. However, the very wide grades can easily lead to demixing or segregation, so that it is difficult to effectively control the quality of stone delivered.

7.5.5 Stability of toe

It is certainly not necessary to extend the armour layer over the full water depth down to the seabed. The Shore Protection Manual gives some rules-of-thumb, indicating that the armour layer should extend down to about one wave height below still water level. The armour layer should then be supported by a toe, for which the same manual gives an indicative stone weight of 10% of the weight of the regular armour.

This approach is not very satisfactory, since one can imagine that the use of heavier stones in the toe allows a higher position of the toe, whereas the use of smaller stones in the toe will lead to a toe with a lower position.

Gerding (Van der Meer, d'Angremond and Gerding [1995]) did useful work on this subject for his master's thesis. He investigated the relation between the unit weight of toe elements, toe level, and damage (Nod). His findings were confirmed by the thesis work of L. Docters Van Leeuwen [1996], who also varied the rock density ρ_r. The result was:

$$\frac{H_s}{\Delta D_{n50}} = \left(0.24 \frac{h_t}{D_{n50}} + 1.6 \right) N_{od}^{0.15} \qquad\qquad (7.22)$$

Critical values for N_{od} are:

N_{od}	Character of damage
0.5	Start of Damage
1.0	Acceptable Damage
4.0	Failure

These values apply for a standard toe, with a height of 2 to 3 D and a width of 3-5 D. The validity range is:

$$0.4 < h_t/h < 0.9$$

$$3 < h_t / D_{n50} < 25$$

A definition sketch is given in Figure 7-5

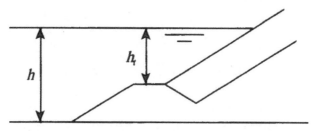

Figure 7-5 Definition sketch toe stability

7.5.6 Breakwater head

The head of a breakwater is relatively vulnerable since, owing to the curvature, the armour units are less supported and/or less interlocking. In general, damage occurs on the inner quadrants, which is understandable if one looks at the 3D shape.

Therefore, the head of a breakwater is often reinforced either by:

- using larger size armour units or by
- increasing the density of the armour units or by
- reducing the slope.

An idea of the required measures can be obtained from the recommended K_D values given in the Shore Protection Manual.

Neither of the structural solutions is ideal: larger and heavier blocks pose construction problems; a reduced angle of slope may cause a hazard to navigation.

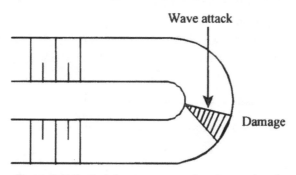

Figure 7-6 Typical damage pattern breakwater head

7.5.7 Stability of crest and rear armour

As long as the crest of the structure is so high that it prevents considerable overtopping, the armour units on the crest and the rear slope can be much smaller than the armour on the front slope. In this context, it must be mentioned that in the

Netherlands the 2% run-up level is used for dikes with an inner slope consisting of grass covered clay.

The size of stone on the inner slope of a high-crested breakwater is often determined by waves generated in the harbour basin by wind or passing ships. Only in the vicinity of the harbour entrance, must one take into account waves penetrating through the entrance.

In many cases, however, the crest will not be so high, and in the design conditions considerable overtopping will take place. In some cases, the crest will even be below still water level under such conditions.

This means a reduction in the direct attack on the armour on the outer face, and at the same time a more severe attack on the crest and the inner slope. In this way, there is a relation between the choice of crest level and the material on the crest and the inner slope. The reduction of the attack on low crested breakwaters is discussed in Section 7.5.8.

There are insufficient data available in the literature to indicate the critical crest level that requires continuation of the armour layer over the crest along the inner slope to a level that is well below still water level. This is certainly the case for submerged breakwaters, but it is certainly not required if the crest level exceeds the significant run-up level. For crest levels between these extremes an answer may be found in engineering judgement and specific model tests. In a limited study for a breakwater with tetrapod armour layer, De Jong [1996] concludes that the worst condition for the rear armour exists when the freeboard R_c divided by the nominal diameter is between 0 and 1 ($0 < R_c / D_n < 1$).

Another student, G. Burger [1995] analyzed data of Van der Meer and Vidal. He reached the conclusion that the inner slope is relatively safe when the dimensionless freeboard R_c / D_{n50} is larger than 4. When the crest is lower than this value, wave attack on the inner slope necessitates the use of relatively heavy blocks. When the crest is submerged equally (freeboard $< - 4 R_c / D_{n50}$), the wave attack is reduced so much that the damage decreases rapidly for all sections (crest, front slope and rear slope).

7.5.8 Stability of low and submerged breakwaters

Essentially two cases can be discerned: a crest above still water level and a submerged crest.

Low crest

Van der Meer derived a reduction factor for the armour size D_n. This reduction factor is:

$$\frac{1}{1.25 - 4.8R_p^*}, \text{ for } 0 < R_p^* < 0.052 \tag{7.23}$$

With:

$$R_p^* = \frac{R_c}{H_s} \sqrt{\frac{s_{op}}{2\pi}} \tag{7.24}$$

Application of this formula leads to a reduction in block size of up to 80% of the original value if the crest is at still water level. This is equivalent to a weight reduction of about 50%.

Research to distinguish damage to the front slope, crest and rear slope of low crested breakwaters is still in progress. For example, Burger [1995] concluded that Vidal recommends a much lower reduction of block weight for low crested breakwaters than Van der Meer suggests in the above formulae. Great care is therefore required when it is decided to reduce armour weight of a low-crested breakwater. General rules are not available.

An exception is made for De Jong [1996]. He gives a special stability formula for a Tetrapod armour layer on a low-crested breakwater. This formula is valid for the armour units on crest and front slope:

$$\frac{H_s}{\Delta d_n} = s_{om}^{0.2} \left\{ (2.64 k_\Delta + 1.25) + 8.6 \left(\frac{N_{od}}{\sqrt{N}} \right)^{0.5} \right\} * \left(1 + 0.17 e^{-0.61 \frac{R_c}{D_n}} \right) \tag{7.25}$$

Submerged crest

When the crest is submerged, the wave attack is no longer concentrated on the slope, but rather on the crest itself. Van der Meer re-analyzed the data of Givler and Sorensen [1986] with the following result:

$$\frac{h_c}{h} = (2.1 + 0,1S) \, e^{-0.14 N_s^*} \tag{7.26}$$

in which:

h_c = height of the crest, measured from the bottom
h = water depth, measured from the bottom
S = damage level as defined earlier
$N_s^* = \dfrac{H_s}{\Delta D_{n50}} s_p^{-1/3}$
s_p = local wave steepness

7.6 Future developments

In view of the relatively large scatter in the results of the stability formulae and of the large influence of factors that are not included in the formulae (shape of foreshore and others), it is still good engineering practice to test a final design in a physical model. In such a model test, the structure must be exposed to wave loads higher than the design wave. Preferably, the test should be continued until failure of the structure, so that the safety margins can be judged. Such model tests are quite labour-intensive, and therefore quite costly.

To overcome this problem it is possible to continue the kind of work started by Van der Meer. By generalizing model investigations, it should be possible to improve and refine the design formulae. This will certainly be done.

Another development is a change in the type of model studies. With the increasing capacity of computers it will be feasible to solve the Navier Stokes equations and calculate the water movement in front of and within rubble mound structures. At the moment, this is already feasible in a 1D approach, by solving the long wave equations. Van Gent describes this method in his PhD thesis. The model ODIFLOCS is available at Delft University of Technology. Van Gent has calculated the reshaping of berm breakwaters by using this model.

The next step will be a 2D approach, and progress is being made in this direction. This opens the possibility of studying the behaviour of rubble mound structures with the aid of a PC or workstation. For this purpose, development of a 2D model is not sufficient. There must also be a clear insight in the hydrodynamic and soil mechanical forces that determine stability or the loss of it.

8 DYNAMIC STABILITY

Design of berm breakwaters is gaining more and more attention. The term berm breakwater refers to the presence of an extended quantity of stone on the seaward side of the breakwater. This volume is sufficient to allow for a certain deformation without endangering the integrity of the structure as a whole. This chapter gives some empirical data on the deformation of such berm breakwaters. It is not necessary to learn formulae by heart, but it is good to try and understand the structural principles

8.1 Introduction

In Chapter 7, the stability of blocks on a slope was studied on the condition that the units would be stable In principle, no movements were permitted. We have seen that this requires the use of heavy blocks. It is not always possible to obtain such heavy blocks of quarry stone because of geological limitations of the quarry. Casting the blocks in concrete is complicated because very large units are rather sensitive to breaking.

Whether it would be possible to allow slight movements of the armour stone, so that the shape of the outer slope can adapt itself to the prevailing wave conditions has therefore been studied. It is evident that a steep slope will tend to become gentler. To maintain the overall function of the breakwater, including the required crest level, one has to provide extra material. That is why this type of breakwater is often called a berm breakwater, since the extra material is placed in a berm on the outer side of the structure.

The application of extra material is only feasible when the cost of this material is not too high. This is the case when the quarry is not too far away from the construction site of the breakwater. Concrete units are never used in berm breakwaters, they are too costly and their sensitivity to abrasion is too high.

An added advantage of a berm breakwater is the fact that a wider gradation of material can be used. This eliminates the need for expensive sorting operations at the quarry. The wider gradation, plus the fact that the maximum stone size is limited, makes it easier to bring the stone demand for the design into accordance with the yield curve of the quarry. In the simplest form, the quarry yield is split into a maximum of three categories: filter material, core material and berm material.

Allowing the waves to reshape the outer slope eliminates the need to construct this slope at a specific angle. The contractor can dump the stone by truck and level it with a bulldozer, leaving the slope to assume the angle of internal repose. This again represents an important saving on construction cost. However, one must ascertain that a sufficient volume of material is used per running metre of cross section.

8.2 Seaward profiles

Model tests on slopes that are not statically stable have indicated that a typical profile is formed according to Figure 8-1.

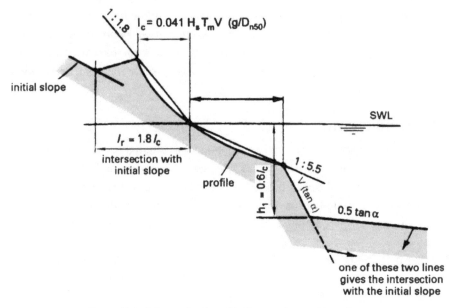

Figure 8-1 Schematized profile for sand gravel beaches

The characteristic dimensions can be expressed in terms of wave parameters:

$$\ell_c = 0.041 H_s T_m \sqrt{\frac{g}{D_{n50}}} \tag{8.1}$$

$$\ell_s = \ell_r = 1.8 \, \ell_c \tag{8.2}$$

$$h_t = 0.6 \, \ell_c \tag{8.3}$$

As can be seen from Figure 8-1, the intersection point of the profile with the still water level determines the position of the newly formed slope. From this point, an upper slope is drawn at 1:1.8 and a lower slope at 1:5.5. The horizontal distance ℓ_c determines the position of the crest on the upper slope, the distance ℓ_s determines a transition point on the lower slope. The actual slope in the zone of wave attack is a curved line through the three points. Below the lower transition point, a very steep slope develops at the angle of natural repose φ. If the original slope was already steep, the steep lower slope continues until the bottom is reached. If the original slope was gentle, the steep part continues until a level h_t below SWL. From the newly formed crest, the equilibrium profile connects to the original slope at a distance ℓ_r (the run-up length).

The position of the intersection point with SWL is not known in advance, but can be found easily when one realizes that the volume of erosion should be equal to the volume of accretion.

An example of slope development based on the formulae is shown in Figure 8-2.

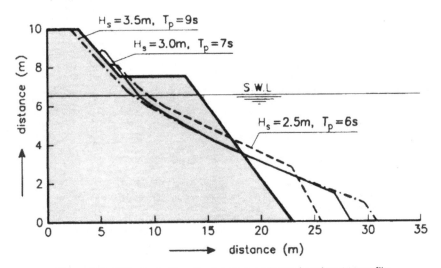

Figure 8-2 Influence of wave climate on a berm breakwater profile

The designer of a berm breakwater can change the width and height of the berm by trial and error in such a way that the core is always protected by at least a double or triple layer of armour material. The trial and error work is made easier by the use of the software package "BREAKWAT" of WL | Delft Hydraulics.

In principle, two types of initial cross-section are used, one with a berm at crest level, and one with a berm slightly above MSL. In the latter case, the chosen berm level is at such a height that trucks can safely drive over the berm.

8.3 Longshore transport of stone

When a statically stable breakwater loses armour units from the cross-section, it is not relevant in which direction the units are moving, since they are already accounted for as damage. When armour units of a berm breakwater are moving out of their places, it is assumed that they find another, more stable position within the same cross-section. This assumption is not correct when the wave approaches the structure at an angle. In that case, the armour unit may be transported for some distance along the breakwater. Another unit, originating from a profile a little further upstream may fill its place. This process cannot continue indefinitely, since there will be a section that is eroded continuously so it is no longer possible to maintain an equilibrium profile.

This is why one should not accept considerable longshore transport along a berm breakwater. Burcharth and Frigaard [1988] did some research on this and they state that longshore transport remains within reasonable limits if the armour size used for berm breakwaters is not too small. They recommend the following limits:

Trunks exposed to steep waves:

$$\frac{H_s}{\Delta D_{n50}} \leq 4.5 \tag{8.4}$$

Trunks exposed to oblique waves:

$$\frac{H_s}{\Delta D_{n50}} \leq 3.5 \tag{8.5}$$

Breakwater Heads:

$$\frac{H_s}{\Delta D_{n50}} \leq 3 \tag{8.6}$$

Van der Meer has carried out similar tests and concludes that the number of stones transported per wave along the breakwater $S(x)$ is at its maximum for wave angles between 15 and 35 degrees. According to Van der Meer [1992]) the transport is:

$$S(x) = 0 \quad \text{for } H_o T_{op} < 105 \tag{8.7}$$

and for $H_o T_{op} > 105$:

$$S(x) = 0.00005 \left(H_o T_{op} - 105 \right)^2 \tag{8.8}$$

in which:

$$H_o = \frac{H_s}{\Delta D_{n50}} \quad \text{and} \quad T_{op} = T_p \sqrt{\frac{g}{D_{n50}}} \tag{8.9}$$

8.4 Crest and rear slope

One of the design principles of a berm breakwater is the simplification of the cross-section. Therefore, the armour on the crest and the rear slope is the same as on the front slope. Since we have seen in the previous sections that $H_s/\Delta D_{n50}$ should be in the order of 3 to 3.5, this applies for the armour on the rear slope as well.

From model investigations by Van der Meer and Veldman [1992], it appeared that the crest level determines the damage to the inner slope, with a slight influence of the wave steepness as well.

Start of damage:

$$\frac{R_c}{H_s} s_{op}^{1/3} = 0.25 \tag{8.10}$$

Moderate damage:

$$\frac{R_c}{H_s} s_{op}^{1/3} = 0.21 \tag{8.11}$$

Severe damage:

$$\frac{R_c}{H_s} s_{op}^{1/3} = 0.17 \tag{8.12}$$

8.5 Head of berm breakwater

It is evident that wave attack on the round head of a berm breakwater shows similarities to wave attack by oblique waves. As stated in Section 8.3, it is recommended that the value of $H/\Delta D$ should be limited to less than 3.

Furthermore, it is a wise measure to supply an extra buffer of armour stone at the round head when longshore transport is expected. The extra quantity in the buffer can be created by increasing either the height or the width of the berm. However, the buffer must not become an obstruction to safe navigation.

9 STABILITY OF MONOLITHIC BREAKWATERS

The title of Chapter 9 may be a little misleading. Although it refers to monolithic breakwaters, in practice it deals mainly with vertical wall breakwaters or even caisson breakwaters. The subject is so full of uncertainties that it makes no sense at all to know formulae in a quantitative way. However, it is important to understand the difference between static, quasi-static and dynamic loads, and their effect on the stability.

9.1 Introduction

The problem of the stability of monolithic breakwaters has not yet been solved in a satisfactory and generally accepted way. Research efforts are under way, but have not resulted in a generally applicable theory or formula. Nevertheless, monolithic breakwaters are being built, and designers do use practical formulae. In this chapter, we will discuss a theoretical approach and a practical method developed in Japan. As the stability is a joint effect of wave load and subsoil resistance, some soil mechanics will be discussed as well. In addition to the stability of the monolithic breakwater, some other aspects of wave structure interaction will also be discussed.

Because of the intense interest in many countries, rapid development of the knowledge of monolithic breakwaters must be expected, comparable with the evolution of rubble mound breakwaters between 1988 and 1993. For the reader this means that the most recent sources of literature must always be consulted in addition to this textbook.

9.2 Wave forces and their effects

9.2.1 Quasi static forces

In Chapter 5, the formula for the pressure distribution under a wave according to the linear wave theory has been given. On the basis of this formula, Sainflou [1928] developed a method to calculate pressures exerted on a vertical wall by non-breaking waves. Rundgren [1958] carried out a series of model experiments and concluded that Sainflou's method overestimates the wave force for steep waves. Rundgren then used and modified the higher order approach proposed by Miche [1944]. This Miche-Rundgren method gives satisfactory results for steep waves, whereas the original Sainflou-method is best suited for long and less steep waves.

The main and most important aspect of the Miche-Rundgren approach is the definition of a parameter h_0, which is a measure for the asymmetry of the standing wave around SWL. This leads to the pressure diagrams shown schematically in Figure 9-1.

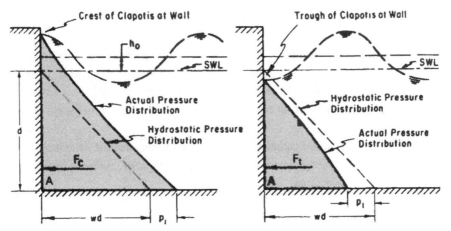

Figure 9-1 Schematic pressure distribution for non-breaking waves (from: Shore Protection Manual)

In this figure wd and p_1 are given by:

$$w d = \rho g h \quad \text{and} \quad p_1 = \left(\frac{1+r}{2}\right)\frac{\rho g H_i}{\cosh\left(2\pi h/L\right)} \tag{9.1}$$

The Shore Protection Manual gives design graphs for the calculation of h_0 as a function of wave steepness, relative wave height (H/h) and reflection coefficient. It also gives graphs to calculate integrated pressures and resulting turning moments for crest and trough of the wave.

This leads to a relatively simple load diagram (Figure 9-2), in which the horizontal hydrostatic forces on the front and rear wall have been omitted because they eliminate each other. For stability, one must consider the resistance against translation and the resistance against rotation. *Here it is stressed that the resistance against rotation cannot be taken simply as the sum of the moments around point A. Long before the structure starts to rotate, the pressure under point A has reached a value that leads to failure of the subsoil or failure of the corner of the structure.*

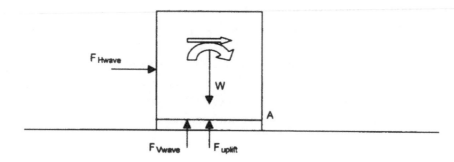

Figure 9-2 Load and equilibrium diagram

Since these formulae have been derived for regular monochromatic waves, it is necessary to combine them with spectral theory and arrive at a statistical distribution of wave forces and overturning moments. It can then be decided what frequency of exceedance is acceptable during the lifetime of the structure. In this way, the design loads can be established.

The loads defined so far are called quasi-static forces, because they fluctuate with the wave period of several seconds and do not cause any dynamic effects. Inertial effects need not be taken into account.

9.2.2 Dynamic forces

In Section 9.2.1, we restricted ourselves to the forces exerted by non-breaking waves. However, when waves are breaking, we know that impact or shock pressures occur in the vicinity of the water surface. The duration of these pressures is very short, but the local magnitude is very large. The quasi static pressures are always in the order of $\rho g H$, but the impact pressures can be 5 to 10 times higher. An example of a pressure record is given in Figure 9-3.

Many researchers have studied the phenomenon in the laboratory and none have come up with a satisfactory explanation that can predict the occurrence and the magnitude of a wave impact as a function of external parameters. Bagnold [1939] was the first of these researchers. He found that the impact pressure occurs at the moment that the vertical front face of the breaking wave hits the wall, and mainly when a plunging wave entraps a cushion of air against the wall.

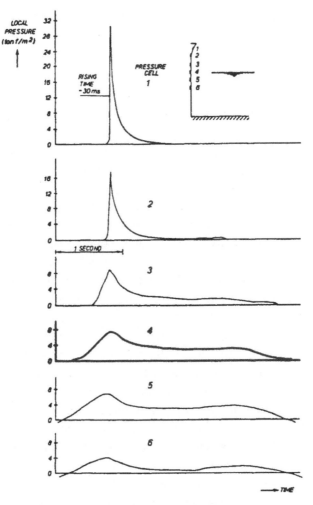

Figure 9-3 Example of a pressure record

Apparently, the deceleration of the mass of water in the wave crest, combined with the magnifying effect of the air cushion, causes the high pressures. Two models can be used to describe and calculate this effect:

- The continuous water jet hitting a plane yields a pressure:

    $p = \frac{1}{2} \rho u^2$ (u is the water velocity in the jet)

- The water hammer effect, resulting in:

    $p = \rho u c$

    in which:

    u = the water velocity in the conduct

    c = the celerity of sound in water (1543 m/s)

The water velocity in the crest of the breaking wave is equal to the wave celerity (in shallow water: $\sqrt{gh}$)

Substitution of reasonable figures leads to a water velocity in the order of 10 m/s and impact pressures:

Continuous jet: 55 kPa (5.5 mwc) [1]

Water hammer: 16,000 kPa (1600 mwc)

In reality, we know that the impact pressures reach values between 50 and 150 mwc. Measurement of the impact pressures in a model is complicated because the short duration of the load requires a very stiff measuring system to provide proper data. Moreover, the compressibility of the water (influenced by entrained air) is an important factor because it determines the celerity of the compression wave in water. Uncertainties about model conditions endanger scaling up to prototype figures.

Minikin [1955 and 1963] has given a method to calculate wave impact pressures, but his method overestimates impact pressures and does not lead to satisfactory results.

From all experiments, however, it has become clear that the duration of the wave impact is short and the area where the impact takes place at the same time is small. This means that the wave impact forces cannot be used for a static equilibrium calculation. The dynamic effects must be taken into account, including mass and acceleration of the breakwater in conjunction with its elastic foundation and the added mass of water and soil around it. Preliminary analysis has shown that it is specifically the momentum connected with the breaking wave that determines the stability or loss of stability of the breakwater. Care must also been taken of potential resonance phenomena when the loading frequency coincides with the resonance frequency of the structure as a whole or of some individual members of the structure.

A sound method of design would establish a physical relation between the impact pressure, the hydraulic parameters and the structural parameters. On the basis of this, one should establish the exceedance curves of certain loads during the lifetime. Taking into account the response of the structure, one can then determine the probability of failure of the structure during its lifetime. Unfortunately, the physical description of wave impacts is insufficient to start this approach.

The most important lesson that can be learned from this section is the uncertainty that is connected with wave impact forces and their effects on the stability of monolithic breakwaters. It is therefore good engineering practice to try to avoid the exposure of monolithic breakwaters to breaking waves. In this context it is good to remember that even if no breaking waves are expected at the location of the breakwater, the breakwater cross-section itself may induce them, specifically when the monolith is place on a high mound of stone(see Figure 9-4).

---

[1] mwc stands for metres water column.

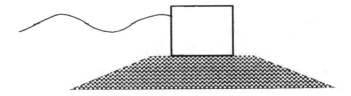

Figure 9-4 Changes to incoming wave front induced by high mound breakwater

It can also be concluded that the risk of local impact pressures increases for structural elements that entrap breaking waves. If water can escape sideways from the impact area, the pressures remain low (compare free jet), while if water cannot escape, the local pressures may become quite high (compare water hammer). In this way, certain features of monolithic breakwaters are relatively vulnerable (Figure 9-5).

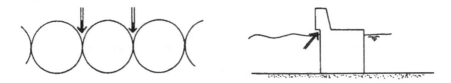

Figure 9-5 Risk of local impact forces

9.2.3 A working compromise: the Goda formula

While the uncertainties around the design of vertical breakwaters have reduced the number of such breakwaters in Europe and the USA, in Japan, construction has continued with varying satisfaction. Goda analyzed many of the successful and unsuccessful structures and came up with a practical formula that can be used to analyze the stability of a monolithic breakwater. From a theoretical point of view, one can object that Goda is not consistent in his definition of design load and risk. In practice, the safety factors he proposes are apparently adequate, as long as one realizes that conditions with breaking waves should be avoided as much as possible. If this is not possible, extensive model investigations must be carried out, followed by a dynamic analysis of the structure and the foundation. In such cases one must take into account all inertial terms.

Goda [1992] has summarized his work in an article published in 1992 for the short course on design and reliability of coastal structures. This article is added to this book as Appendix 4. Pending further theoretically based developments, the Goda formula can help to establish a preliminary idea about the stability of a monolithic breakwater.

9.2.4 Influencing the forces

It has been shown that the quasi-static forces and the dynamic forces have a tendency to translate and rotate the structure, resulting in the displacement of the structure and/or damage to the foundation and the bottom corners.

The effect of the external forces can be reduced by changing the direction of the horizontal force or by spreading the force in space and in time.

The first effect is easily understood if one realizes that the water pressure is always acting normal to a plane. When the front wall of the monolith is tilted, it means that the wave force is no longer horizontal, but directed towards the foundation. This reduces the horizontal component and strengthens the vertical component of the force. Altogether, the likelihood of sliding is reduced and the overturning moment is also reduced (Figure 9-6).

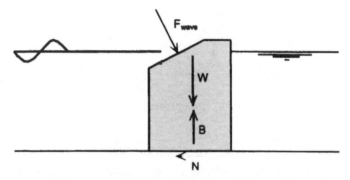

Figure 9-6 Hanstholm caisson

Another method involves the creation of a chamber in front or on top of the structure, so that the point of application of the force is spread over two walls, and a time lapse is created between the two forces. This reduces the maximum instantaneous force, although the duration is prolonged. Jarlan [1961] first applied such idea (Figure 9-7), partly to reduce forces, partly to reduce the reflection. In Japan, a large number of similar ideas have been developed and brought into practice. In a number of cases, the idea is combined with power generation (Figure 9-8). Many of the designs have been described by Tanimoto and Takahashi [1994]. Two examples of their typical designs are given in Figure 9-9 and in Figure 9-10.

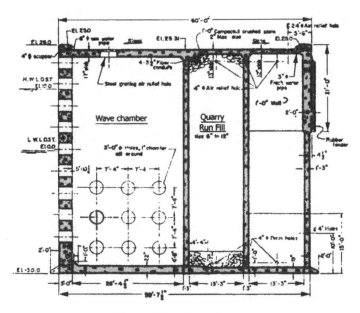

Figure 9-7 Jarlan caisson

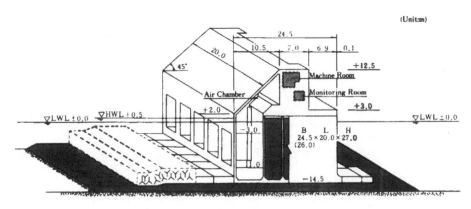

Figure 9-8 Breakwater with a wave power generating system

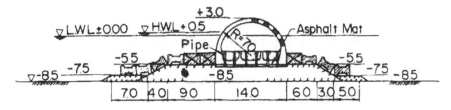

Figure 9-9 Possible cross-section of semi-circular caisson breakwater for extremely high breakers

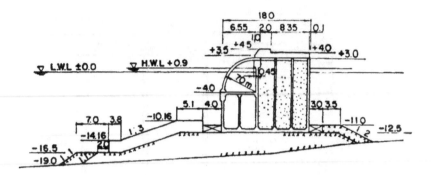

Figure 9-10 Cross section of curved-slit breakwater at Funakawa Port

Many other designers followed the ideas of Jarlan. One of these designs is given in Figure 9-11.

Figure 9-11 Honey wall breakwater

9.3 Scour

Due to the standing wave or clapotis in front of a vertical wall breakwater, the orbital velocities just above the seabed become quite high at the location of the nodes ($\frac{1}{4}L \pm \frac{1}{2}L$) in front of the structure. This can lead to unwanted scour. In this case, material that is necessary from a soil-mechanical point of view will be eroded. This can lead to overturning of the wall towards the sea. The pattern of erosion has been investigated in various models. The results differ largely, depending on the grain size of the seabed (Figure 9-12). The differences must probably be attributed to the fact that finer material is getting in suspension, and coarser material remains on the seabed. In view of the uncertainties it is recommended that the seabed in front of the wall should be protected over the length that is essential to ensure soil mechanical stability, with a minimum of 10 to 15 m.

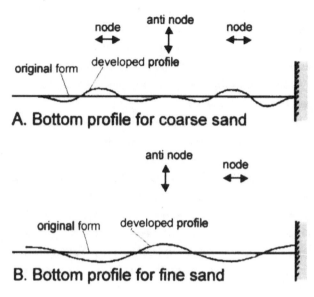

Figure 9-12 Erosion pattern depending on the grain size

9.4 Foundation

The hydraulic forces exerted on the caisson, plus the weight, determine what the local pressures in the interface between the caisson and the foundation will be. It will be clear that these pressures must not lead to soil mechanical failure. Because the foundation is flexible to a certain extent, it is necessary to verify whether the mass-spring system formed by caisson (mass) and foundation (spring) gives rise to resonance phenomena. Depending on the outcome of this investigation, one may decide that a static stability analysis is sufficient (as is often the case). Soil-mechanical failure is nevertheless one of the most likely failure modes.

Even if after analysis it is decided a quasi-static approach is justified, the cyclic effect of the load may not be overlooked. In any case the load will cause an increase in the total stress level (σ) and initiate compression of the subsoil. In first instance this will lead to a higher stress level in the ground water (p). Depending on the permeability of the soil, the excess water will drain and gradually, the effective stress (σ') will increase. This is all in accordance with one of the basic laws from soil mechanics:

$$\sigma = p + \sigma' \tag{9.2}$$

Because of the cyclic character of the load, it is possible that drainage of excess water is not complete when the next loading cycle starts. In this way, the water pressure may gradually increase due to rocking of the caisson. Eventually, this will lead to a condition in which the effective stress σ' becomes very low or even zero. A low effective stress will greatly reduce the resistance against sliding; while an effective stress equal to zero leads to liquefaction or the formation of quicksand. This is the main reason that care is recommended when designing monolithic breakwaters in areas that are sensitive to liquefaction: soil consisting of fine, loosely packed sand as in the SW part of the Netherlands.

It is possible, but expensive to use preventive methods against liquefaction. Soil replacement and compaction of the subsoil are the most commonly used methods. Widening the base of the caisson is also an effective measure.

Because of the possibility that high ground water pressures may occur under the corners of the monolith, large vertical gradients are also likely. It is therefore necessary to cover a fine-grained subsoil with an adequate filter. Because of the large gradients, it is recommended that the filter be designed as a geometrically impervious filter. Filter rules have been treated extensively by Terzaghi. The theory has also been covered in other textbooks during the curriculum (Schiereck [2001]). The most important features are summarized in Appendix 5.

A granular foundation layer may also be required if the structure is placed on an uneven hard seabed. In this case, it is the function of the foundation layer to flatten the seabed and to avoid pressure concentrations and an unpredictable support pattern for the structure. Alternatively, one may create pre-designed contact areas in the bottom of the structure, so that the bending moments in the floor plate can be calculated.

To create a perfect homogenous contact plane between the foundation and the structure, a grout mortar is sometimes injected. This technique has been developed in the offshore industry for the foundations of gravity platforms, but the use has spread to other coastal engineering projects. To avoid loss of grout, a skirt is provided along the circumference of the bottom of the caisson. This skirt (usually a steel sheet) penetrates into the foundation and creates a chamber that can be filled with the grout mortar.

10 WAVE–STRUCTURE INTERACTION

There is a strong interaction between a wave and a wave damping structure like a breakwater. This interaction is visible in front of the structure (reflection), on the slope of the structure (run-up) and behind the structure (overtopping and transmission). This chapter summarizes these interactions. It is not necessary to learn all the formulae by heart, but one should be aware of the order of magnitude of the effects.

10.1 Introduction

Even if a breakwater structure is stable under the action of waves, there is interaction between the structure and the wave field near the structure. We discern various phenomena that lead to different wave patterns in the vicinity of the structure:

* Wave reflection
* Wave run-up
* Overtopping
* Wave transmission

Wave reflection plays a role in front of the structure, wave run-up takes place on the slope of the structure, and overtopping and transmission are important for the area behind the structure. Before using any of the expressions given in this chapter, it is useful to analyze which phenomenon influences the design problem in question. Too often, formulae for run-up or overtopping are used when the designer wishes to address wave transmission.

10.2 Reflection

In Chapter 5, we have already seen the formula for a standing or partially standing wave.

The wave motion in front of a reflecting structure is mainly determined by the reflection coefficient r.

If 100% of the incoming wave energy is reflected, one can safely assume that the reflection coefficient $K_r = H/H_i = 1$. This is generally valid for a rigid vertical wall of infinite height.

The reflection coefficient for sloping structures, rough or permeable structures, and structures with a limited crest level is smaller.

Postma [1989] has investigated the reflection from infinitely high rock slopes. He found a clear influence of the breaker parameter ξ, and of the "permeability" P as defined by Van der Meer (compare Figure 7-3).

For a first estimate, Postma proposes the use of a simple formula:

$$K_r = 0.140 \, \xi_{op}^{0.73} \tag{10.1}$$

For a more accurate approach, he gives the formula:

$$K_r = 0.081 \, P^{-0.14} \cot\alpha^{-0.78} s_{op}^{-0.44} \tag{10.2}$$

This formula can only be used within the validity range of the various parameters as given below:

$$
\begin{aligned}
0.1 &< P &&< 0.6 \\
1.5 &< \cot\alpha &&< 6 \\
0.004 &< s_{op} &&< 0.06 \\
0.7 &< \xi_{op} &&< 8 \\
0.1 &< K_r &&< 0.8 \\
0.03 &< h/L_{op} &&< 0.3 \\
0.09 &< H_{si}/h &&< 0.23 \\
2 &< H_{si}/D_{n50} &&< 6
\end{aligned}
$$

10.3 Run-up

Standard case: smooth, impermeable

Wave run-up is the phenomenon in which an incoming wave crest runs up along the slope up to a level that may be higher than the original wave crest. The vertical distance between still water level SWL and the highest point reached by the wave tongue is called the run-up "z". From this definition, it is clear that we can only speak of run-up when the crest level of the structure is higher than the highest level of the run-up (Figure 10-1). Run-up figures are mainly used to determine the probability that certain elements of the structure will be reached by the waves. Run-up can be indirectly used to estimate the risk of damage to the inner slope of the structure.

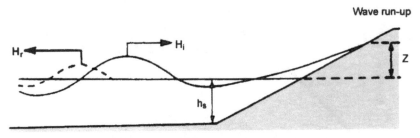

Figure 10-1 Definition of wave run-up

In the Netherlands, research on run-up has always attracted a lot of attention (Battjes [1974], Anonymous [1998]). This was mainly because of the need to assess the required crest level for dikes and sea walls. Most of the data have been collected and published by the "Technische Adviescommissie voor de Waterkeringen, TAW" (Technical Committee for sea defences). Since most of the research effort was directed to run-up on dikes with a slope protection of asphalt or stone revetment, most results are valid for smooth, impermeable cover layers on the seaward slope.

Since the inner slopes of many dikes in this country are covered with grass, it is not acceptable for a large percentage of the incoming waves to reach the crest and subsequently cause damage to the inner slope. Therefore, in most cases, the 2% run-up is given: the run-up level that is exceeded by 2% of the incoming waves. It is assumed that the grass on the inner slope of a sea dike can withstand this condition. The run-up on a smooth impermeable slope is then expressed as:

$$z_{2\%}/H_s = 1.6\xi_{op} \tag{10.3}$$

with a maximum of 3.2, in which:

$z_{2\%}$ = run-up level exceeded by 2% of the waves

H_s = significant wave height

ξ_{op} = breaker parameter for deep water and peak period

The run-up level can effectively be reduced by designing a berm at still water level, by increasing the roughness of the surface, or by increasing the permeability of the structure. Waves approaching the structure at an angle will also lead to reduced run-up levels. This reduction is expressed in terms of reduction factor γ. The effective run-up is than calculated by multiplying the value for smooth impermeable slopes by the correction factors. In the sections below, some values will be presented for the respective cases.

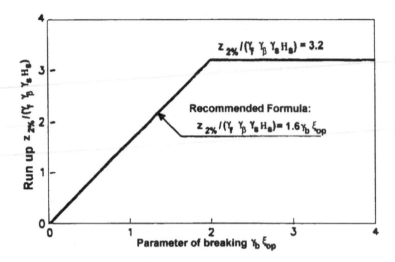

Figure 10-2 Wave run-up on a smooth impermeable slope

Rough slopes, impermeable/non-porous

Run-up levels on rough, impermeable/non-porous slopes can easily be calculated by applying the correction factors indicated in Table 10-1 below. The factors are valid only for $\xi_{op} < 3$ to 4. Above the value of 4, the reduction is no longer applicable.

Structure	γ
Smooth, impermeable (like asphalt or closely pitched concrete blocks)	1.0
Open stone asphalt etc.	0.95
Grass	0.9 – 1.0
Concrete blocks	0.9
Quarry stone blocks (granite, basalt)	0.85 – 0.9
Rough concrete	0.85

Table 10-1 Correction factors for roughness

Rough slopes, limited permeability/porosity

Even for rough slopes with a limited permeability/porosity, it is possible to apply the general run-up formula and to make corrections as long as $\xi_{op} < 3$ to 4. Correction factors for such cases are given in Table 10-2.

Structure	γ
One layer of stone on an impermeable base	0.55 – 0.6
Gravel, Gabion mattresses	0.7
Rip-rap rock, layer thickness n > 2	0.5 – 0.55

Table 10-2 Correction factors for roughness and permeability/porosity

Rubble mound (armoured) breakwaters

Special investigations have been done for slopes on an actual breakwater. It appears that the permeability P, as defined by Van der Meer, again plays a role.
Van der Meer and Stam [1992] propose a set of formulae:

for $\xi_m < 1.5$: $\qquad$ $z_{u\%}/H_s = a\,\xi_m$ $\hspace{3cm}$ (10.4)

and:

for $\xi_m > 1.5$: $\qquad$ $z_{u\%}/H_s = b\,\xi_m^c$ $\hspace{3cm}$ (10.5)

These formulae are valid for breakwaters with an impermeable or almost impermeable core ($P = 0.1$). For structures with a permeable core ($P = 0.4$), the relative run-up is limited to a value:

$$z_{u\%}/H_s = d \hspace{5cm} (10.6)$$

in which:
$z_{u\%}$ $\quad$ = run-up level exceeded by u % of the incoming waves
a, b, c, d = parameters according to Table 10-3.
ξ_m $\quad$ = breaker parameter for deep water, mean period

run-up level $u\,(\%)$	a	b	c	d
0.1	1.12	1.34	0.55	2.58
1	1.01	1.24	0.48	2.15
2	0.96	1.17	0.46	1.97
5	0.86	1.05	0.44	1.68
10	0.77	0.94	0.42	1.45
Sign.	0.72	0.88	0.41	1.35
Mean	0.47	0.6	0.34	0.82

Table 10-3 Run-up parameters for rubble covered and permeable lopes

The formulae are presented graphically in Figure 10-3.
The reader is reminded here, that extensive breaking of waves will result in a spectrum that hardly shows a clear peak. A lot of energy has shifted to the lower frequencies. Definition of the wave period becomes difficult, but remains important because of the influence on the run-up. In the near future, results of research into this wil become available.
No separate data are available for run-up on slopes covered with concrete armour units.

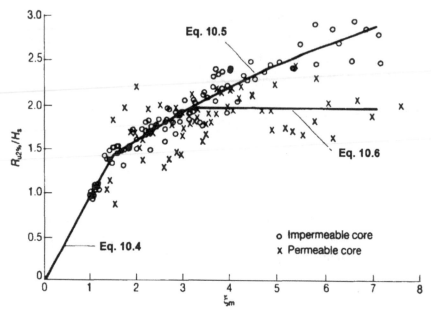

Figure 10-3 Relative 2% run-up on rock slopes

10.4 Overtopping for rubble mounds

Overtopping is defined as the quantity of water passing over the crest of a structure per unit of time. It therefore has the same dimensions as the discharge Q [m³/s]. Because this quantity of water is often a linear function of the length of the structure, it is expressed as a specific discharge per unit length [m³/s/m].

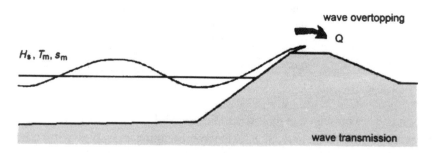

Figure 10-4 Typical wave overtopping

The study of overtopping quantities is again a subject that was initiated in relation to the stability of the inner slopes of grass covered dikes. When designing breakwaters, the quantity of overtopping may be important to determine the capacity of the drainage facilities required for port areas directly protected by the breakwater, or to assess the risk to people or installations on the crest of the breakwater. Although the assessment of these sometimes-subjective risks on the basis of model experiments is difficult, it is possible to derive a trend. Earlier investigations from Japan and Italy (Gerloni et al. [1991]) have been collected and interpreted by Van der Meer [1993]. Their results are given in Table 10-4.

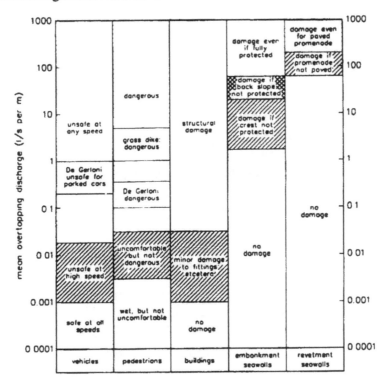

Table 10-4 Overtopping discharges and their effects

Like the conditions governing wave run-up, the quantity of overtopping is also largely influenced by the nature of the outer slope of the structure. An added effect is the influence of the shape and nature of the crest (presence of a crown wall). Unfortunately, various model investigations (Bradbury and al. [1988], Owen [1980], Van der Meer and Stam [1992], De Waal and Van der Meer [1992] and Aminti and Franco [1994]) were carried out with varying structures. Therefore it is not possible to derive a generally applicable formula for overtopping. The reader is advised to study the available literature in detail and to select the most promising approach for his specific problem. For a first approximation, one may use the formula of Bradbury et al (1988), which is valid for a structure without a crown wall, and similar to Figure 10-4:

$$Q^* = a(R^*)^{-b} \qquad\qquad (10.7)$$

with:

$$Q^* = \frac{Q}{\sqrt{g H_s^3}} \sqrt{\frac{S_{om}}{2\pi}} \tag{10.8}$$

and

$$R^* = \left(\frac{R_c}{H_s}\right)^2 \sqrt{\frac{S_{om}}{2\pi}} \tag{10.9}$$

in which:

R^* = dimensionless crest freeboard
R_c = crest freeboard relative to SWL in m
H_s = significant wave height
S_{om} = deep water wave steepness, based on mean period
Q = specific discharge in m³/s/m
Q^* = dimensionless specific discharge

Relevant values of the coefficients a and b depend on the structural details as given in Figure 10-5. The values for *a* and *b* are given in Table 10-5.

Type of structure	a	b
Section A	$3.7 * 10^{-10}$	2.92
Section B	$1.3 * 10^{-9}$	3.82

Table 10-5 Overtopping coefficients

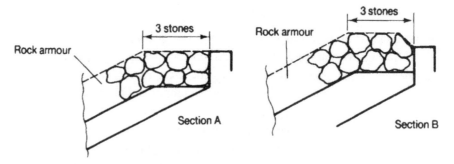

Figure 10-5 Overtopped rock structures with low crown walls

10.5 Overtopping and transmission for vertical walls

Goda [1985] has investigated the transmission of vertical wall breakwaters placed on a rubble berm. He relates the relative freeboard R_c/H to the transmission coefficient $K_t = H_t / H_i$ and finds a slight influence of the height of the berm (see Figure 10-6).

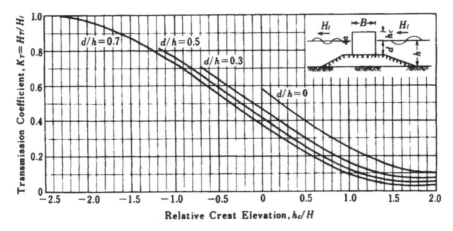

Figure 10-6 Wave transmission for a vertical breakwater

Overtopping has been investigated by Franco et al [1994]. They describe the unit discharge q as:

$$\frac{q}{\sqrt{gH_s^3}} = a \exp(-b\frac{R_c}{\gamma H_s}) \qquad (10.10)$$

in which:

q = unit discharge (m³/m/s)

a, b = experimental coefficients

γ = geometrical parameter

For a rectangular shape:

$a = 0.192$

$b = 4.3$

$\gamma = 1$

It must be kept in mind that a vertical face breakwater causes a lot of spray when hit by a wave. The spray may also be blown over the breakwater. This effect is not included in the above formula.

10.6 Transmission by rubble mounds

Wave transmission is the phenomenon in which wave energy passing over and through a breakwater creates a reduced wave action in the lee of the structure (Figure 10-7). This will certainly happen when considerable amounts of water are overtopping the structure. Wave transmission is also possible, however, when the core of the structure is very permeable and the wave period is relatively long. It is specifically the influence of these two factors that for a long time has prevented the derivation of an acceptable formula for wave transmission by rubble mound breakwaters. In Section 10.5, the simpler case for a monolithic breakwater was presented.

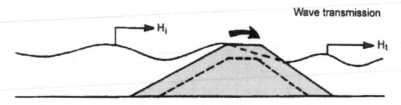

Figure 10-7 Typical wave transmission

Many authors (Seelig [1980], Powell and Allsop [1985], Daemrich and Kahle [1985], Van der Meer [1990]) have investigated the effects of wave transmission. This has resulted in the diagram presented in Figure 10-8. It must be noted that the transmission coefficient can never be smaller than 0 or larger than 1. In practice, limits of about 0.1 and 0.9 are found (Figure 10-8).

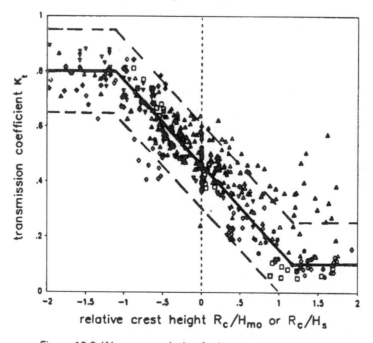

Figure 10-8 Wave transmission for low crested structures

It is remarkable that for $R_c = 0$, which represents a structure with the crest at SWL, the transmission coefficient is in the order of 0.5. This means that a relatively low structure is already rather effective in protecting the harbour area behind the breakwater. In combination with the requirements for tranquillity in the harbour, the designer can decide on the minimum required crest level.

Eventually, Daemen [1991] (see also Van der Meer and d'Angremond [1991]) in his MSc thesis was able to produce an acceptable formula that relates the transmission coefficient to a number of structural parameters of the breakwater. To account for the effect of permeability, Daemen decided to make the freeboard R_c of the breakwater dimensionless dividing it by the armour stone diameter. This eliminates much of the scatter that was present in previous approaches. For traditional low crested breakwaters the Daemen formula reads as follows:

$$K_t = a \frac{R_c}{D_{n50}} + b \tag{10.11}$$

with:

$$a = 0.031 \frac{H_i}{D_{n50}} - 0.24 \tag{10.12}$$

$$b = -5.42 s_{op} + 0.0323 \frac{H_i}{D_{n50}} - 0.0017 \left(\frac{B}{D_{n50}} \right)^{1.84} + 0.51 \tag{10.13}$$

in which:
$K_t = H_{st}/H_{si}$ = transmission coefficient
H_{si} = incoming significant wave height
H_{st} = transmitted significant wave height
R_c = crest freeboard relative to SWL
D_{n50} = nominal diameter armour stone
B = crest width
s_{op} = wave steepness

The use of the Daemen formula is complicated when it is decided to use a solid crown block, or to grout armour stones with asphalt into a solid mass. Therefore, another MSc student, R.J. de Jong [1996] (see also d'Angremond et al. [1996]), reanalyzed the data and came up with a different expression. He choose to make the freeboard dimensionless in relation with the incoming wave height:

$$K_t = a \frac{R_c}{H_{si}} + b \tag{10.14}$$

with:

$a = -0.4$, and

$$b = 0.64 \left(\frac{B}{H_{si}} \right)^{-0.31} \left(1 - e^{-0.5\xi} \right) \tag{10.15}$$

The factor 0.64 is valid for permeable structures; it changes to 0.80 for impermeable structures.

11 DESIGN PRACTICE OF BREAKWATER CROSS-SECTIONS

The design of a breakwater is not only an application of theory, but is also largely based on experience, therefore this chapter attempts to combine a set of simple rules into a design manual. However, it is felt that a set of design rules can never satisfactorily solve a design problem so it is essential to look into cross-sections of existing breakwaters, while reading this chapter. To facilitate this process, some examples of breakwaters in the Netherlands have been given in one of the appendices. Studying this chapter with some degree of success is only possible if one analyzes the examples at the same time and wonders why the specific designs have been followed.

11.1 Introduction

In the foregoing chapters we have used terms such as 'armour layer' and 'core', we have discussed the stability of concrete armour units, and we have seen the principles of a berm breakwater. It is now time to discuss some practical rules for the design of cross-sections. Many of these rules are simply rules-of-thumb, and only gradually has an experimental basis been created on which acceptance or rejection of these rules can be based. The present chapter is thus a mixture of research, experience and plain engineering judgement.

It must be kept in mind that natural rock is usually obtained by blasting, and that the size of the stone obtained (yield curve) can only be influenced to a limited extent. It is much easier to increase the percentage of fine material than the percentage of coarse material. Any material that is blasted must be handled in the quarry, whether or not it is used in the breakwater. If it is not used, it must be left somewhere, and in many countries, the deposition of waste material requires special licences that are difficult to obtain. It is therefore an element of good design to try to use all materials produced by the quarry.

In principle, we can follow two options.

- The first one is to split the quarry production into two or three categories (filter material, core and armour). This almost automatically leads to a berm breakwater. The gradation of the stone categories is rather wide, (D_{n85}/D_{n15} up to 2.5 or even 3).
- The second option includes the classification of the quarry output into a larger number of categories, each with a narrower gradation (D_{n85}/D_{n15} up to 1.5). In this way, it is possible to select the proper stone size for a specific function. It leads to a more economical (and thus more economic) use of material.

The berm breakwater with its larger volume has the advantage when the production cost is low and the quarry is located near the site of the breakwater. When quarry stone is more costly and the quarry is at greater distance, it is more economical to build a multi-layered breakwater. However, in this case also it is advantageous to keep the design of the cross-section as simple as possible.

11.2 Permeability/porosity and layer thickness

11.2.1 Permeability/porosity

When rock or concrete blocks are placed in the cross-sections of a breakwater, it is important to have an idea of the permeability/porosity and the layer thickness. The permeability/porosity is important because it determines at least part of the hydraulic response of the structure, and it influences the stability (P in the Van der Meer formula). During construction, it is important because the porosity determines the bulk density. Quarry stone is often paid for per ton of material to the quarry operator. When the contractor is paid per m³ for placing the material in the cross-section, as is often the case, to make a proper cost estimate it is essential to know the bulk density of the material.

The volumetric porosity n_v is defined as follows:

$$n_v = 1 - \left(\frac{\rho_b}{\rho_r} \right)$$

(11.1)

in which:

ρ_b = bulk density as laid
ρ_r = density of rock

Determination of the bulk density is not simple because of the errors made at the boundaries of the measured volume. Preliminary data can be found in the Shore protection Manual (Anon. [1984]), the CUR/CIRIA Manual (Anon.[1991]), or in the more recent MSc thesis of Bregman [1998].

Table 11-1, with data from the Shore Protection Manual and the CUR/CIRIA Manual, indicates porosity levels between 35 and 40% for quarry stone placed in thin layers. Bulk handling may lead to a porosity that is up to 5% higher than the values in Table 11-1. A wider gradation, however, may lead to lower porosity. Because of the uncertainties in the determination of the porosity and the bulk density, it is recommended that some in situ tests to ascertain the actual values should be carried out. It is emphasized that special placement of quarry stone (with the longest dimension either parallel or perpendicular to the slope) has a big effect on both layer thickness, and stability.

In Table 11-1 the porosity of concrete units is also given. Here again, the method of placement may cause large differences in porosity and in stability.

Type and shape of units	Layer thickness n	Placement	Layer coefficient k_t	Porosity n_v	Source
Smooth quarry stone	2	Random	1.02	0.38	SPM
Very round quarry stone		Random	0.80	0.36	Cur/Ciria
Very round quarry stone		Special	1.05 – 1.20	0.35	Cur/Ciria
Semi-round quarry stone		Random	0.75	0.37	Cur/Ciria
Semi-round quarry stone		Special	1.10 – 1.25	0.36	Cur/Ciria
Rough quarry stone	2	Random	1.00	0.37	SPM
Rough quarry stone	> 3	Random	1.00	0.40	SPM
Irregular quarry stone		Random	0.75	0.40	Cur/Ciria
Irregular quarry stone		Special	1.05 – 1.20	0.39	Cur/Ciria
Graded quarry stone		Random	-	0.37	SPM
Cubes	2	Random	1.10	0.47	SPM
Tetrapods	2	Random	1.04	0.50	SPM
Dolosse	2	Random	0.94	0.56	SPM
Accropode	1	Special	1.3	0.52	Sogreah
Akmon	2	Random	0.94	0.50	WL

Table 11-1 Thickness and porosity in narrow gradation armour layers

11.2.2 Layer thickness and number of units

For armour layers, it is important to know the effective layer thickness for single, double or triple layers of material. This is essential information when designing and constructing a cross-section. The crest level is determined on the basis of the required protection on the lee side of the breakwater. Given this required crest level, one must know at what level the core must be finished. If the crest of the core is too high, this indicates that too much material has been used, which will probably not be paid for by the client. If the crest of the core is too low, and the client still wants the given crest level, this means that extra armour stone has to be used. Since armour stone is generally more expensive than core material, this again will cause a financial loss to the contractor.

The effective layer thickness is discussed extensively in the Shore Protection Manual [1984]. The thickness of a layer t (in m) is calculated as:

$$t = n\,k_t\,D_{n50} \tag{11.2}$$

in which n is the number of stones across the layer.

In the American literature, k_Δ is often used instead of k_t and r instead of t.

Values for k_t are also given in Table 11-1.

It is also important to know the number of armour units N required to cover a certain area A. This number is:

$$N = n\,k_t\,A\left(1 - n_v\right)D_{n50}^{-2} \tag{11.3}$$

Although Table 11-1 gives values of k_t and n_v for concrete armour units as well, it is emphasized that those values can fluctuate considerably. This depends on the interpretation of the qualification "random placement". It is possible for instance to place cubes much more densely than indicated in the table. This will improve the stability, but it will also influence the reflection, the run-up, the overtopping, and the transmission. Therefore, care is required when data from inexperienced researchers are used.

11.3 Berm breakwater

In principle, the cross-section of a berm breakwater consists of two materials, core material and armour material. The armour material is the coarsest fraction of the quarry yield, the core material is the finer fraction. The armour material is located in a berm along the outer slope of the breakwater. The quantity of armour material is chosen in such a way that after a series of storms envisaged during the lifetime of the breakwater, the core be always covered by at least a double layer of armour material. This applies to the front slope, the crest, and the exposed part of the inner slope, i.e. to a little below low water level.

Crest level

In Chapter 8 (Dynamic Stability), it was explained that the crest level determines the stability of the inner slope. Given a reasonable ratio between wave height and nominal stone size, the minimum crest level follows from the accepted level of damage rather than from the functional requirements. When designing the cross-section it is good engineering practice to create a safe level at which trucks, bulldozers and cranes can work without much interference from waves.

Filter layer

On the front slope, the incident and reflected waves create a complicated pattern of orbital velocities and pressure fluctuations. This will cause larger stones to sink slowly into the seabed, unless the latter consists of rock. The same hydraulic conditions will enhance the risk of scour and erosion of the seabed just in front of the breakwater. Eventually, both phenomena together will lead to the loss of material and potentially to the loss of the stability of the entire breakwater. Therefore, to prevent the formation of a scour hole close to the structure, it is good engineering practice to provide a filter layer under the toe of the breakwater and to extend this filter some distance in front of the toe. In the case of a berm breakwater, it is also wise to apply such a filter in the area that will be covered by armour stone after reshaping of the seaward slope.

Protection of the seabed may consist of a granular filter, or a geotextile with a cover of quarry stone. Filter rules are beyond the scope of this book; the reader is referred to the CUR Manual (Anon. [1995]) on the use of rock in hydraulic engineering or to Schiereck [2001]. A brief summary of filter rules is given in Appendix 5.

Slopes

Since wave action will reshape the profile anyway, it makes no sense to construct the cross-section to a particular slope. It is generally accepted that the material should be left to find its natural angle while it is being deposited by barge, dump truck and/or bulldozer.

This leads to two frequently used basic cross-sections, one with the berm at crest level and one with the berm just above MSL. These two cross-sections are presented in Figure 11-1 and Figure 11-2.

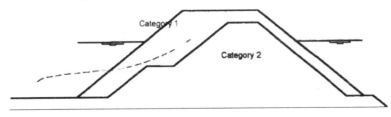

Figure 11-1 Berm breakwater with high berm

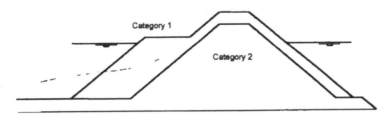

Figure 11-2 Berm breakwater with berm at MSL

11.4 Traditional multi-layered breakwater

11.4.1 Classification

Although there are some standard cross-sections for traditional multi-layered breakwaters, numerous variations can be made, which makes classification quite difficult. The first, and most logical classification is done by crest level. A second type of classification is by type of crest, and a third type refers to the kind of armour units.

High / low crest

In this context, a high crest is considered so high that the inner slope is not severely attacked by the action of overtopping waves. The inner slope is designed for waves generated by passing ships or waves locally generated in the harbour basin. A low crest is so low that the inner slope is severely attacked by waves passing over the crest. To protect the inner slope, usually the same armour type is used on the inner slope as on the seaward slope. The crest level is generally determined on the basis of the functional requirements (wave tranquility on the lee side). It must be clear that selection of too high a crest level leads to excessive use of material because the volume of the structure is proportional to the square of its height. Selection of a low crest level may also have serious consequences for the construction method and the maintenance. A low crest may lead to the exclusive use of floating equipment.

Crest design

A second type of classification is possible with respect to the nature of the breakwater crest. This can consist of armour units or of a solid cap block. If the crest consists of armour units, the crest is usually inaccessible to people or equipment. This means that maintenance is only possible by using floating equipment. If the crest is formed by a solid block it is common practice to design it in such a way that it can be used as a road, both during construction and subsequently for maintenance or other purposes.

Rock or concrete armour units

A third type of classification is possible on the basis of the type of armour material. Since the maximum size of quarry stone is limited, it is not uncommon to reduce the seaward slope to obtain sufficient stability. (Note: the inner slope can be steeper!). If concrete armour units are used, it can easily be demonstrated that a steep slope of say $1:1\frac{1}{2}$ (cot $\alpha = 1.5$) leads to the most economic design.

11.4.2 General design rules

To explain the various design rules, a definition sketch of a multi-layered cross-section is given in Figure 11-3. It indicates the main elements of such a breakwater and their respective names.

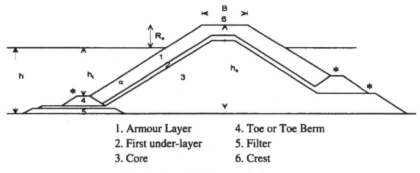

1. Armour Layer	4. Toe or Toe Berm
2. First under-layer	5. Filter
3. Core	6. Crest

Figure 11-3 Definition sketch of cross-section

Tolerances

Regardless of the construction element under consideration, it must be clear that considerable size deviations can be expected. Not only deviations between size of the design and actual breakwater, but also deviations in size from one location to another location in the breakwater. The construction of mounds of stones with units up to 1 or 2 m diameter has an inherent inaccuracy in the order of magnitude of the size of the stones. The design of the structure must take into account these tolerances. At different levels and locations, berms must be provided in which deviations in the actual measurements can be taken into account. These locations have been marked with an asterisk in Figure 11-3 through Figure 11-8.

Armour layer

It is evident that the armour layer must be able to withstand the wave attack during design conditions. The severity of the stipulated conditions follows from economic considerations. In general, the armour is placed in a double layer ($n = 2$), since this allows a few armour units to be displaced before the underlying material is exposed.

The armour layer consists of concrete units or quarry stone. In the case of quarry stone, it is generally the heaviest fraction of the quarry yield curve. This has a narrow grading ($D_{85}/D_{15} < 1.5$). If quarry stone is used, it is possible to reduce the slope in order to improve the stability. When concrete armour units are used, it is more economical to increase the block weight if this is needed for stability.

Crest

If the crest consists of loose armour units, its width (B) must be at least 3 stones, or in the form of a formula:

$$B = n\,k_t\,D_{n50} \tag{11.4}$$

where $n = 3$, and k_t is taken from Table 11-1.

If the crest is formed by a concrete cap block, it must be ensured that the layer on which the block is placed or cast in situ is wider than the cap-block. It is never possible to fill voids under the cap-block after it has been placed. To protect the cap-block, it is recommended that armour material be placed at the seaward side to the full height of the breakwater. Parapet walls extending above the level of the armour units will be loaded heavily and in many cases, such walls have suffered extensive structural damage. To ensure that the armour layer shields the cap block properly it is recommended that a horizontal berm is maintained in front of the cap that is least as wide as one armour unit (see Figure 11-4).

First under-layer

The layer directly under the armour layer is called the first under-layer. It is obvious that the units forming this layer must not pass through the voids in the armour layer. In the literature, one finds a rule that the weight of units in the under-layer should not be less than 1/10 of the weight of the armour proper. This seems a very strict rule if compared with the filter rules of Terzaghi. These rules allow a ratio of 4 to 5 in diameter between two subsequent filter layers. However, one must remain a bit on the conservative side because of the consequences of failure of the filter mechanism. The filter must therefore be "geometrically impermeable". It is recommended that the weight ratio of subsequent layers of quarry stone be kept between 1/10 and 1/25. (D_{n50} ratio between 2 and 3). For more information, one is referred to Appendix 5. It is noted that in this context, choosing finer material for the first under-layer influences the notional permeability parameter P in the Van der Meer formula. This leads to the need for heavier armour material.

A second consideration for the selection of a certain size for the material in the first under-layer may be the stability during construction. Depending on the construction sequence, the first under-layer may be exposed to a moderate storm during construction.

If the armour units are concrete blocks, the first under-layer is the heaviest fraction of the quarry yield curve. When the armour units are quarry stone, the first under-layer is composed of an intermediate fraction of the quarry yield. It is generally a narrow grading.

Toe berm

The toe berm is the lower support for the armour layer. In traditional literature, one finds a weight recommended that is equal to the weight of the first under-layer. With the most recent data that are presented in the chapter on stability formulae, the designer can find a balance between the level of the toe berm and the size of the stone in the berm.

Core

In most cases, the material of the first under-layer is such that the core can be situated directly under it. Again assuming a weight ratio between the first under-layer and the core of between 1/10 and 1/25, this means that the core material is a factor 100 to 625 lighter than the armour material. This means that usually it is not necessary to apply a second under-layer between the core and the first under-layer.

For the core, a material called "quarry run" or "tout venant" is usually used, indicating that it is meant to represent the finer fractions of the quarry yield curve. It must be noted that under no circumstances can overburden (degraded or weathered rock) be mixed with the quarry run. Quarry run generally has a wide ($1.5 < D_{85} / D_{15}$ < 2.5) to very wide ($D_{85} / D_{15} > 2.5$) grading.

The use of large units in the core is not a problem from a stability point of view, although it does have a negative influence on wave transmission and sand tightness.

Filter

Specifically under the seaward toe, large pressure gradients that tend to wash out material from the seabed through the structure may exist. Even extension of the core material under the toe berm may not guarantee the integrity of the structure as a whole. Loss of material in this region is an important threat to the stability of the armour layer. There are ample examples in the literature that show that substandard filters have initiated the failure of a breakwater.

It is therefore recommended that a geometrically impermeable filter should be placed under the seaward part of the breakwater. This filter may consist of a number of layers of granular material, or of a geotextile or other mattress. The pressure gradients under the centre of the structure and under the inner toe are generally much lower. Here, the quarry run may often act as a filter of sufficient quality. Care must be taken, however, when land is reclaimed directly behind the breakwater. Internal reflection may then again cause filter problems at the inner boundary of the breakwater. In such a case, special investigations are required to determine a satisfactory solution (reverse filter).

Since the layers of a granular filter are constructed at a considerable depth under water, it is necessary to make any separate layer thick enough to guarantee the presence of that particular material at any location. It will also be useful if the presence of the required material can be ascertained by inspection. In practice, this means that no layers thinner than 0.5m should be designed. This may lead to a relatively thick filter bed if a granular filter is used.

Scour protection

Just in front of the breakwater, the seabed may be eroded owing to a concentration of currents, or to a partially standing wave. Since the loss of bed material directly in front of the toe may cause a soil mechanical stability problem, it is recommended that a blanket should be placed in front of the breakwater as scour protection. The width should be determined on the basis of local conditions, but should not be less than 5 to 10 m for a rubble mound breakwater and 10 to 15 m for a vertical wall breakwater.

11.4.3 Standard cross-sections

In Figure 11-4, Figure 11-5, Figure 11-6, Figure 11-7 and Figure 11-8,examples of standard cross-sections based on the considerations given in Section 11.4.2 are given. These cross-sections show the changes of the leeward slope for increasing crest level and thus for decreasing overtopping and transmission. Examples of cross-sections with a concrete cap-block are also given. This feature is rarely seen in low-crested or submerged breakwaters, probably because of the difficulty of placing the block.

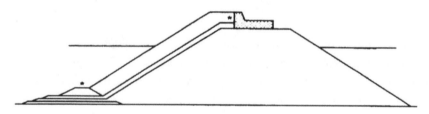

Figure 11-4 Rubble mound breakwater – light overtopping (with cap block)

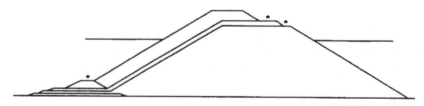

Figure 11-5 Rubble mound breakwater - light overtopping

Figure 11-6 Rubble mound breakwater - moderate overtopping

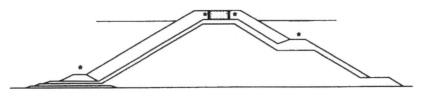

Figure 11-7 Rubble mound breakwater - moderate overtopping (with cap block)

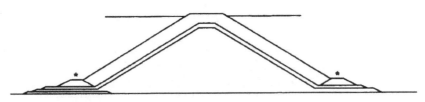

Figure 11-8 Rubble mound breakwater - severe overtopping

It must be noted that owing to the relation between local water depth and local significant wave height, the cross-section (including the size of the armour units) will vary considerably along the alignment of the breakwater. This gives the designer an added opportunity to match the quarry output to the over all demand of the project.

In the shallow water close to the shore, the standard design with a granular filter under the toe, will be difficult to construct. Owing to the thick filter bed, the level of the toe berm becomes too high. The problem can be solved by dredging a trench for the toe, by replacing the granular filter by geotextile, or by modifying the toe berm. These four solutions are sketched in Figure 11-9, Figure 11-10, Figure 11-11 and Figure 11-12.

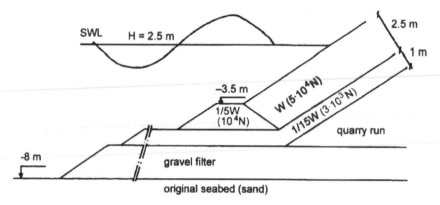

Figure 11-9 Shallow water, dredged trench

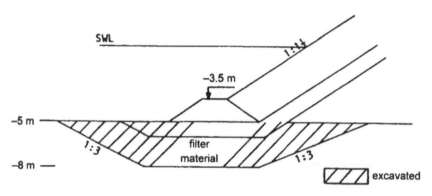

Figure 11-10 Shallow Water dredged trench gravel filter

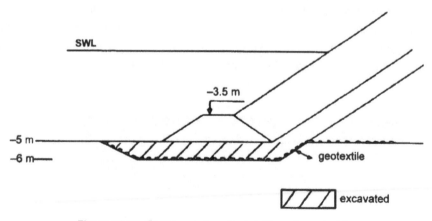

Figure 11-11 Shallow water, dredged trench geotextile

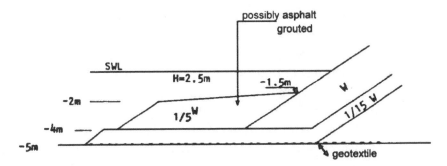

Figure 11-12 Shallow water (no excavation, geotextile and increased berm)

Although the dredging of a trench seems expensive and a rather academic solution, it is not. In many cases, the bearing capacity of the subsoil is insufficient to create a safe foundation for the heavy load presented by the breakwater. This will be demonstrated by low safety coefficients in slip circle calculations, specifically when the soil is compressible and impermeable (consolidation time). In such cases, it is good engineering practice to apply soil improvement. Placing the toe in the dredged trench creates the intended soil improvement.

Although the figures presented in this chapter give a good impression of possible cross-sections, the reader is recommended to study cross-sections of actual breakwaters as well. This applies to both successful designs that have survived, and to the unsuccessful examples that failed. When studying cross-sections in handbooks, it must be kept in mind that sometimes for the sake of simplicity, essential features are not shown. In Appendix 6, a review of all alternative cross-sections that were generated during the design process of the Europoort breakwaters is given. In the same Appendix, sketches of the breakwaters built in IJmuiden and Scheveningen are also given.

11.5 Monolithic breakwaters

In spite of the complexity of calculating the stability and the structural strength of monolithic breakwaters, the international commission on the study of waves (Anonymous [1976], commissioned by PIANC) has established a number of design recommendations. It is stressed here that these data are relatively old, and that a PIANC working group is currently preparing a new report on vertical wall breakwaters.

The Commission on the Study of Waves distinguishes two limit states:

- A limit state for the use of the breakwater characterized by a wave height H_u with a reasonable return period
- A limit state for the rupture of the breakwater characterized by a wave height H_r that is an extreme wave height

For establishing a preliminary design of a vertical breakwater the commission recommends a first approximation for the cross-section as follows (Figure 11-13):
- Wall that presents a free height that is at least $1.5H_r$ below low water (follows a recommendation by the XVIIIth International Navigation Congress, Rome 1953)
- Wall, the thickness of which is at least 0.8 times the free height
- A toe protection against undermining the thickness of which is at least 0.15 times the free height. (This places the seabed at least $1.72\ H_r$ under LW)
- Crest rising to an elevation of 1.3 to 1.5 times H_u above HW on the sea side and 0.5 times H_u on the harbour side
- Parapet wall the thickness of which is about $0.75\ H_u$
- Scour protection extending at least $2.5\ H_u$ in front of the wall, with a minimum of 10 to 15m

This leads to a basic cross-section as sketched in Figure 11-13. This sketch can only be used as a first approximation in a design and its stability must be verified thoroughly for every new application.

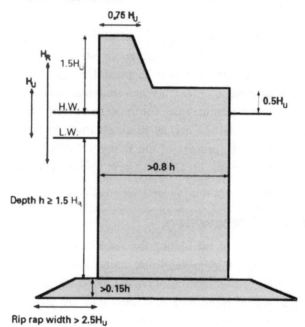

Figure 11-13 Recommended cross-section according to PIANC (1973)

12 DESIGN PRACTICE FOR CLOSURE DAMS

Chapter 12 treats the design of closure dams. Contrary to Chapter 11, little attention is given to the cross-section of the closing dam. The main issue in this schapter is the closing operation itself. The reader must attempt to understand the relation between narrowing the cross-sectional profile and the tidal motion.

12.1 Closing an estuary, creating final gaps in the tidal channels

A detailed plan showing the construction phases during the closure of an estuary entrance comprising several channels and tidal flats has not yet been presented. In this chapter some examples of a hypothetical closure will demonstrate various options. A number of alternatives will be outlined and their relations with some historic cases will be discussed. Data on flow velocities and discharges has been taken from mathematical calculations. Only data that is relevant to the decisions in the design is presented.

The example assumes a tidal estuary that has to be closed along a fixed alignment. The longitudinal profile of the total closure (see Figure 12-1)consists of (from left to right):

- a foreshore, 250 m wide, 0.5 m lower than mean sea level.
- a secondary gully of 200 m width, with an average depth of 4 m below mean sea level (MSL),
- a tidal flat 300 m wide, at a level of about MSL,
- the main gully 250 m wide, with an average depth of 6.5 m below MSL and the greatest depth along the bank.

The longitudinal profile of the closure gap is thus 4000 m^2 at high water and 1800 m^2 at low water. The tide is a semi-diurnal sine wave with a range of 3 m. In all calculations the tidal range is taken as constant; neap-spring variation is ignored. The storage area of the basin is 20 km² at high water and 5 km² at low water.

Three main options will be studied:

a) dam sections first across the shallows and then closing the gullies (Section 12.2).

b) dealing with the gullies first and closing the shallows later (Section 12.3).

c) simultaneous closures (Section 12.4).

Each option may have several alternatives.

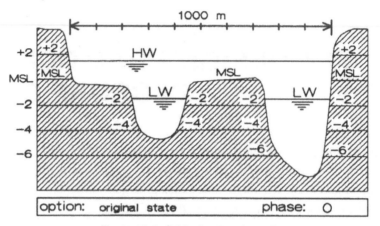

Figure 12-1 Original state, phase 0

12.2 Blocking the shallows first

Dam sections across the shallows will create two gaps. Many possibilities are created, but some are unattractive. For example the secondary gap is very shallow for caissons, while for a vertical closure, the total length of the two sills (only 450 meters) is rather short (this will be clarified when describing options b and c). To obtain the required data for a well-considered decision it is necessary to use a mathematical model. Horizontal closure by tipping quarry stone in both gaps is a very good possibility, but for the purpose of this example, a combination of placing caissons and tipping quarry stone will be considered.

Some construction phases are presented in Figure 12-2, Figure 12-3 and Figure 12-4. The programme reads:

Two gaps are created by damming the shallows. Bottom protection is provided for in the remaining gaps. Then, both closure gaps are slightly diminished in sectional profile by creating sills. For a caisson closure abutments are also made. These are concrete structures or sheet-pile walls that shape the vertical sides of the closure gap. Next, caissons are positioned in the main gap and finally the secondary gap is closed by tipping quarry stone.

The programme is summarized in Table 12-1.

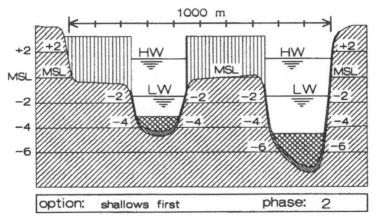

Figure 12-2 Shallows first, phase 2

phase	action	foreshore	sec. gully	tidal flat	main gully
0	original state	250 m; -0.5	200m; -4	300m; msl	250m; -6.5
1	bottom protection + shallows	dammed	200m; -3.5	dammed	250m; -6
2	partial sills in both gaps	dammed	200m; -3	dammed	250m; -4.5
3	final sill, abutments	dammed	200m; -2.5	dammed	190m; -4.5
4	first caisson in place	dammed	200m; -2.5	dammed	128m; -4.5
5	sec. caisson in place	dammed	200m; -2.5	dammed	66m; -4.5
6	third caisson in place	dammed	200m; -2.5	dammed	closed
7	narrowing on sec. sill	dammed	100m; -2.5	dammed	closed
8	further narrowing	dammed	50m; -2.5	dammed	closed
9	last gap	dammed	10m; -2.5	dammed	closed

Table 12-1

As the two gaps are blocked virtually in succession, there may be considerable imbalance in the tidal system between the gaps. To prevent this, the tipping into the secondary gap has to occur simultaneously with the placing of the caissons. The effect of this procedure is that the flow conditions during the sinking of the last caisson are too high. It is always necessary to check whether the flow conditions during the last phase are acceptable.

Checking on these conditions gives the data in Table 12-2. At the moment when two caissons are being placed (phase 5) the maximum discharge via the secondary gap has doubled and reaches about the same magnitude as the main gap originally had, while on the contrary, discharge via the main gap has halved. The secondary channel will have to accommodate the doubled quantities, for which its profile will be barely adequate. The scouring of a gully across the shallow from the main to the secondary channel is the likely consequence of the imbalance.

(The values in Table 12-2 have been calculated for the same tide variation at sea; spring/neap variations are not included. Maximum flow does not necessarily coincide with maximum discharge, neither do the maxima of the two gaps always coincide).

| U in m/s | | secondary gap | | | | main gap | | | |
| Q in m³/s | | during ebb | | during flood | | during ebb | | during flood | |
phase	situation	U_{max}	Q_{max}	U_{max}	Q_{max}	U_{max}	Q_{max}	U_{max}	Q_{max}
0	original	1.09	915	1.07	940	1.09	1810	1.07	1825
1	bp+dams	1.33	1010	1.27	1045	1.33	2070	1.27	2085
2	sills	1.67	1065	1.57	1135	1.67	1935	1.57	1995
3	abutment	2.12	1090	1.94	1215	2.12	1790	1.94	1865
4	1 placed	2.71	1305	2.39	1505	2.57	1385	2.26	1470
5	2 placed	3.57	1550	3.00	1875	3.19	820	2.69	895

Table 12-2

(phase 2 is pictured in Figure 12-2)

The positioning of the last caisson takes place in the situation of phase 5 during HW-slack, because at LW there is insufficient water depth. At the end of the flood period the flow diminishes as follows:

- 30 min. before slack: $u = 1.50$ m/s,
- 20 min. before slack: $u = 1.20$ m/s,
- 10 min. before slack; $u = 0.70$ m/s.

For a safe sinking operation, these values are far too high. Consequently, the programme has to be adapted. Instead of using ordinary caissons, caissons equipped with sluice gates can be used. The programme then reads as indicated in Table 12-3.

phase	action	foreshore	sec. gully	tidal flat	main gully	sluice gate
4	first placed, opened	dammed	200m; -2.5	dammed	128m; -4.5	56m; -3.5
5	sec. placed, opened	dammed	200m; -2.5	dammed	66m; -4.5	112m; -3.5
6	third caisson placed	dammed	200m; -2.5	dammed	0m	112m; -3.5
7	narrowing on sill	dammed	100m; -2.5	dammed	0m	112m; -3.5
8	further narrowing	dammed	50m; -2.5	dammed	0m	112m; -3.5
9	last gap in sec.	dammed	10m; -2.5	dammed	0m	112m; -3.5
10	close sluice gates	dammed	dammed	dammed	0m	closed

Table 12-3

(phase 4 is pictured in Figure 12-3)

This time the positioning of the last caisson takes place in phase 5, with the sluice gates of the two other caissons opened (assumed to have 56 m effective width each and a floor thickness of 1 m). Then, at the end of the flood period the flow conditions appear to be acceptable:

- 30 min. before slack: $u = 0.75$ m/s,
- 20 min. before slack: $u = 0.50$ m/s,
- 10 min. before slack; $u = 0.20$ m/s.

At first, the balance between the gaps will be better than in the former scheme because the gates provide a flow profile while tipping starts in the other gap. Although the secondary gap is closed before the gates close, the main channel does not exceed its original discharge. The balance can still be improved by closing a number of gates simultaneously with the tipping. However, that worsens the conditions for the tipping. The flow conditions are given in Table 12-4.

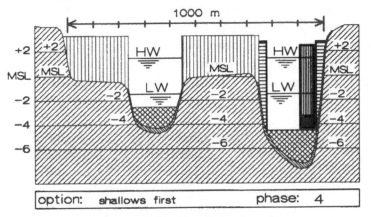

Figure 12-3 Shallows first, phase 4

U in m/s		secondary gap				main gap **			
Q in m³/s		during ebb		during flood		during ebb		during flood	
phase	situation	U_{max}	Q_{max}	U_{max}	Q_{max}	U_{max}	Q_{max}	U_{max}	Q_{max}
5	1+2 open	2.60	1260	2.30	1445	2.32	1460	2.06	1580
6	3 placed	3.35	1480	2.85	1775	2.82	965	2.40	1095
7	100m gap	3.87*	830	3.40	1040	3.67	1155	3.03	1360
8	50 m gap	3.78*	410	3.57	535	3.95*	1220	3.36	1485
9	10 m gap	3.62*	80	3.58	105	4.05*	1245	3.58	1560

* means critical flow ** via the sluice gates

Table 12-4 Flow and discharge conditions

(phase 7 is pictured in Figure 12-4)

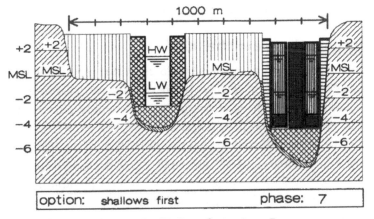

Figure 12-4 Shallows first, phase 7

Critical flow also occurs during the ebb in the sluices. Per tidal cycle this lasts for nearly 2 hours in phase 8 and for 2.5 hours in phase 9. This can probably be prevented by providing the third caisson with sluice gates. This was not required for the sinking operation. If sluice gates are provided for the third caisson the conditions for the tipping of the stone are better, but the imbalance between the gaps increases. Whether doing this is worth the extra expense, depends on the savings on stone tipping and bottom protection. The flow and discharge conditions for three sluice caissons are given in Table 12-5.

U in m/s Q in m³/s		secondary gap				main gap **			
		during ebb		during flood		during ebb		during flood	
phase	situation	U_{max}	Q_{max}	U_{max}	Q_{max}	U_{max}	Q_{max}	U_{max}	Q_{max}
7	100m gap	3.35*	720	2.80	875	3.14	1570	2.63	1805
8	50 m gap	3.55*	385	3.09	480	3.51	1695	2.91	1980
9	10 m gap	3.49*	80	3.15	100	3.81*	1780	3.15	2120

Table 12-5 Flow and discharge conditions

Critical flow in the sluices now occurs during the final operation and lasts for half an hour only. Maximum flow velocities in the secondary gap reduce by about 10%.

A historic case of the use of the above system of closing a tidal basin was the construction in of the "Brouwersdam" between 1965 and 1972. The damming the "Brouwershavensche Gat" created Lake Grevelingen. The dimensions of the channels and the basin were large. The total dam alignment was 6 km long. In this case the minor gully was closed by sinking 12 sluice-caissons, each 68 m long, on a sill levelled at 10 m below MSL. The main gully was closed by gradual closure with 50 kN concrete cubes. The profile of this gap was about 13000 m². Unlike the hypothetical example, the gradual closure was a vertical closure, dropping the cubes by means of a pre-installed cableway. As in the hypothetical example, the flow condition forms a limiting factor for the progress of the gradual closure.

12.3 Blocking the main channel first

In this section the same estuary as above is closed by first reducing the profiles of the gullies. Next the main gully is completely blocked. The secondary gully will then be further reduced and finally the total remaining profile will be blocked. This option is worth consideration if it leads to a cheaper closure operation. When the channels are blocked first the obvious disadvantage is that the bottom of the shallows also has to be protected against scour. Cost savings on the other items need to compensate for this expensive extra. Such savings may result from a possible reduction in the dimension of the protection in the gully and from cheaper caisson design. A major saving would result if caissons without sluice gates were used, while another saving could be obtained by using two caissons only.

A determining factor for the decision to omit sluice gates is the positioning of the last caisson. The flow conditions will be best when the cross sectional area of the flow profile is as large as possible at that moment. This is the case when there is no sill in the secondary gap. Assuming the same dimensions of the caissons, the determining total flow profile is the original profile, reduced by bottom protection over the full length, by the abutments on both sides of the main gully, by the foundation sill and by the caisson(s) already placed. The HW-slack period then is characterised by:

- 30 min. before slack: $u = 0.80$ m/s,
- 20 min. before slack: $u = 0.60$ m/s,
- 10 min. before slack; $u = 0.25$ m/s.

Positioning will not be a problem. However there is a substantial imbalance between the two gullies. The maximum flood via the secondary gully is 1975 m^3/s, which is more than twice the original. A sill in this gully, up to the level -3m, brings the discharge down to 1860 m^3/s. The effect is small and flow velocities in the main gully increase up to:

- 30 min. before slack: $u = 0.95$ m/s,
- 20 min. before slack: $u = 0.65$ m/s,
- 10 min. before slack; $u = 0.35$ m/s.

Sinking is possible, but further raising of the sill is not acceptable. The construction of the sill up to -3m in the secondary gully can best be done concurrently with the foundation sill in the main gully. After that, caissons can be placed at short intervals to limit the duration of the imbalance.

Saving on the number of caissons depends on the flow conditions for creating the smaller gap for the two caissons, as the narrowing of the gully has to be done by pumping sand. The flow profile available is the original profile reduced by bottom protection and by the island for the abutment on the shallows. The longitudinal profile then consists of the following:

- 250 m foreshore, 200 m secondary gully and 250 m shallows, all provided with bottom protection
- an island section on the shallows of 50 m length
- an adjoining island in the main gully along 75 m, leaving a gap length for two abutments of 25 m each
- two caissons of 60 m each and 5 m extra (see Figure 12-5)

In the secondary gap, bottom protection will be present and some of the sill construction may exist. The calculations show that the maximum flow velocities in the gap are 1.70 m/s during ebb and 1.60 m/s during flood, which is no problem for the construction of the island.

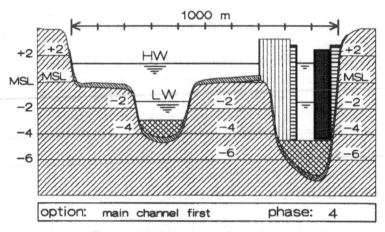

Figure 12-5 Main channel first, phase 4

The conclusion is that blocking the gullies first can be achieved by closing the main gap with two simple caissons, while a restricted sill is present in the secondary gully. The remaining flow profile consists of two 250 m-long shallow sections with a partly blocked gap in between. The construction phasing up to this moment thus reads as per Table 12-6.

phase	action	foreshore	sec. gully	tidal flat	island	main gully
0	original state	250m; -0.5	200m; -4	300m; MSL	none	250m; -6.5
1	bottom prot. + island	250m; MSL	200m; -3.5	250m; +0.5	125m	175m; -6
2	sills in both gaps	250m; MSL	200m; -3	250m; +0.5	125m	175m; -4.5
3	sill, abutments	250m; MSL	200m; -3	250m; +0.5	150m	125m; -4.5
4	first caisson placed	250m; MSL	200m; -3	250m; +0.5	150m	65m; -.5
5	sec. caisson placed	250m; MSL	200m; -3	250m; +0.5	-	closed

Table 12-6

Flow velocities and discharges are given in Table 12-7.

U in m/s Q in m³/s		secondary gap **				main gap			
		during ebb		during flood		during ebb		during flood	
phase	situation	U_{max}	Q_{max}	U_{max}	Q_{max}	U_{max}	Q_{max}	U_{max}	Q_{max}
0	original	1.09	915	1.07	940	1.09	1810	1.07	1825
1	bott. prot. + island	1.61	1155	1.52	1215	1.61	1695	1.52	1720
2	sills	2.01	1175	1.85	1295	2.01	1525	1.85	1600
3	abutment	2.42	1355	2.19	1525	2.29	1205	2.07	1285
4	after 1st	3.06	1585	2.68	1860	2.73	705	2.40	770
5	after 2nd	3.98*	1860	3.37	2310	closed	0	closed	0

** the central 200 m section only (the shallows falling dry during low tide).

Table 12-7

The next step could be a horizontal closure by tipping quarry stone from either one side or from both sides. This creates high flow velocities in the secondary gap. The situation is comparable with the former option, except for the sluice gates. In this case the flow velocities will increase to about 4.5 m/s. Therefore, it is more appropriate to try to reduce the profile of the secondary gap, maintaining the flow across the shallows. Dumping quarry stone from dump-vessels will be impossible because of draught restriction. However, vertical closure by means of a temporary bridge (to be installed in the previous period) or a cableway is possible. The length is considerable (700 m), but 500 m of the bridge crosses shallow water so the cost of the foundation is limited.

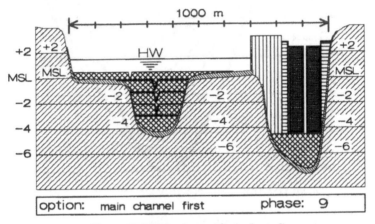

Figure 12-6 Main channel first, phase 9

Another method, which involves a difficult operational procedure, takes into account the fact that during LW the dam section across the shallows falls dry for several hours. The first step is to bring the sill level up in two layers to above LW, (from -3 to -2 and then to -1). The water level in the basin will not follow the sea level and the relation between the levels of the sea, the basin and the sill determine the operational possibilities. The equipment to be used is a shallow-draught crane vessel and dump trucks with hydraulic cranes, approaching via the drying dam sections. The operational period per tidal cycle (the work-window) is short and the production is low, but the equipment is available on the market and investment in a bridge or cableway can be avoided. Gradually, layer by layer, the sill will be raised. For every layer, the determining moment exists when the last 10 m has to be placed. That missing part of the layer gives a dip in the sill level, which is subjected to higher (critical) flow.

The crane vessel can operate when anchored near the gap during the periods that flow velocities are lower than 2 m/s. For the layer from −3 m to −2 m (phase 5 to 6) this averages 2 hours during HW and 1 hour during LW. For the next layer (phase 6 to 7), up to −1 m, this is 1.5 hours at HW and three quarters of an hour during LW. After that, dump trucks may start to assist during the LW-period, because the water level at the seaward side will fall lower than the sill level. At the start of the next layer up to MSL (phase 7 to 8), the basin level falls below that level during about 1 hour (b-b in the figure). When finishing the layer, the water levels are less than 0.5 metres above the sill level for two hours. Though it is risky, trucks can operate in water as long as it is less than 0.5 metres deep.

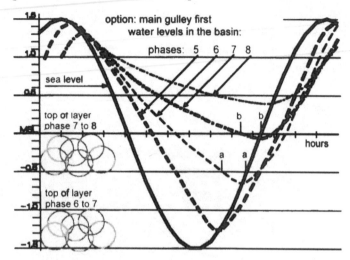

Figure 12-7 Main channel first, water levels in the basin

The construction phasing thus continues for the 700 m section vertically (see Table 12-8):

phase	action	foreshore		sec. gully		tidal flat
6	first layer	250m; MSL	97m; -2	6m;-2	97m; -2	250m; +0.5
7	first layer	250m; MSL	97m; -1	6m; -2	97m; -1	250m; +0.5
8	level foreshore	250m; MSL	97m; MSL	6m; -1	97m; MSL	250m; +0.5
9	level tidal flat	222m; +0.5		6m; MSL	222m; +0.5	250m; +0.5
10	level + 1	347m +1		6m; +0.5		347m; +1
11	final layer	dammed		6m; +1		dammed

Table 12-8

Table 12-9 gives the flow velocities for the various levels of the sill.

U_{max} in m/s		deepest part		deepest but one		deepest but two		deepest but three	
phase	situation	ebb	flood	ebb	flood	ebb	flood	ebb	flood
5	after 2nd	3.98*	3.37	2.34*	2.85*	1.80*	2.50*		
6	up to -2	4.22*	3.43	3.81*	3.84	2.28*	3.03*	1.94*	2.32*
7	up to -1	3.82*	3.38	3.28*	2.68*	2.38*	3.15*	2.02*	2.32*
8	up to MSL	3.27*	2.92	2.50*	2.91*	2.06*	2.32*	not applicable	
9	up to 0.5	2.32*	2.67	1.98*	2.32*	not applicable			
10	up to + 1	1.86*	2.18*	1.05*	1.55*				
11	up to HW	0.88*	1.55*	high water free					

* means limited by critical flow condition.

Table 12-9

In the table the deepest part represents the dip in the sill mentioned above. Although in all sections of the sill critical flow limits the flow velocity, in the dip this is only true for the ebb. Besides, ebb is determining for sill levels up to about MSL, above that flood flows are higher.

Considering the above results, it appears that the maximum flow velocity in the secondary gap during the raising of the sill is not very much less than would have been the case if a horizontal closure had been designed (4.22 m/s instead of about 4.50 m/s). This is due to the fact that the limiting critical flow condition for the –2m sill level under these circumstances is about the same as the normal flow condition for a narrow gap with a 3 m tidal range. The determining condition occurs for the dip in the low sill in the 200 m gap. In that stage of the process, the 500 m shallow sections are too elevated to be useful. This example proves by its exception that the general rule that vertical closure leads to smaller flow velocities than horizontal closure is not always applicable. From a financial point of view, neither the difficult operational procedure and small production capacity, nor the investment for a bridge or cable-way, can compete with the dump trucks operating from two sides for horizontal closure.

A rather critical point in the phasing of the closure is the situation near the island after the caissons have been placed. The main gully is blocked and the discharge of the secondary gully is more than doubled. An easy way for the water to pass through the gap is by following the main gully, circling across the shallows around the island and to return into the main gully. Scouring a short-cut like that is a typical example of the consequence of the imbalance. It would cause a disaster and has to be prevented.

A historic case of a closure in which first gullies were blocked by caissons and then the shallows were closed, is the closure of the Schelphoek, one of the major dike breaches in the south west of the Netherlands (1953). The situations of the breaches and the closure alignment are given in Section 2.5 showing the development of erosion gullies. The picture of the situation after 20 weeks shows the two gaps that had been shaped, typically suited for caisson placing, while the long overland stretches were protected by mattresses. After the caissons had been placed the overland sections were constructed by horizontal closure. A large number of shallow concrete units was placed every tide in such a short sequence that the overland flow could not scour a short-cut.

The vertical closure procedure, using dump trucks driving on the sill and using hydraulic cranes to level the cobbles, was executed during the closure of the Markiezaatskade (1983), one of the secondary dams required in the Delta Works. The dam closed a shallow tidal basin of about 20 km^2 with a tidal range of about 4 meters. The final gap was 800 m long and had a basic sill level of 2 m below MSL.

In that case, the vertical closure was an advantage as the full 800 m had the same low sill level. The quarry stone dam was constructed in 1 or 0.5 m thick layers.

12.4 Closure over the full dam length

The logical option to close over the full length of the alignment is a vertical closure or a combined closure. The principle is to first partially block the gullies by sills, until one level exists over the full length and then continue either vertically or horizontally until the 1000 m long gap above that sill is closed. This differs from the option to close one of the gullies first in that there are no caissons, that the imbalance will be smaller and that flow conditions are more favourable due to the longer weir. The first two phases of construction are:
- a bottom protection along the full alignment,
- sills to be dumped up to the level of –2.5 m (draught of the vessels permitting).

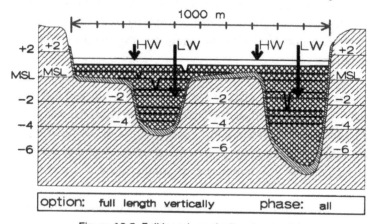

Figure 12-8 Full length vertically, phase: all

Then, horizontal or vertical closure has to be selected. With a horizontal closure one final gap is created. While proceeding to that gap the flow conditions on the lowest part of the sill will be determining. As that sill level is -2.5m, the situation of the former options is again created. The same final stage is reached at the cost of a bottom protection over the full length. The bottom protection in the shallow areas is therefore superfluous so the full-length option is possible only if vertical closure is used. The next phases therefore are:
• a layer bringing the level of the sills up to –1m,
• further layers of 0.5 m thickness up to HW.
The procedure is identical to the last phases of the former option. The difference is that the weir length is 1000 m instead of 700 m.
The phasing of the closure thus reads (Table 12-10):

phase	action	foreshore	sec. gully	tidal flat	main gully
0	original state	250m; -0.5	200m; -4	300m; MSL	250m; -6.5
1	bottom prot. + sill (-3.5)	250m; MSL	200m; -3.5	300m; +0.5	250m; -3.5
2	sills dumped (-3)	250m; MSL	200m; -3	300m; +0.5	250m; -3
3	sills dumped (-2.5)	250m; MSL	200m; -2.5	300m; +0.5	250m; -2.5
4	sill by trucks (-1)	250m; MSL	200m; -1	300m; +0.5	245m; -1
5	up to MSL	445m; MSL	5m; -1	300m; +0.5	250m; MSL
6	up to +0.5	445m; +0.5	5m; MSL	300m; +0.5	250m; +0.5
7	up to +1	445m; +1	5m; +0.5	300m; +1	250m; +1
8	up to HW	445m; +1.5	5m; +1	300m; +1.5	250m; +1.5

Table 12-10

The influence on the flow conditions appears from the lists below (to be compared with the table in the former option for the 700 m long weir):

U in m/s Q in m³/s		secondary gap				main gap			
		during ebb		during flood		during ebb		during flood	
phase	action	U_{max}	Q_{max}	U_{max}	Q_{max}	U_{max}	Q_{max}	U_{max}	Q_{max}
0	original	1.09	915	1.07	940	1.09	1810	1.07	1825
1	protect. + sill	1.78	1230	1.66	1310	1.78	1535	1.66	1635
2	sills -3	2.06	1180	1.90	1310	2.06	1475	1.90	1635
3	sills -2.5	2.48	1110	2.23	1305	2.48	1385	2.23	1630
4	sill -1	2.99*	710	3.49*	1135	2.99*	870	3.49*	1390

Table 12-11

The discharge quantities in the remaining gap diminish with the progressing construction of the sill. The determining flow conditions are those in the narrow dip of every layer. The flow velocities in the various parts of the 1000 m gap are listed in Table 12-12. The phase with the determining flow conditions is phase 4, shown in the graph.

Comparing the phases 4 to 8 of this option, with phases 7 to 11 of the former option, it appears that the maximum flow velocity is considerably less (3.62 m/s and 4.22 m/s) as consequence of the longer gap length. All flow conditions reach the critical flow stage, except for in the last dip in the sill, which occurs as consequence of the construction of the layer up to the level of -1 m. In the dip the flood flow is still sub-critical. The magnitude is slightly less than the critical flow above the −1 m elevated sill, which is due to the low discharge-coefficient used for this narrow gap.

Even more important is the fact that the time window for the equipment to operate on the sill is much longer than in the former option. This is due to the lower low-water level in the basin for the comparable level of the sill. On the other hand, an operational difficulty of the option under consideration is that the layer construction has to advance over 500 m from each side instead of 350 m.

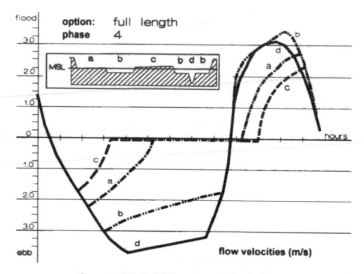

Figure 12-9 Full length vertically, phase 4

| U$_{max}$ in m/s | | deepest part | | deepest but one | | deepest but two | | deepest but three | |
phase	situation	ebb	flood	ebb	flood	ebb	flood	ebb	flood
4	sill -1	3.62*	3.09	2.99*	3.49*	2.24*	2.74*	1.68*	2.27*
5	up to MSL	2.97*	2.87	2.33*	3.05*	1.98*	2.32*	not applicable	
6	up to 0.5	2.32*	2.64*	1.95*	2.41*	not applicable			
7	up to +1	1.90*	2.05	1.11*	1.55*				
8	up to HW	0.88*	1.55*	high water free					

Table 12-12

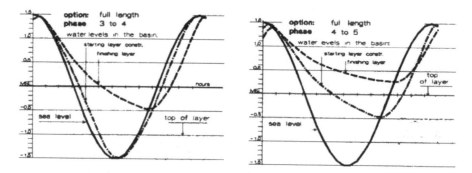

Figure 12-10 Full length, phase: 3 to 4 and 4 to 5

An alternative way to avoid this problem of 500 m driving distance, is to prepare an approach-road towards the centre of the sill via an artificial island on the tidal flat. The construction can then advance from four sides each section being only 250 m long. This solution must be assessed against the extra cost of installing transhipment facilities. The total sill length is reduced only by the width of the construction island.

With certain dimensions the method is even changed and the first option with the final gaps in the channels, using vertical closure instead of caissons is followed.

An example of a closure by constructing sills in the gullies and providing one level over the full length of the alignment is the damming of the River Feni in Bangladesh in 1984/85. Owing to shallow water effects the tides were very variable,. Spring ranges were double the neap ranges, while low water levels were always about the same (see Figure 5-6, influence of MSf). During spring tides the tidal wave entered the estuary as a tidal bore. Therefore, conditions during neaps were much more favourable than during springs and the last layers of the vertical closure had to be forced in a major effort over the neap tide period. During that closure day a neap-tide-safe profile had to be constructed on top of the sill, which had to be heightened up to a spring-tide-safe profile within a week's time.

The tide on the closure day rose from -0.5 m to +2.65 m (range 3.15 m), while the starting level of the sill was +0.70 m. On the closure day an embankment was constructed up to the level +3 m by piling up jute bags filled with clay. In total 1,000,000 bags were positioned by hand by 12,000 Bangladeshi people, in a time of five hours. In order to minimise the hauling distance the bags had been stored in 12 stockpiles along the alignment of the dam, which reduced the total length of the gap to about 1000 metres. Trucks tipping clay raised the profile to the spring tide safe profile.

12.5 Cross section of closure dams

In damming activities sometimes a choice can be made between a rather permeable initial closure dam or a profile provided with an impermeable (which in practice generally means with very low permeability) core or structure. The choice will depend on various circumstances. The main consideration is that in a permeable profile the phreatic pressure decreases gradually but the flow in the full profile has to remain within limits to ensure micro stability. A watertight core leads to a high pressure difference over the core, which endangers macro stability. Moreover, locally along the boundary of the core or structure, the high pressure difference may lead to micro instability.

In several closure designs made for the Delta scheme in The Netherlands, an asphalt mat was used to protect the bottom. Although permeable mats composed of graded stone with an asphalt binder can be made, these mats were practically impermeable. Between the sea and the basin the phreatic level in the soil underneath the mat gradually diminished according to an extended flow net.

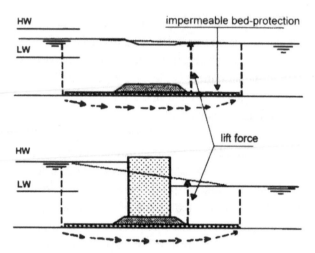

Figure 12-11 Uplift of impermeable bed protection

Since the protected length (*L*) is large, *i* is small and so is *u* (Darcy). After construction of the dam profile in the centre of the protected area, the difference in the water level increases and thus u also increases. As shown in Figure 12-11, the phreatic level underneath the mat just downstream of the dam results in a lift force. The relatively thin mat, meant to prevent scour, suddenly gets another task: to counteract the upward pressure by its own weight. If lifting should occur, the mat might tear and the water would break out. The seepage length (*L*) would then be about half the original and in a two-dimensional case *u* would double. In practice there would be one location which would give way first and the flow would converge on this point. The flow would start a piping process and after that the whole dam profile would collapse. Consequently, the mat had to be much thicker than in cases in which a permeable mat has been designed.

In the design process the various combinations of water levels and soil parameters for the subsoil and the structural details have to be considered. Permeability coefficients have to be tested or assumed. Determining conditions in the various construction phases have to be concluded. Next, groundwater flow nets can be drawn and analyzed. In three-dimensional situations (beside a caisson or at the end of a sheet pile wall), the flow net is difficult to draw and great care must be given to these details. A problem may arise if infill of fines gradually plugs the soil. If this occurs the permeability changes and the pressures start building up. The reverse, (erosion of fines), may lead to higher flow velocities that worsen the erosion and finally lead to a scour pipe in which seepage changes into piping.

Generally, the most critical stage is the closure of the final gap. This is a situation in which a bottom protection against scour has been laid on which in a short period of time an initial dam or structure is made. Differential water levels attack the freshly-made profile. With caissons in particular, the stability of the foundation bed and of the side connections is important.

A typical example of piping actually occurred during the closure of the River Feni in Bangladesh. During preparatory operations, stockpiles of jute bags filled with clay had been stored along the initial closure alignment. These stockpiles had been built on geotextile sheets strengthened with bamboo grating. During the closure operation sufficient bags would be carried from the stockpiles into the longitudinal initial dam profile to block the flow. Immediately afterwards this profile would be strengthened by a heavy clay profile placed behind it. The procedure required the removal of the bamboo gratings but time did not allow the complete removal and the clay profile covered several remnants of these bamboo canes. During high water, the groundwater then found its way underneath the clay along these bamboo canes and several wells gave artesian water. The problem was solved by dumping some extra clay on top of the wells. This appeared to be sufficient to block the flow and prevent the escalation of the piping.

If piping occurs, several measures can be taken to prevent escalation. The best measure is to block the inflow on the upstream side. However, although the outflow point is known where the water enters is generally unknown. The next measure is to ballast the well as in the above example. But sometimes, water may break out time and again. In such cases, the well can be covered by a permeable sheet that is sufficiently ballasted with cobbles. Owing to the permeability, the upward pressure will remain low and further ballasting is possible. The flow will carry soil particles that become stuck in the sheet and gradually the flow is blocked while the pressure increases. Ballast and soil on top will ultimately stop the flow. In shallow water and above water, a method is to construct a ring-shaped wall of sufficient height around the well. Within this ring the water level will rise until the high level at the entrance of the pipe is reached. Then the flow stops and any sort of material can be dropped onto the well and block it.

12.6 Final remarks

The calculations and considerations relating to the closure options for the above example demonstrate that the change in conditions during the closure operations depends on the method adopted. Furthermore, it corroborates the statement, made in the preface, that there is no single correct solution to this design problem. Without further details about availability and costs of materials and equipment no conclusions can be drawn on the basis of the above calculations.

Moreover, various other considerations may influence the decisions, such as: "Is it possible to build a large dry-dock for the construction, immersion and float-out of the caissons? Is there any social-economic reason why labour intensive low-level technology is preferred above a high-skilled approach? How skilled is the available labour force and in what sort of operations are they experienced? Are there any restrictions on the import and use of equipment or materials from outside the country and what is the taxation situation?"

But even the technical arguments have not all been considered fully. What about operational conditions and time periods? Are there seasonal changes in sea level or tidal amplitudes, periods with storm surges or cyclones, monsoons restricting operations because of waves and swell, limiting the execution of part of the works to specific work-windows? What happens when, due to unforeseen set-backs, the critical operation will exceed that time window?

No attention has yet been paid to the impact of extreme conditions on the design. Is sufficient data available to make an analysis of the probability of the occurrence of such an event? In the considerations included in the above example, not even the tide has been varied for springs and neaps.

Some operations are more vulnerable to extreme conditions than others and so consequently the measures to be incorporated in the design may take different proportions of the costs. These have to be included in the assessment of the various options in order to make a proper choice. In the event of a failure, the options for repair depend largely on the closure method used. These considerations are difficult or sometimes impossible to assess, and yet it is all part of the design process.

It is impossible to provide a single recipe for the closure of all dams. The solution is always site specific. The objective is to illustrate the basic technical problems likely to be encountered and the application of scientific principles to their solution, substantiated by the wealth of practical experience involved.

13 CONSTRUCTION METHODS FOR GRANULAR MATERIAL

Chapter 13 is meant to give students an impression of the various available construction methods. The chapter requires careful reading and an attempt to understand the background.

13.1 Introduction

Although for a first approach theoretical and analytical considerations are indispensable to designers of closure dams and breakwaters, the method of construction is equally important. Based on theoretical considerations, a preliminary design is put on paper. For a breakwater this leads to the details of the cross-section, the crest level, the outer slope, the weight of armour and sub-layers, generally starting the calculation from top down to bottom. For a closure it is the materials needed to cope with the tremendous change in flow-conditions during the progressive stages of horizontal and vertical closure that influence the design. Next, an analysis has to be made of the conditions in all construction phases and of the equipment required to operate in those conditions, from start to finish and from bottom to top. The main problems will be the stability (or vulnerability) of every construction phase, the accessibility of the work front and the safety of the equipment. In order to optimize the feasibility, alternatives have to be generated by making small or even radical changes to the design, in order to find an acceptable compromise between design requirements and the possibilities of the available equipment and materials.

Closure dams and breakwaters have one important aspect in common: their massive character. Construction of the structure requires a tremendous amount of material that must be acquired, for instance from a quarry, transported to the location of the structure, and then placed within the profile along the alignment. This is a logistical problem that covers much more than the handling of the material alone. It concerns opening up the various working sites, mobilizing and maintaining equipment, mobilization and accommodation of personnel, and last but not least, the actual handling of millions of tons of material. Optimizing the solution of this logistical problem may create savings that are much larger than the little extra cost of improving a sub-optimal design of the cross-section.

It is impossible to make a complete description of all options for the construction of closure dams and breakwaters in a book like this. In practice, consultants and contractors have a special and very well documented department to do this kind of work in the tender and pre-tender stages. This book attempts to give students an idea of possibilities and problems. Reference is made to the CUR/RWS Manual No.169 (Anonymous [1995]), where more details are given, and from which some information has been included here.

In this chapter, we follow the materials like sand and blocks from source to destination. Various materials are used to create structural elements, such as the geotextile plus bamboo plus quarry stone used to make a bottom-mattress. They have to be obtained, brought to the site, reworked or stored, transported to the actual site and placed by appropriate equipment in a variety of wave and flow conditions. This mainly concerns:

- Bottom fixation or bottom protection: Mattresses made of geotextiles, branches or bamboo, sunken or placed onto the bottom in open water and ballasted with quarry stone (or alternatively: granular filters)
- Shore-connections and initial dam sections: Sand-fill or quarry-run dams with embankment protection, made by dredging or by dry earth-moving equipment
- Abutments: Sheet-piled walls or a caisson to provide a vertical connection with the large caisson units placed afterwards
- Breakwater core and dumped sills: Quarry stone, "all in" or with selected gradation, transported and positioned by dump trucks or dump vessels
- Cover layers and armour: Selected quarry stone or artificially made concrete units, placed by hydraulic equipment and cranes

Caissons and monolithic structures have such specific problems of execution that they will be treated in a separate chapter (see Chapter 14). A more thorough knowledge is required to satisfy some aspects of the above items. Details of the construction and use of mattresses (13.3), the provision and handling of quarry stone (13.5), the use of rolling and floating equipment in various conditions (13.6) and finally some very specific techniques and ancillary equipment (13.7) are given in the next sections. Some details of construction equipment are given in Appendix 8.

13.2 Scour prevention by mattresses

Every closure of a watercourse leads to a situation in which the flow accelerates, circles around dam heads or crosses over materials with different hydraulic roughness. Each of these results in increase of the capacity to erode. In the sand closure process the scour is accepted since its magnitude is limited because of the restrictions to the flow velocities which offer the possibility to apply the method. For all other methods and dependent on the resistance of the bottom material against erosive action, scour holes may develop, which can endanger the stability of the closure dam. This has to be prevented by the placing of bottom protection means at all relevant locations. These do not prevent the scour completely but shift its bearing towards less vulnerable locations and may reduce the scour depth. Scour prevention therefore is part of most closure processes and generally one of the first actions in practice.

Generally speaking the scour resistance of the bottom material is difficult to predict. Rock and stiff clay will be very resistant, soft clay is rather resistant, peat may stand the attack quite long and then suddenly break out in large lumps. The behaviour of sand has been investigated intensively and several formulae have been derived to predict the scour hole development. Since in practice a sandy subsoil is but seldom homogeneous, the actual scour may still deviate from the predicted values.

In short, scour holes can be expected at places where:

  i.   the flow velocity increases in course of time
 ii.   the flow distribution over the vertical changes
iii.   the flow is not saturated with sediment
 iv.   the turbulence intensity increases

These aspects occur in closure processes tor instance:

- when diminishing the gaps profile (item i)
- at the end of a stone protection as consequence of change in roughness (item ii)
- due to reduction of the discharge quantity in the approach gulley (item iii)
- around dam heads, structures and obstacles (item iv)

Of course combinations of these four often occur.

In one-directional flow, the scouring process creates a hole which is characterised by its steep slope at the upstream side, its depth and its gentle downstream slope. In tidal areas, where the flow changes direction in every tidal cycle, the shape of the hole will be different. The reversing flow smooths the hole out slightly by which the slopes become more gentle. The development of the hole goes quickly in the beginning but gradually slows down. By creation of the hole, the bottom topography adapts itself to the flow's eroding capacity and in the end an equilibrium state is reached. The depth of the scour hole develops in an exponential relation with time. In many cases the equilibrium state is reached so quickly that the intermediate stages are of no importance. However, if a number of caissons are placed one after the other in a couple of weeks time, the flow pattern changes stepwise in short periods and so does the scouring capacity. The scour hole then develops as a summation of intermediate successive stages (see Figure 13-1).

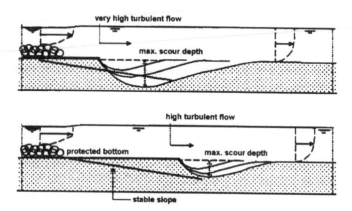

Figure 13-1 Development of scour hole

The development of scour holes in itself is not dangerous. Only in cases where they come too close to either the closure dam or adjoining dams or structures. will they endanger stability of these structures. Then. uncontrolled scour should be prevented. A scour protection by a bottom mattress or a filter layer will be required. Since the costs of these protections in closure works generally are considerable, minimizing the dimension is important. However, the installation has to be done in advance of the determining situation (construction phasing) and a too short protection may give a large risk. The longer the protected area, the further away the hole develops and since on that spot the attack will be less, for instance because of spreading of the flow or diminishing of the turbulence intensity, the equilibrium depth of the hole will be less. Both aspects, further away and lesser depth, improve the stability consideration of the endangered structure. Usually, as a first approximation, a protected length of about 10 times the water depth is considered safe. For detailed engineering in case large costs are involved, the optimization requires physical model investigation in a hydraulic laboratory.

For the stability considerations the dam in the closure gap and the joining bottom protection act as a total structure. Consequently groundwater flows and potential head differences will build up over this protection. Therefore the protection has to:

- be flexible to follow changes in bottom topography
- be well connected to the bottom, leaving no room for piping
- be sufficiently heavy to prevent flapping in the flow

- be extra ballasted at its end to prevent turning up when the tide turns, be impermeable for the soil material underneath
- be stable in all flow conditions in all relevant construction phases
- be permeable for water to prevent high pressures underneath (sometimes the requirement for an impermeable part is combined with the scour protection, in spite of the pressure increase, see Figure 12-11).

13.3 Construction and use of mattresses

As soon as artificial structures for breakwater or dam construction disturb the governing flow and wave attack, fixation of the bottom by taking protective measures becomes necessary. A granular filter can be used but sometimes more coherent structural materials are required. The structure has to cover a rather large area and needs to be sand-tight but permeable to water. Moreover, it has to be stable under all prevailing flow and wave conditions.

Usually, mattresses, which are floated to the site or rolled onto a sinkable cylinder, then stretched out onto the bottom and subsequently ballasted, are used. Each mattress overlaps its predecessor. In the old days thick willow mattresses were the only suitable structures (see Appendix 9). However, since the introduction of geotextiles, adapted structural designs are used. The sand-tightness, formerly obtained by the thick layer of osiers can now be acquired by using a geo-textile sheet.

This modern version of bottom protection is based on splitting up the functions:

1. Strength is obtained by using geotextile of the desired material
2. Sand-tightness is obtained by using geotextile of the correct mesh
3. Floatation is obtained by adding willow or bamboo bundles (as the old version)
4. Rigidity during the sinking operation is obtained by adding one willow or bamboo grating (as compared to two in the old version)

The items 1 and 2 are usually combined in one sheet, but sometimes, when the bottom material is very fine grained (silty), a double sheet is required that consists of a non-woven (felt type) sheet and a strong woven sheet.

The items 3 and 4 are not required for a sheet, which is unrolled from a cylinder straight onto the bottom.

In all cases a considerable quantity of ballast is needed to fix the sheet in position. This is usually provided after the sheet has been initially ballasted during the placement operation. Initial ballasting is used to sink the sheet when the protection is floated in place. In addition, this keeps the sunken or unrolled sheet on the bottom after sinking. This initial ballast has to be immediately followed by part of the final ballast to secure the protection against the next flow attack.

The final ballast should remain stable on the sheet and not roll or slide away. This is not only a matter of the size of the ballast material (Shields) but also a matter of friction. A bottom protection, generally drawn nicely horizontally on the design-drawings may in practice cover an undulating or sloping area. The above mentioned items 3 and 4 therefore have another important function in providing fixation for the ballast material. This is a complication for the unrolling system. Though the sheet is thin and can simply be rolled-up, some sort of structure has to be added to provide resistance. This is usually achieved by the mechanical fixation of the initial ballast to the sheet (by pins or glue) before it is rolled up onto the cylinder. Such a cylinder is therefore large and heavy.

A very special type of bottom protection was designed for the Eastern Scheldt Storm Surge Barrier. The protection used in this case had to be of the granular-filter type and needed to be laid in a single operation per mattress. The three-layer filter was wrapped up in geotextile sheets that were kept in the correct shape by a steel mesh (like gabions). Special equipment was built to handle these mats. This type of mattress was chosen for that part of the sill where loss of bottom material would cause structural failure. Geotextile alone was not considered safe enough over the design life time of 200 years.

Mattress-type of scour protection has a number of advantages and disadvantages when compared with granular filters:

Advantages are:
- Limited construction height
- Applicable on steep slopes

Disadvantages are:
- Difficult to remove
- Presence of structural joints that tend to be vulnerable
- Vulnerable for mechanical damage
- Life time is restricted (ultra violet radiation for geotextiles, pile-worm (teredo) for willow or bamboo)

13.4 Construction of granular filters

The use of granular filters is more common in breakwaters than in closure dams. This is a consequence of the different working conditions that are present. Breakwaters are generally constructed in places with a relatively flat bottom, whereas a scour protection for a closure dam will often be extended along the steep slopes of the gullies. The material that is meant to form the granular filter tends to roll down the slope in that case.

The granular filter is composed of several layers of filter material, of which the finest material is used to provide a sand-tight filter over the virgin bottom material. Subsequent layers are stepwise coarser to prevent underlying layers from washing out, until such grain size is obtained that the upper layer can withstand the external forces by waves and currents.

The construction of such filter is complicated since the filter has to be built up layer after layer. To ensure the required sand-tightness, it is unacceptable that layers are interrupted locally. The presence of each (designed) layer must be guaranteed and proved by the contractor. In view of the tolerances on one hand and the measuring accuracy on the other hand, a layer thickness of at least 0.5 m or $2D_{n50}$ is recommended when the filter material is barge-dumped.

The dumping itself is done by barges that are able to produce a controlled flow of material. These are either split barges or side dumping barges that create a sort of curtain of falling material. By hauling the barges slowly sideways, an even layer can be applied. It is necessary to control both, the discharge of material and the lateral velocity of the barge. This control is often easier at breakwater locations than in potential closure locations, where the current velocities tend to be higher in the construction stage.

The advantages of a granular filter are:
- Self-healing after minor damage (for instance by a dropped anchor)
- Absence of structural joints
- Simple to be removed (dredging)
- Towards the end of the area to be protected, they can be faded out gradually

Disadvantages are:
- Absence of structural coherence, they disintegrate on steep slopes
- The total construction height is considerable because of the minimum thickness per layer and the large number of layers

13.5 Providing and handling of quarry stone

It has been indicated earlier which data must be collected before can be decided to open a quarry in a particular rock formation. Before starting the actual quarrying operations, some requirements must be met:
- Access must be ascertained
- Required permits must be available
- Protective measures against damage to human interests and ecology must be operational
- Accommodation must be available for personnel
- Maintenance facilities for equipment must be available
- Supplies of fuel, spare parts, explosives, etc. must have been arranged

The planning of the quarrying operation is mainly based on the expected fragmentation curve. The blasting and quarrying must be done systematically, following a pre-determined mining plan. During the blasting, bench floors that can be used for sorting the material, loading and transport are created. The width of the bench floor must provide adequate working space for these purposes.

In most cases, it will be difficult to obtain the larger fractions. In the beginning, this does not appear to be a problem, since in the first phases of the project, only finer material for filters and core is used. The larger fractions are required in the later phases of the project, when armour layers and breakwater heads are under construction. Then, however, it is too late to modify the blasting scheme and to obtain the required percentage of armour stone. It is therefore recommended that larger fractions should be produced right from the start of the operation, and efforts to obtain these should not be postponed to the last stage of quarrying. The consequence is that all stone gradations must be sorted and stored separately, right from the beginning, even if some categories of stone are not required until later. This leads automatically to the need for a large stockpile area where stone can be stored until it is used.

It is recommended that the classification of stone in the quarry is facilitated by providing sample stones per category and by frequently using a weighbridge to check the weight of sample stones.

More details about the quarry are given in Appendix 2.

The transport of material from the quarry to the work site can be done in four different ways:

- By road
- By rail
- By water (either inland or sea)
- By a combination of methods

It is impossible to indicate a preferable mode of transport as the choice depends on local conditions, available facilities and the extra investments required. In general, transport by water is far cheaper (4 to 5 times per ton kilometre) than transport by road or rail. This is valid only if a waterway of sufficient width and depth is available, or can be made available at little extra cost. Rehandling of material due to change of transport mode will create extra cost, which must be compared with the savings.

It is not certain that delays in the transport chain will coincide with delays in the quarry operation. This is a second reason to provide a stockpile area at the quarry site that has sufficient capacity to cope with irregularities in the production and delivery of stone.

13.6 Use of rolling and floating equipment

At the location of the dam or the breakwater, a relatively large construction yard is required. Space has to be provided for offices, accommodation, workshops, etc. In general, a stockpile for quarry stone and other construction materials is also required to act as a buffer when supply and discharge are not in balance. When concrete armour units are used, a concrete mixing plant is required as well as a block casting area and a storage area for the armour units.

In principle, there are three methods to bring the material into the designed profile:

- By floating equipment
- By rolling equipment
- By a combination of both

It is evident that for detached breakwaters or for dam sections not yet connected to the mainland floating equipment or transhipment is the most logical choice.

With regard to the method of construction, the major differences between breakwater construction and closure dam construction are caused by the differences in the cross-section of the profile to be made. A breakwater generally consists of a core covered by several layers of different materials of various sizes (see for instance Figure 11-3). The cross-section of a closure dam is much more uniform, and typically, cover layers are not required. However, different materials may be used in the longitudinal profile for different dam sections because of the increasing flow velocities at the work front when the gap narrows. Different longitudinal cross-sections may also be required for the breakwater, usually because the exposure is in a different wave-environment (for instance in deeper water or near the breakwater head).

The main problem of breakwater construction is to build out core and cover layers consecutively in such a manner that every part which is not yet stabilized by its cover is not damaged by the environmental conditions (weather) during construction. All damage, which occurs during construction, has to be repaired according to the prescribed layer profile, as the functioning of the breakwater depends on the filter-design rules. Therefore, it is necessary to consider tolerances and to maintain a very strict position control during the construction of the breakwater.

Environmental conditions (storm surge) may also damage a closure dam. Although the repair may require extra material to compensate for the loss, construction generally simply continues on top of the remains. In some cases continuation implies a completely different design. For instance a planned vertical closure may have to be completed horizontally or by using an obsolete vessel as a caisson. Basically, there is no need to keep to the original design, as long as the gap gets closed. There, strict adherence to the tolerance and positions is not as essential as it is in breakwater constructions.

Several aspects of using rolling equipment are identical for breakwaters and dams. However, breakwater construction, with its layers, is the more complex operation. For floating equipment, there is a distinct difference because of the environmental conditions. For breakwaters, the main problem is operating in wave conditions: "how to place the material dangling from a moving floating crane, on the right spot". For closure dams, the flow characteristics determine the operations: "how to keep position or anchor in the flow or how to operate in the short moment of turning tides". Of course, in some cases waves and flow may occur at the same time.

13.6.1 Rolling equipment

If rolling equipment is used, a dam is built out with a work front in several phases, for filters, core material, first under-layer, toe, etc. The crest of the dam is used as main supply road. It has a minimum width of 4m for one lane traffic. For two-lane traffic the crest width must be at least 7m. As an alternative, one can create passing places. Since it is virtually impossible to drive over the armour units (they are too large), the access road to the work front is often created on the crest of the core or at the crest of the first sub-layer. The level of this crest must be high enough above HW to guarantee the safety of equipment and personnel working there.

In this way, the full length can be constructed according to the design, except for the armour units on the crest. These units can be placed in the final stage, working backwards from head to mainland (see Figure 13-2). A picture of such breakwater under construction is given in Figure 13-3.

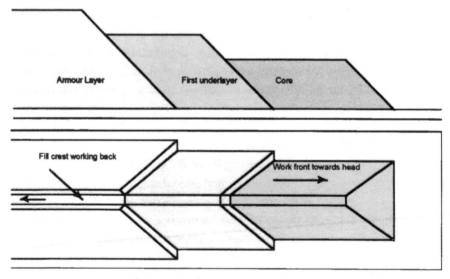

Figure 13-2 Subsequent work fronts

Figure 13-3 A breakwater under construction

If a concrete crest block is used instead of a crest of loose armour units, one may position this cap-block building out the dam. The cap block can than be used advantageously to provide a better quality access road to the work front.

The filters and the core are often placed by bulk dumping. The armour units are always placed individually by crane to avoid the risk of breakage or misplacement. The method used to place the first under layer depends on its size and the local conditions. For the material that is placed individually, nowadays the crane is fitted with electronic positioning equipment to place the units in the pattern prescribed in the specifications.

A disadvantage of the dry construction method is the fact that all material must be transported over a rather primitive road with limited capacity (see Figure 13-4). This becomes ever more important as the length of the dam increases. When the crane at the work front prevents direct dumping, one may consider the use of a gantry crane. The required width and height of the work-road over the crest may lead to a much bulkier design.

Figure 13-4 Trucks waiting on the breakwater

Figure 13-5 Use of cheap local equipment

The major advantage of the dry construction method is the potential use of cheap local equipment (see Figure 13-5) and the independence of working conditions at sea (fog, waves, swell, currents). For execution of a closure, obviously, rolling equipment can only be used for the horizontal method or for finishing a combined closure horizontally. The crest height of the closure profile is taken at a storm safe level. The material size is identically for the full profile as the flow attack is at the work front, over the full height. In order to improve access for the rolling equipment, the crest layer is filled-out with small size material; a possibility that does not exist for the breakwater as it would disturb the layer characteristics. In exceptional cases the method is used on breakwaters, on condition that the fines are later removed.

13.6.2 Floating equipment

When floating equipment is to be used, a work harbour is required from the start of the operations. In this harbour, the barges can be loaded.

Split barges or preferably a side-dumping vessel may be used to place filter layers of a limited thickness. Bulk dumping can be used for the construction of the core, with bottom door barges, split barges, and tilt barges or flat deck barges and with bulldozers pushing the material over board. Intermediate layers may be applied along the slopes by using side-unloading barges.

As soon as the structure reaches a level higher than 3 to 4 meter below HW, the use of these barges and vessels becomes impossible. If one continues to use floating equipment, floating cranes or crane platforms are needed to finish the upper part of the profile. The use of cranes may also be necessary to trim the slopes of the core that are constructed in bulk dumping operations.

The main advantage of the "wet" construction method is the possibility to start construction at more than one work front, or to build detached breakwaters. Another advantage is that one can bring in the greater part of the material independently of the limited transport capacity over the crest. When material is transported by barge from the quarry to the site, it is an advantage to use the supplied material without stockpiling.

A disadvantage is the dependence on working conditions at sea and the need for a working harbour right from the beginning of the works.

For closure dams, floating cranes can be used for horizontal closure by positioning them at the downstream side of the closure gap, in the area protected from the flow behind the dam head (see Figure 13-6).

For vertical closure, floating cranes are not very adequate. Dump barges are the best tools but they can only provide the lower portion of the dam profile. The top section, crossing the waterline, has to be made in a different manner. This is either a continuation of the horizontal system, making the total closure method a "combined method" (see Section 13.6.3). or a vertical continuation, using very special equipment like a cableway or a bridge (see 13.7.2) Dumping creates a layered build-up. Depending on the way the dumping is executed, this results in a triangular or a trapezoidal profile (Figure 13-7 and Figure 13-8).

Figure 13-6 Positioning the floating cranes

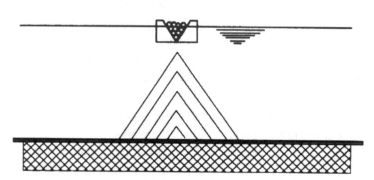

Figure 13-7 Profiles in line-dump

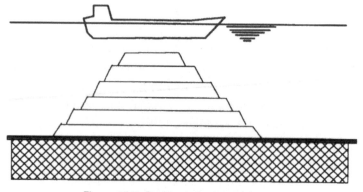

Figure 13-8 Profiles in horizontal layers

In closure dam design, the creation of a sill by dump vessels not only has a financial advantage but also the possibility to create extra stability in the flow attack. Instead of tipping from a centre line, vessels can dump horizontally layer by layer in a trapezium-shaped profile. The side slopes thus depend on the individual dimension of every successive layer (Figure 13-8). This is an important stabilizing factor for the flow crossing over the sill. This may also be an advantage from the point of view of soil mechanics.

13.6.3 Combination of floating and rolling equipment

A construction method in which dry and wet equipment is used in combination is often preferred. The main reason is that generally the unit transport cost over water is lower than by dump truck. As the largest part of the cross-section is the lower part of the profile, dump vessels are used for the bulk of the material in these sections. This is specifically true for deep-water dams, as dumping will not reach higher than about 4 metres below high-water level. The profile accuracy is low; both for the finished level and the width and for a breakwater, some reprofiling of the side slopes will have to be done later. This can be done, simultaneously with the placing of the next layer for stabilization, by a crane operating from above water, after further tipping by dump trucks has provided access (see Figure 13-9).

In the combined method of closure dam construction, the dumping seldom creates a sill sufficiently high to reach the critical flow stage. But even if this is so, the horizontal continuation of the closure on top of it may change the levels such that the flow will change back to a sub-critical flow. In all cases, the flow velocity in the last part of the gap will be much higher than that occurring during the dumping process of the top layer. This layer therefore needs much heavier units (stone), certainly in the centre section of the gap, than when the vertical closure system is continued.

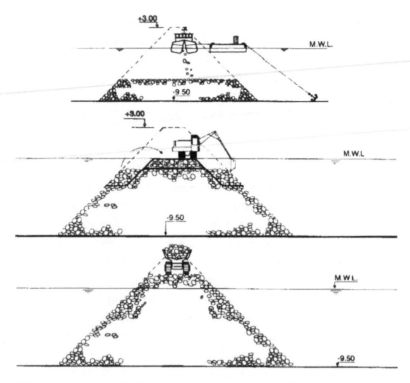

Figure 13-9 Construction of a closure dam by successively using waterborne equipment, grab and dump truck

13.7 Very specific techniques and ancillary equipment

13.7.1 Closure by hydraulic filling with sand only

In a number of cases, a tidal basin can be closed just by pumping sand into the gap. The principle is simple. As long as more sand is pumped into the closure gap than the flow erodes away, the gap narrows. Due to the development of dredgers with very high capacities (5000 to 8000 m3 sand per hour), this has become a realistic option. Bulldozers are needed to spread the sand-water mixture over the fill and shape the desired profile of the dam. In addition, they prevent the erosion of gullies on the fill slope and compact the deposited sand. The equipment, used to control the fill process has improved likewise.

Thus the main question is how much capacity is needed. A very high capacity can be attained by using many delivery lines from various dredgers. However, to keep the fill under control, for these operations every delivery line needs a specific width. More capacity means a much wider fill-profile, which does not improve progress. So there is a practical limit to the supply. Of course, the gap can be approached from two sides.

A sand closure is a horizontal closure, with a progressively narrowing gap, in which the flow increases until the very last gap can be blocked. The flow in the gap has a sand transporting capacity that is related to the flow velocity to the power of the order of four. This means that when the flow increases from 2 m/s to 2.5 m/s, the transport of sand is multiplied by a factor 2.5 and when it reaches 3 m/s it is multiplied by a factor 5.

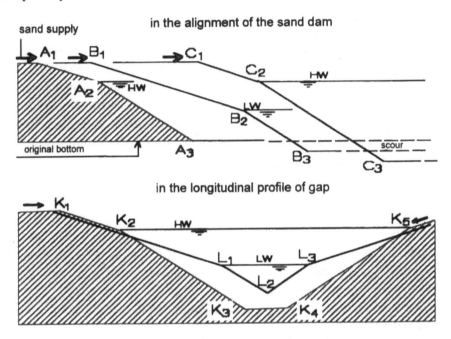

Figure 13-10 Closure by pumping sand

The normal process of building a sand fill dam is as follows. A sand water mixture is pumped by the dredger via a delivery line to the fill to a sufficient height above HW (point A1 in the figure). The mixture runs down the slope to the waterline (A2) and into the water where the sand settles. The sand creates a slope above water (A1-A2) which is much less steep than the submerged slope (A2-A3). In the ebb period, going from HW to LW, the progression, as shown in the figure, is from A1-A2-A3 to B1-B2-B3. Progress seems small, as the delivery point moves very little (A1-B1).
All the sand goes into the toe circle. During flood however, the line B1-B2-B3 shifts to C1-C2-C3. The nett progress per tidal cycle is A1-C1. Scour will erode the original bottom in front of the dam and enlarge the profile to be made. In addition, part of the supplied sand will also be taken by the flow and carried outside the alignment area. The further the dam extends the higher the rate of flow in front of it and the smaller the nett progression per tidal cycle.
When hardly any progress is being made, the final blocking has to be enforced. This is a special operation, which starts at the moment of HW, which is shown in the

figure by the line K1-K2-K3-K4-K5 for two fills approaching the gap from both sides. Progress during the ebb may even seem to be negative, as the line K1-K2 retreats but the size of the gap at LW is very small (L1-L2-L3). Then, just before slack water, the profile has to be blocked. This requires the temporary interruption of hydraulic transport by taking recourse to earth moving plant, thus obviating the erosive action of the hydraulic transport by water. Bulldozers and cranes will have to shift as much sand as possible into this LW-gap to shape a tiny ridge in the triangular profile (L1-L2-L3). This sand is taken from the slope above water (K2-L1). The ridge being ready, the flow is blocked and all the sand supplied by resumed pumping will settle in the profile. The ridge, protruding above LW has to be heightened and widened to stay ahead of the rising outer water. The volume required increases tremendously with the rising level, as the length of the ridge also increases. To create a stable profile over the full length (K2-K5), that is able to stand the head difference of the full tidal range, in a few hours, requires a very skilled fill procedure and a sufficient sand delivery capacity as well.

Thus the principal questions are; how large a gap can be closed in the last tidal cycle and what will the sand transporting capacity of the flow in that gap be. The last operation in the gap takes place in one tidal cycle, so for a semi-diurnal tide this is about twelve hours. The twelve hours before that is the last tidal cycle in the normal fill procedure in which some narrowing progress is still to be made. Consequently, the possibility of closing is determined by the operations during the last day. The capacity required thus is determined by:

- the normal process followed to attain the size of gap that can be closed in one tidal cycle
- the ebb-phase during which the tiny ridge can be shaped in the LW-slack period
- the flood-phase in which stable profile must be built before the next HW

With that capacity laid down, the progress in the days or weeks before the closure can be calculated by phase-wise determination of the nett progress per cycle. Summing-up these phases gives a reasonable approximation of the total time needed for the construction of the dam.

Several sand closures have been constructed. In these cases it appeared that, depending on the grading of the sand, the maximum flow velocity that could be accommodated was in the order of 2 to 2.5 m/s. According to the flow formulae in gaps (Section 5.2.3), these velocities occur for a head difference of about 0.30 m, for rounded sand dam heads. The gap size for which this boundary condition exists can be calculated by using a mathematical model of the closure procedure.

For instance, in the mathematical example "case 1" of Section 16.2 (Figure 16-5) the flow reaches about 6 m/s in the final stages (4 and 5), when the head difference is slightly more than half of the tidal range (1.5 m). The 2.5 m/s is already reached at stage 2, which is for a gap dimension of 4000 m2. This gap size is far too much for the final day's operation. Nevertheless, the sand-closure of the "Krammer" (in the rear of the Eastern Scheldt basin) in 1987 closed a basin of about 55 km2, which had a tidal range of 2.70 m. However, by that time, the storm surge barrier in the entrance of the estuary was operational and the tidal range was artificially reduced to 0.60 m during the closure. Thus, to enable the closure flow velocities were kept under 2.5 m/s. This shows that if the tidal range is smaller than 0.6 m, even very large basins can be closed by pumping sand.

Basins with larger tidal ranges can only be closed by sand pumping if the storage area of the basin is much smaller. An example of that is the sand closure of the Wohrdener Loch in northern Germany near Meldorf in 1978 (Figure 13-11). The tidal range during the neap tide on closure day was 3.20 m. The storage area was 10 km2, the grading of the sand about 350 µm, the total installed dredge capacity 8000 m3 per hour and 14 bulldozers and 8 hydraulic excavating machines were busy at the fills. The length of the gap at the waterline during HW (K2-K5) (Figure 13-10) in that case was 120 m. The capacity to strengthen the ridge during the flood phase appeared to be the determining factor.

Figure 13-11 Final closure using sand only

A few conditions determine the possibility to close with sand only:
- the tidal range or the storage area of the basin has to be sufficiently small
- large quantities of good quality sand must be available nearby
- high-capacity dredgers have to be used
- A well-organized fill-procedure using cranes, backhoes and bulldozers is required

If the original bottom in the gap has a resistance against erosion comparable to the sand used for closing, protection against scour is not relevant. Scour is acceptable unless it endangers stability of structures in the proximity. A considerable volume of sand is carried by the flow outside the desired profile. This reduces progress but is part of the method. The lost sand is not considered a permanent loss. Actually, instead of providing an expensive bottom protection, scour is compensated by using an extra quantity of sand. The many machines operating at the fill require that there is good bearing subsoil. Sand closures using very fine or silty fine sand are impracticable or impossible. Therefore very fine or silty fine sand is not suitable for the construction of sand closures.

13.7.2 Use of a temporary bridge or a cable way

When closing vertically, the flow pattern will be completely different. Building up layer after layer, the flow will reach a stage in which critical flow starts. Stability of the profile depends on its shape, which in turn depends on the way the dam is built.

Dump vessels or dump trucks cannot create the layers to bring the sill above the waterline in the vertical closure system. Therefore different equipment is required. The most common method is to use a temporary bridge or a cableway. Of course this equipment has to be installed as the first part of the full closure operation. Generally, the method requires a high initial investment but it may be an alternative to a caisson closure in cases where the provision of a caisson construction dock is expensive. The advantage of vertical closure over combined or horizontal closure, which is obvious in the classical system of sinking mattresses, is much less determining when using modern equipment.

However, when a cable or a bridge is installed, it seems obvious that this should be used for the entire profile of the closure dam. This means that the profile is built-up in the line system (see Figure 13-6 and Figure 13-12). Using the return cable or by tipping at both sides of the bridge also makes a two-line dump possible, but these two tops are relatively close. The disadvantages of this type of operation are:

- Steep side slopes with a risk of sudden instabilities in the build-up. This is specifically true in the final stage for the downstream slope owing to seepage flow through and overflow over the top.
- Irregular very sharp crested top, which determines the critical flow.
- A low production rate and a high in-situ unit cost (manufacturing and handling)
- The largest portion of the profile must be placed in the final stage of closure

The advantage is that any place in the alignment can be reached at any moment, for instance for repair of an unstable section.

Nevertheless, it is always worthwhile to consider the construction of the lower portion by dump vessels.

Figure 13-12 Closure with a cableway

13.8 Minimizing risks during construction

It has been mentioned several times that attention should be paid to the stability of the cross section during construction. For closure dams, this may be evident, because the closure dam itself is a temporary structure that will be replaced and protected by the final structure as soon as possible. Nevertheless, one must take into account that the closure dam must withstand the hydraulic loads that are expected during the period of exposure.

In a similar way, one must consider the various construction phases of a breakwater. Considering Figure 13-13, it will be clear that the work front cannot withstand the design storm. Therefore, one must consider what risks are threatening the structure during the construction phases and find ways to reduce these risks. Common methods are to:

- Select a specific construction period
- Reduce the exposed length of the vulnerable part of the structure
- Construct protective bunds

Specific construction period

Sometimes it is possible to reduce the risks by limiting the construction period to the summer season (or to a particular monsoon season). If the whole structure can be completed in this calm period, the risk can be much smaller. Otherwise it may be possible to interrupt construction during the rough season. When the work fronts are well protected it will be possible to resume construction the next year or season.

Reduce exposed length

In other cases, an option may be to keep the distance between the work fronts as small as possible. If damage occurs, it will be restricted to a small stretch of the structure. Since the contractor is still present with all his equipment, repair of the short damaged section need not to pose great difficulties.

Protective bunds

Instead of the work sequence as sketched in Figure 13-13 (left), it is possible to dump the first under-layer before the core material. This method can be compared with the construction of reclamation bunds around a confined reclamation area in dredging. The disadvantage of the method is that more material is required for the first under-layer than strictly necessary on the basis of the theoretical two-layer design. In Figure 13-13, the traditional method is compared with the alternative.

It is clear that the alternative method (on the right hand side of Figure 13-13) provides a better protection of the core material during construction phases. It is also clear at the same time that it requires more expensive first under-layer material and that the construction method is a little more complicated. Depending on the availability and cost of material versus the cost of handling, one can save some of the extra material by double handling.

Construction sequence as in:

1. Filters
2. Core
3. First under-layer
4. Armour layer
5. Crest (working back)

Construction sequence providing protective bunds:

a) Filters
b) First under-layer (part)
c) Core
d) First under-layer (part)
e) Core
f) Armour layer (part)
g) First under--layer (part)
h) Armour layer (part)
i) Crest (working back)

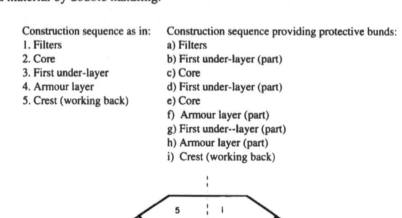
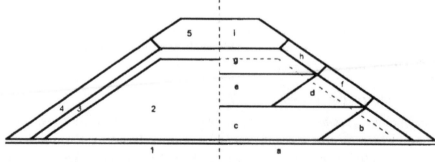

Figure 13-13 Varying construction sequence

14 CONSTRUCTION METHODS FOR MONOLITHIC STRUCTURES

Like Chapter 13, this chapter gives a description of practical problems encountered when using caissons or building monolitic breakwaters. It should be read as an example rather than be considered as absolute facts.

14.1 Introduction

Both, in breakwater constructions and in closure dams, it can be an advantage or even essential to use very large monolithic elements. As far as breakwaters are concerned, the monoliths may consist of loose elements that are assembled to form the final monolithic structure. For closures, the large elements are generally of the caisson-type.

Caissons have been widely used in both closure dams and monolithic breakwaters. Although there may be some structural and operational differences, the basic principles are identical. The structural differences may be due to the different load pattern or to the fact that in closure dams, the caissons are designed to allow discharge until gates are closed. The operational differences may be due to the fact that in closure dams the current is usually the main constraint during the placement of the caissons, while waves are the main constraint for breakwaters.

Because of the small differences, here no distinction is made when discussing the constructional aspects of caissons for the two types of application. The reader is referred to Chapter 9 for a discussion of wave loads on monolithic breakwaters.

14.1.1 Caissons, closed or provided with sluice gates

Caissons are used in breakwater construction in order to create a near vertical or vertical wall, which minimizes the cross-sectional profile. This is most effective in deep water and is a choice requiring a small but expensive volume. The resulting reflection of the waves is the main disadvantage. The design will always consist of a rather large number of identical caissons, placed in line. The length of each depends on the handling and stability during transport and sinking and only very seldom on the desire to minimize the number of sinking operations.

In closure designs caissons are large, artificially made structures or vessels used to block a final closure gap in one major effort or in a minimum of major steps. Therefore, they are designed to be as large as is feasible from a constructional point of view. In emergencies existing ships and pontoons or the like have been used, sometimes after adaptation to fit the gap dimensions. In normal circumstances, caissons are specifically designed for the purpose. Generally they are made of concrete, have a box shape and are self-floating. Three different typical systems can be distinguished:

- The final gap is closed in a single operation by placing one or several caissons simultaneously
- Several identical units, which together fit into the gap, are made. They are placed one after the other in a period of several days, during which for each successive caisson the positioning conditions will be more severe
- Several units are used, a number (or all) of which are provided with sluice gates. Each unit is placed with its gates closed, but after stabilizing of the caisson, the gates are opened. As soon as all caissons are rigidly in place, all the gates are closed at a suitable moment.

Which system is used depends on circumstances and conditions. They are progressively more expensive.

In this chapter, the following subjects will be discussed:
- monolithic breakwaters assembled from small units
- monolithic breakwaters consisting of large units constructed in-situ
- monolithic breakwaters consisting of prefabricated large units, floating and non floating
- monolithic breakwaters consisting of prefabricated large floating units i.e. caissons

Specifically for the caissons, the identical aspects of the use in breakwater and closure design are given in detail. The construction, the transport, the foundation bed and abutments, the floating stability and sinking, and the stability after placement are described. A few aspects typical of the use of caissons in one or other of the applications are then discussed.

14.2 Monolithic breakwaters

14.2.1 Monolithic breakwaters constructed by assembling small units

The oldest vertical wall breakwaters were composed of rectangular blocks of natural stone. These blocks were sawn in the quarry and placed in the breakwater according to a pattern comparable with the present brickwork techniques. The blocks were connected with dowels to ensure the monolithic behaviour of the structure.

This technique is still used, although the blocks are often cast in concrete nowadays, and steel reinforcement and cement mortars are used to connect them (Figure 14-1 and Figure 14-2)

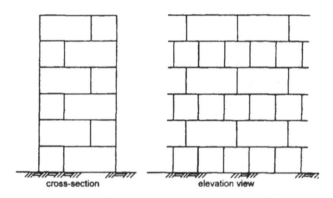

Figure 14-1 Typical block wall

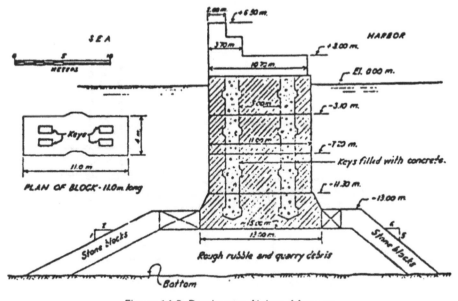

Figure 14-2 Breakwater Algiers, Morocco

14.2.2 *Monolithic breakwaters consisting of large units constructed in-situ*

The most common example of in-situ construction of large monolithic units is the construction of sheet pile cells. The main problem in constructing this type of structure is the closure of the slots between the individual piles. The workability during pile driving may also cause problems.

The cells are filled with soil or stones. It must be assumed that owing to overtopping and spray, the fill material is saturated with water over its full height. Depending on the type of fill material, cyclic loading and wave impact forces may cause liquefaction of the fill material, which results in relatively high ground pressures on the sheet piles. In such cases, poor connections between the piles cause a serious complication.

14.2.3 *Prefabricated large units*

Prefabricated large units can be transported in different ways. The most elementary way is to use the buoyancy of the elements. In that case, we speak of caisson type structures. Because of their specific importance for both dams and breakwaters, they are treated in a separate Section 14.3.

However, it is not necessary that the prefabricated unit is fitted with a watertight bottom. It is possible to place circular or rectangular rings on a foundation bed and fill them with quarry run to act as a monolithic breakwater. The units can then be brought into place by using separate floats or lift barges. It is also possible to roll them out over the crest of the placed units and lower them into position with a gantry crane. This method was used in Hanstholm (Denmark) and Brighton (see Figure 14-3).

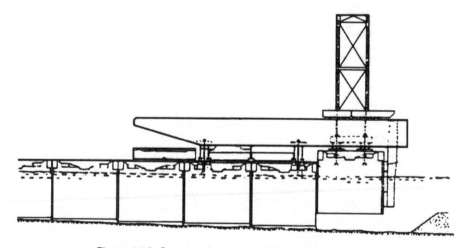

Figure 14-3 Construction method Brighton breakwater

14.3 Caissons

14.3.1 Building yard

It is possible to construct caissons in different ways. The main differences lie in whether construction is completed in the dry, or only the bases of the caissons are constructed in the dry. In this case when their buoyancy is sufficient they can be launched and construction can be completed in a floating condition.

Whether only the lower part is constructed in the dry or the whole caisson makes little difference to the initial stage of construction. The construction yard must be kept dry until the structures are ready to float. For this purpose, one may use the following facilities:

- Dry-dock at a shipyard
- Slipway
- Lift deck
- Dredged special purpose dock

The first three facilities are common features of a shipyard. They can be used if available at an affordable cost. The disadvantage may be that the space is too small to construct the required number of caissons in a limited period. Then, one may consider floating out the caissons long before they are completed and finishing the upper part and the superstructure in the floating condition.

The last option, a specially dredged large dock is a common solution in the Netherlands. The dock is kept dry by deep wells, and the closing dike can easily be breached by a dredge when the construction of the caissons has been completed. All caissons for the closures in the Delta project were constructed this way (Figure 14-4).

Figure 14-4 Dredged dock for construction of caissons at Neeltje Jans construction island

The advantage of such a special purpose dock is that it can be built close to the site where the dam is to be constructed. Moreover, the size of the dock can be made to comply with the specific requirements.

The construction site for the building of caissons is fitted out like any construction site for a large concrete structure.

14.3.2 Transport

After completion of the caissons, they have to be transported to the site of the dam or breakwater. It is essential that sufficient depth and keel clearance is available throughout the route from the dock to the site. Tugboats with sufficient power to overcome currents and to maintain a reasonable speed tow the caissons.

The force required to attain a specific speed depends on three parameters. One is the submerged cross-section of the caisson in the tow-direction and its drag coefficient (see Section 5.2). Next the friction force along the bottom and the sides, in fact the length of the caisson, has an influence. Third, the motions in the water that are caused by turbulence and waves have an impact on the speed. The acceleration of such a large mass requires a large force. Generally speaking, the drag coefficient for caissons will be in the order of about 4. It should be kept in mind that the pulling force is determined by the square of the relative velocity of the caisson in relation to the water. To calculate the actual speed over the ground, the relative velocity has to be compensated for the direction and velocity of the water flow. To generate the pulling force, often tugboats are used. As a rule of thumb, about 10 HP is required for every kN pulling force (1 HP (horsepower) = 0.75kW).

Finally, apart from the force in the tow direction, it is necessary to have lateral control to counteract yawing and power to slacken the tow. In consequence of these requirements, the total power needed is double the power calculated above. The worst case, is probably the moment when the caisson passes over the shallowest part of the transport route. Owing to the high return current under and alongside of the caisson in the narrow profile, the drag coefficient increases. The speed therefore drops and sideways control becomes more difficult. Increasing the power used (full ahead) brings a risk of dipping the caisson and touching the ground. This happens because the pressure underneath the caisson is less than the hydrostatic pressure (see Section 5.2).

Proper attention must also be paid to the stability of the caissons. This means that adequate calculations of the metacentric height must be carried out (see Section 14.3.4).

14.3.3 Preparation of foundation and abutments

All caisson structures have to be placed onto a prepared bed, which has to fulfil several functions. Before placement of the caisson, it is necessary to:

- bring the bottom to the desired level and smoothen it.
- keep it that way until the caisson is in place (flow and waves).
- provide proper connection between consecutive caissons

After the caisson has been put onto the foundation bed, four aspects are important:
- The load of the caisson should be well spread over the bed, which defines tolerance of the bed level in combination with the structural strength of the caisson. This is in particular the case, when the caissons have a permanent function, such as in breakwaters. Uneven (and unpredicted) supports may lead to failure of the bottom of the caisson by excessive bending moments.
- In closures, flow underneath the caisson will soon reach high values but piping (with scour of bed material) should be prevented.
- In closures, the gap size needs to be longer than the length of the caisson to allow for tolerances and diagonal length during the swing motion; however, outflanking of the flow along the sides has to be blocked immediately.
- The caisson will soon be subjected to high forces caused by waves or hydrostatic head, which will try to either shift or overturn the caisson. Generally, a linear decline of the potential head underneath the caisson is assumed. However, if permeability of the bed is lowest at the downstream side, the upward pressure is higher than average. Moreover, seepage flow in the bed material concentrates along the lower edge.

The last function of the bed is performed by the part that acts as a scour-prevention in front of the caisson. Although the foundation bed and the bottom protection are not necessarily related, they are usually made simultaneously and of the same materials in order to prevent structural inconsistency at the boundary between them.

The first (or only) caisson has to fit to the adjacent dam sections that have been made in a different way. This can be done in various ways. One method is to place a short caisson in advance in order to create a vertical side at its free end. This is linked to the dam section by encircling the other side with the other dam material, (clay, sand, quarry run or the like). The other method is to use a U-shaped sheet pile wall and link the dam to this structure by normal fill. This latter method requires a floating plant to hammer the piles, which may be an expensive operation in choppy water.

14.3.4 Floating stability during transport, positioning and ballasting

The transport of caissons can be an important factor in the design of a caisson. Although caissons are usually designed to withstand forces in their final placed state, it must also be possible to transport the caisson to the site and therefore the draught may be restricted. The dynamic stability of the caisson during transport is another important aspect.

For example, assume a caisson of infinite length, a width of 9 m and a height of 12.5 m. All walls are 0.5 m thick and are made of concrete with density 24 kN/m³ The caisson is made in a dry dock and floated by raising the water level in the dock. Then (per metre length) the characteristics of the caisson are:

weight of concrete (G)	396 kN
centre of gravity above bottom (g_b)	4.80 m
moment of inertia ($I = \dfrac{1}{12}LB^3$)	60.75 m⁴
displacement (V)	39.6 m³
draught of caisson (d_r)	4.40 m
centre of buoyancy above bottom (c_b)	2.20 m
MC = I/V	1.53 m
metacentre height (m_c)	negative 1.07 m

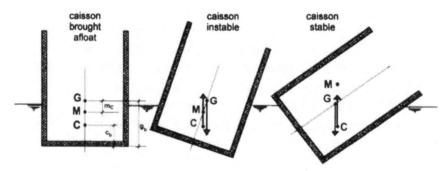

Figure 14-5 Stability of a floating caisson

Since MC + c_b (3.73 m) is smaller than g_b (4.80 m), the caisson is unstable and will tilt as soon as it comes off the ground. Figure 14-5 shows a clockwise tilted position; the centre C of the submerged part is situated to the left of G and the tilting moment will continue. Although the caisson capsizes, it will not be submerged. At an angle of about 45° a stable situation is reached. The submerged part is triangular and the water line is much longer than the 9 m, which increases the I, and thus the MC. The MG (m_c) is positive and quite large; thus stability is very good. M is situated on a new stability axis. This is true as long as the vessel does not get water inside, for instance because of motion or wave action. In the figure, the tilt has been drawn just over its stability point to show the righting moment.

The tilt can be avoided by adding ballast to the caisson. Solid ballast (e.g. sand) is advantageous, but in the example shown below, a 2 m layer of water is pumped in and the data change as tabulated below:

weight of concrete and water (G)	556 kN	increased by 160 kN
centre of gravity above bottom (g_b)	3.85 m	down by 0.85 m
moment of inertia (I)	60.75 m⁴	unchanged
displacement (V)	55.6 m³	increased by 16 m³

draught of caisson (d_r)	6.18 m	increased by 1.78 m
centre of buoyancy above bottom (c_b)	3.09 m	raised by 0.89
MC	1.09 m	decreased by 0.44 m
metacentre height (m_c)	0.33 m	increased by 1.40 m

Consequently, the MC + c_b becomes 0.33 m larger than g_b and stability seems to be marginally reached. However, if the caisson gets a little tilt, the liquid cargo starts moving so the centre of gravity also shifts and the resulting moment is increasing the tilt. In the calculation a correction for the I is needed. The water area inside the vessel has to be subtracted. This can be avoided by using bulkheads to subdivide the caisson into compartments . Using solid ballast avoids the dynamic effects of the cargo (see Figure 14-6).

The sinking of a caisson in a closure gap is usually done by ballasting with water. Therefore, its stability has to be calculated for all stages of that operation. Although the sinking is intended to ground the vessel, a list will, at the very least, result in a position out of place.

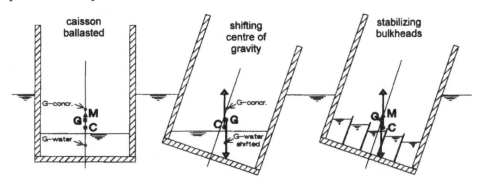

Figure 14-6 Stability of a caisson ballasted with water

The above calculations are valid for situations where the outer pressure around the vessel is hydrostatic (Archimedes). This may not be true for a vessel in a strong current flow, as for instance in a closure gap with too little keel clearance. A flow net around the vessel is then needed to give the outer pressure distribution and enable the determination of the position of the centre of buoyancy.

14.3.5 The sinking operation

Placing

When on site, the caissons must be placed in the required position on the seabed. Adequate tugboat assistance is required to tow the units into position and to keep them in position during sinking. This is generally done at slack water. The question of whether the placing operation should be carried out during HW slack or during

LW slack depends on the draft of the caisson and the available water depth. In breakwater construction in many places the advantage of the LW slack is a reduction in the action that is required.

It is not easy to keep the unit in position during the sinking operation. It is recommended that at least one or two winches should be available to make a connection with the shore or with previously placed caissons. Shortly before it comes to rest on the foundation the unit has a tendency to move horizontally out of control. This is due to the overpressure in the thin layer of water between the seabed and the bottom of the caisson. The problem can be solved by increasing the permeability of the foundation layer or by fixing a skirt or some steel rods in the bottom of the caisson.

In closures, the caisson is intended to block the gap or part of the gap and thus will be positioned transverse in the flow. Since dimensions are generally considerable, even small flow velocities will result in high forces during the manipulating and positioning of the caisson. Therefore placing will always be done during slack water when the tide turns. In practice, there is no instant when the flow velocity is zero. Generally, the tide starts turning near the bottom first and later at the surface and this does not usually occur over the full gap length at the same time. Therefore, slack water is the period (time window) in which velocities are smaller than about 0.5 m/s in either direction. In a tidal cycle there are two such periods, these being during high-water when flood changes into ebb and during low-water with the change from ebb into flood.

Figure 14-7 Placing the final caisson in "Het Veerse Gat"

14.3.6 Work-window of flow conditions during the sinking operation

In closure operations, the working window for the sinking of a caisson depends on the duration of the flow with velocities less than 0.5 m/s, called slack water. If a number of caissons have to be placed one after the other, the duration is different for each new caisson. As indicated before, in a tidal cycle, there are two periods of such slack water, one during high and one during low tide level. A number of considerations determine which slack water period is selected for the sinking operation:

- duration of the time window, which is not the same for the two slack periods.
- available keel clearance on the approach route of the caisson; sailing during high water may be preferred.
- draft and stability of the floating caisson
- stability during the ballasting and sinking operation
- the desired water level in the basin after closure (for the last caisson)
- the way of placing and the equipment used

The last item relates to the fact that caissons are preferably positioned by pushing (or pulling) them against the direction of the current. The advantage is that if something goes wrong, the caisson is pushed back by the flow into the free space, while in the opposite case the caisson may float into the gap and get damaged or cause damage. Thus the procedure starts by bringing the caisson into the gap against the diminishing flow well before the moment of still water. The most commonly used way to bring the caisson into the gap is to position it in advance, head-on on to one side of the gap. A corner is connected to a fixed point on the shore (as a hinge) and it can be swung around that point like a door into the gap. Then, by ballasting, the caisson is lowered and put down just before (or at) slack water. Further ballasting will stabilize its position while the direction of the flow underneath and around the caisson is reversed.

For an operation at high water slack, when the tide turns from flood into ebb, the caisson must be positioned opposite to the flood flow, thus from the basin side towards the open water. Therefore, the caisson has to be sailed into the basin via the gap during a preceding high water period (assuming the fabrication dock to be outside the basin). If the (last or the only) caisson is pushed into the gap by tugs, the tugs are trapped in the basin.

An example of an unusual method of closure that clearly demonstrates these aspects is the closure of the Miele in 1978. This was the main gully in the tidal flat area near Meldorf in the north-west of Germany. The closing method that was originally planned failed and left a closure gap with limited possibilities to enforce a closure before the next winter. The gap was 320 m wide and the bottom elevation was 3.60 m below MSL. The tide had a range of 3.5 to 4 metres. It was decided to try a closure by caissons, by adapting five identical sand transport barges. A new bottom protection had to replace the distorted one.

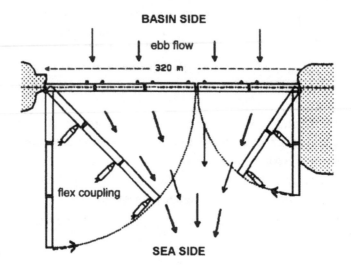

Figure 14-8 Caisson closure in the Miele (Meldorf)

A sill had to be created as foundation for the barges. The limited water depth did not allow the construction of a high sill, or even the use of large stones. Therefore flow velocities had to be kept low and the five barges could not be placed one after the other. The problem was how to put one composite caisson, consisting of five rather fragile steel barges onto a 320 m long sill without risk of breaking, piping underneath, or piping at the ends. The solution found was as follows:

The barges were provided with heavy steel H-profiles underneath, along both sides, to improve longitudinal strength, to improve penetration into the bed material and to reduce the formation of free-spans along uneven bed levels. For stability calculations, it was (pessimistically) assumed that the bed underneath the downstream H-profile would be less permeable than that on the upstream side and determine the upward water pressure. The barges were assembled into two composite caissons, one consisting of two and one of three barges. The connection between the barges was made by using flexible material that allowed each barge to settle independently (within reasonable limits). Much attention was paid to smoothening the sill to avoid high spots that would pierce the bottoms of the barges. The two sets of caissons were positioned near the gap at the two shore ends, where they were connected to a hinge (pole) by steel wire. Both sets of caissons were swung into the gap simultaneously at low water slack. For ease of positioning (to prevent the "doors" from swinging too far) steel tubes had been piled in the alignment of the gap. Tugs on the seaward side had to gently push the caissons against these tubes. Being in line, the caissons had a wide slit where they met. This slit was closed by pushing from the shore-ends and releasing the wire hinges, while the space was divided over the two shore connections. Ballasting was done by pumping water and sand into the barges. Stones were dumped in the shore-end slits and to prevent piping sand was pumped at high capacity along the full length of the caissons via a floating pipeline. As soon as the rising water allowed dump-vessels to sail, stone was also dumped alongside the barges. (After a sand dike had been provided on the basin-side of the caissons, the barges were emptied, refloated, refitted and returned to normal operation. see Figure 14-8)

14.3.7 Number of caissons and/or sluice gate caissons

Generally, several caissons will be placed one after the other over a period of several days. Every caisson blocks part of the profile of the gap and the next caisson will be more difficult to position. Flow velocities increase, the time window diminishes and the turbulent eddies in front of the caisson will be more severe. Although the programme will be planned to work from spring tides to neaps, at least for the later caissons, the last caisson to be placed will determine the design and dimensions of the caissons and foundation bed material. The advantage of this method is that the operational phase in which flow velocities are very high is relatively short. This means that in areas with a limited workable period, for instance because of weather conditions or river discharges, the progress stays within schedule. Moreover, the duration of exposure to high flows with high eroding capacity is small.

For large closures, the last caisson may need unrealistic dimensions in order to achieve a sufficiently long workable period, so the use of caissons provided with sluice gates is a good option. In such cases, the structural components of every caisson (head-walls, diaphragm walls, bottom structure and ballast hold) block a small part of the gap profile. The gates will provide an opening of 80 to 85% of the submerged section of the caisson, which can be multiplied by a discharge coefficient in the hydraulic calculations. Again, the determining conditions are those during placing of the last caisson. These flow conditions determine the dimensions of the total opening provided by the gates. Multiplication by 1.3 gives the total gap size to be blocked by caissons with gates.

15 FAILURE MODES AND OPTIMIZATION

Chapter 15 looks in hindsight at the design and construction procedures for breakwaters and closure dams. An attempt is made to analyze the various ways in which the structures may fail, either during construction or after completion. This analysis requires a systematic approach in which each step is considered critically. The analysis is certainly not complete, and the reader is invited to participate actively. This certainly applies to the optimization procedures discussed in this chapter and elaborated in one of the appendices.

15.1 Introduction

For a long time, the design process used for rubble mound breakwaters was not very analytical. Often, a design wave height was selected on the basis of a limited number of field data. The final design was then tested in a hydraulic model. Usually a geotechnical study completed the efforts.

In the hydraulic model study, the hydraulic stability of the cross section when exposed to the design wave was verified. Although it was evident that this design wave height could be exceeded, waves higher than the design wave were seldom used in the model. Scatter in the model results and inconsistencies in the model procedures were seldom taken into account. Safety coefficients were not commonly used to cope with uncertainties in load or resistance of the structure.

In this way, it could happen that new armour units like the Dolos, with a very good hydraulic performance, were developed. However it was not recognized that the mechanical strength of such units was insufficient to withstand the impact forces under design conditions. In the same way, it was not always recognized that the margin between initial damage and failure of the armour was different for armour layers consisting of traditional quarry stone and the newer artificial armour units.

This caused a reduction of the inherent safety factor of the traditional structure that went unnoticed.

The failure of a number of large rubble mound breakwaters triggered a more analytical approach to the design of such structures. One of the first systematic works on this subject was the CIAD report (Anon. [1984]) on the reliability of breakwater design. It was soon followed by a more comprehensive study by PIANC PTC II Working Group 12 (Anon. [1995]). The PIANC study not only presented methods for probabilistic analysis at levels I, II and III, but also gave recommendations for the values of partial safety coefficients based on a general probabilistic analysis.

In the mean time, PIANC started a new working group to carry out a similar study for monolithic breakwaters. The publication of the report of this working group is expected in 2000.

For closure dams, a systematic probabilistic analysis has been published, as Appendix A in the CUR/RWS Manual on the Use of Rock in Hydraulic Engineering (Anon. [1995]).

15.2 Failure mechanisms

For a good insight into the behaviour and reliability of a structure under design conditions (and excess of design conditions), it is necessary to have a more or less complete idea of potential failure modes or failure mechanisms. A failure is defined as a condition in which the structure loses its specified functionality. This can be connected either to a serviceability limit state or to an ultimate limit state.

For breakwaters, the protective function is the most important one in most cases. Failure is thus related to any damage that leads to unwanted wave penetration into the harbour, followed by further structural and/or operational damage.

The CUR/RWS Manual (Anon. [1995]) gives a general overview of failure modes of rock structures (Figure 15-1). Burcharth [1992] presents a slightly different review more focussed on breakwaters (Figure 15-2).

These overviews are given with some reluctance because it is not yet feasible to give properly defined limit states for each of the failure modes separately in terms of load, resistance and scatter of results.

The same applies to monolithic breakwaters. For this type of breakwater, some failure mechanisms have been assembled in Figure 15-3.

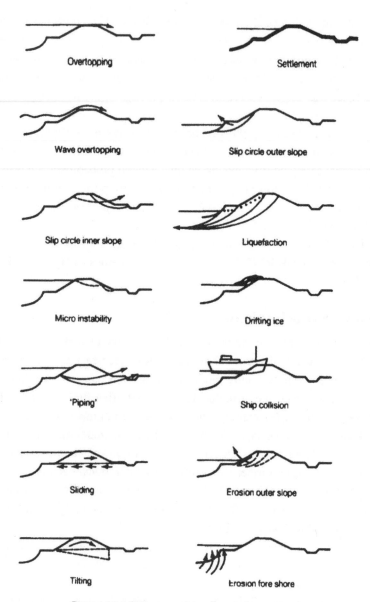

Figure 15-1 Failure modes of rock structures

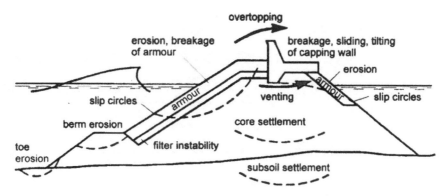

Figure 15-2 Failure modes for a rubble mound breakwater according to Burcharth [1992]

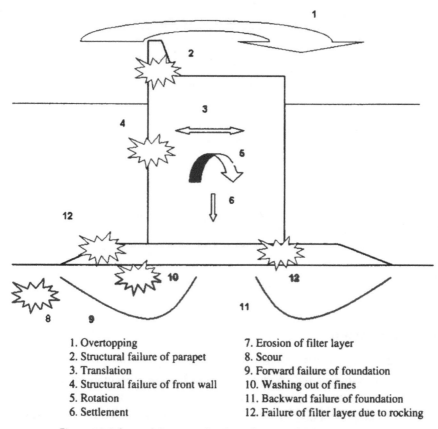

1. Overtopping	7. Erosion of filter layer
2. Structural failure of parapet	8. Scour
3. Translation	9. Forward failure of foundation
4. Structural failure of front wall	10. Washing out of fines
5. Rotation	11. Backward failure of foundation
6. Settlement	12. Failure of filter layer due to rocking

Figure 15-3 Some failure mechanisms for monolithic breakwaters

The CUR/RWS Manual (anon. [1995]) gives an overview of failure modes of rock-fill overflow dams. Failure modes during their construction are shown in Figure 15-4. This list of failure modes is far from complete. In particular, failures caused by three-dimensional mechanisms are easily forgotten.

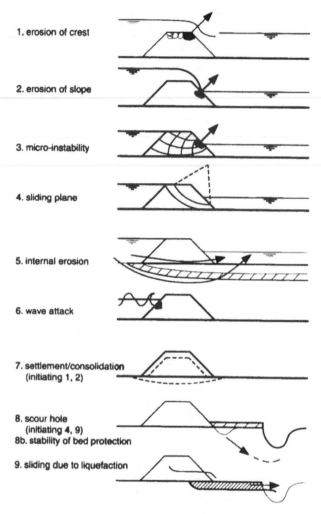

1. erosion of crest

2. erosion of slope

3. micro-instability

4. sliding plane

5. internal erosion

6. wave attack

7. settlement/consolidation
   (initiating 1, 2)

8. scour hole
   (initiating 4, 9)
8b. stability of bed protection

9. sliding due to liquefaction

Figure 15-4 Rock-fill overflow dam failure modes at various construction stages

15.3 Fault trees

A structure can be schematized as a complex system consisting of many components, which may function or fail. A fault tree sketches the systematic relations between failure and malfunctioning of all components in their mutual, interactive relation. Failure of a component may or may not trigger the malfunctioning of a component at a higher level, until eventually the structure as a whole does not perform the functions for which it was meant. Failure of the structure as a whole may also occur if two non-correlated events happen at the same time. Considering these options, one can indicate "AND and "OR" gates, denoting parallel and serial relationships. By quantifying the probability of failure for each component, and by combining the various causes of failure, it is possible to assess the overall probability of failure of the system, be it a breakwater or a closure dam. It is beyond the scope of this book to enter deeply into the theory of fault tree analysis. The reader is referred to more specialized books on reliability theories. However, for illustration, a simplified fault tree and the related calculation of the probability of failure of a breakwater are given in Figure 15-5.

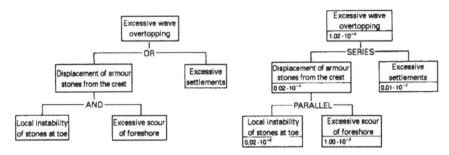

Figure 15-5 Fault tree and probability of failure after CUR/RWS Manual

A complete fault tree analysis reveals the contribution of each failure mechanism to the overall probability of failure for the complete structure.

The probability of failure for each component of the system can be determined by making a preliminary design and assessing the uncertainties in load and resistance (strength) via a reliability parameter Z. This can be carried out at different levels of sophistication. At present, "Level II" methods are most commonly used. Apart from the probability of failure, the Level II calculation specifies the relative contribution of each load and strength parameter. For breakwaters, the uncertainty of the wave climate is often the most important contribution to the probability of failure on the loading side. On the structural side, however, the major contribution is provided by the scatter in the stability formulae and the inaccuracy of the nominal diameter of the armour stone (D_{n50}). In this way, it is possible to make analyses to determine the most promising measures to reduce the probability of failure if necessary.

For closure dams, the CUR/RWS Manual (Anon. [1995]) describes 3 fault trees for the events:

a) Failure of cross-section of rock-fill dam

b) Failure of construction equipment

c) Failure of construction planning

Obviously, these fault trees and failure modes are only used to demonstrate the approach to be taken. For instance, for a sill the individual failure modes will differ but the general characteristics of such a fault tree will hold. The same holds for the transition structures. Examples of such failures are described by Pilarczyk et al (ICOLD 1991).

In many cases, a risk analysis of possible failure modes will not prevent an event from happening. However, by means of this analysis, it should be possible to decrease the probability of its occurrence and/or to limit the consequences of such a failure. One way of achieving this is by diverting some specific construction elements from the critical path in the construction program.

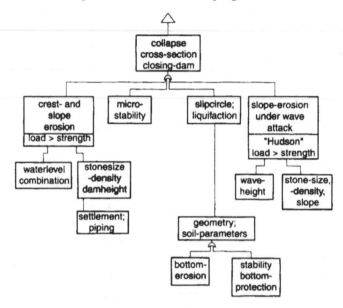

Figure 15-6 Estuary closure dam: fault tree for failure of cross-section

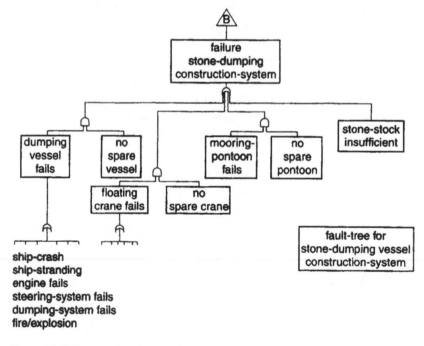

Figure 15-7 Estuary closure dam: fault tree for failure of construction equipment

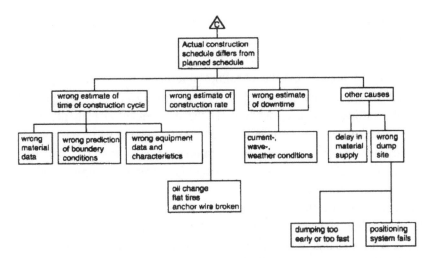

Figure 15-8 Estuary closure dam: fault tree for failure of construction planning

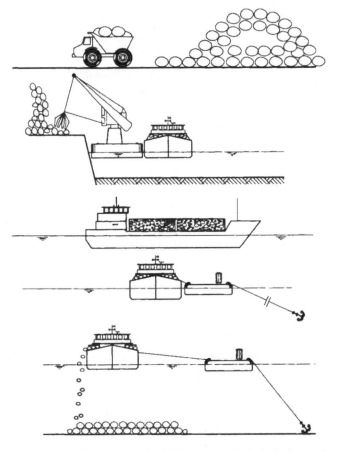

Figure 15-9 Estuary closure dam: equipment utilization **in relation to fault tree**
(Figure 15-8)

The next question is whether the calculated probability of failure for the system is acceptable. After a lengthy study to quantify the probability of failure, it is highly unsatisfactory to make this decision on an irrational basis.

It is then wise to quantify the risk of failure in terms of the product of probability of failure and its consequences (damage). These consequences are not limited to the cost of the failing structure, but include the consequential damage. For a fully destroyed breakwater, the damage thus represents the cost of reconstruction of the breakwater plus the delays in the port operations resulting from this. The risk (being the product of probability and cost of damage) is expressed in terms of cost per unit of time (generally per annum).

It is possible to reduce the risk by strengthening the structure. Usually this can only be done at some extra cost. In this way, the construction cost of the structure increases, but the risk is reduced, mainly due to a lower probability of failure. Since the construction cost is expressed in actual monetary value at the time of construction, it is necessary to capitalize the annual risk due to failure over the lifetime of the structure and calculate its present-day value. The extra construction cost can then be compared with the savings on the capitalized risk.

In practice, the situation is more complicated, because it is not only the risk of failure that has to be accounted for, but also the risk of partial damage, resulting in the need for maintenance and repair. A second complication is that often there are several ways to improve the strength of a structure and it is not always clear what is the best (most economical) way to do this. These macro and micro optimization processes are discussed in Section 15.4.

15.4 Optimization

15.4.1 Micro level

Optimization at micro level can best be explained by considering the deterministic design process. The objective of this process is to make a design that leads to the minimum total cost for a given strength level. To achieve this goal, it is necessary that all material in the structure fulfils its function, and is optimally used.

This can be compared with designing a frame. An attempt is then made to select the members in such a way that all are exposed to a stress level close to the maximum admissible stress. In the same way, an attempt can be made to ensure all elements in a closure dam or a breakwater are close to failure or partial failure when exposed to the design load. In a probabilistic design process, this means that one should avoid a very large contribution to overall failure by a single partial-failure mechanism while other mechanisms make no contribution to the probability of failure. It is wise to distribute the contribution to overall failure over a number of failure mechanisms. In fact, one should base this distribution on considerations of marginal cost. If a construction element is relatively cheap, over-designing it is not so much of a problem. If it is relatively expensive, over-designing leads to too high a cost in comparison with other elements

This means that the designer must attempt to make a balanced design, as can easily be explained when considering the cross section of a rubble mound breakwater. If the crest level designed is so high that no overtopping occurs even under severe conditions, it makes no sense to protect the inner slope with heavy armour stone. For a low crested breakwater on the other hand, it is essential to carefully protect the inner slope.

15.4.2 Macro level

Optimization at macro level can also best be explained by taking the deterministic design process, when only one failure mechanism with simple load and strength parameters is considered. When more mechanisms and parameters play a role, the calculations rapidly become more complicated, and one should be careful not to make mistakes that lead to false conclusions.

Paape and Van de Kreeke [1964] developed the method for rubble mound break-waters as early as 1964. The method is discussed below, and a sample calculation is given in Appendix 7. References to Tables and Figures refer to that Appendix.

The method starts with the assumption that there is a direct relation between one load parameter (the no-damage wave height, H_{nd}) and a strength parameter (the weight of the armour units, W). It is further assumed that the wave climate is known and available in the form of a long-term distribution of wave heights (Table A7-1). The interaction between load and strength is determined on the basis of laboratory experiments, which indicate that damage starts when a threshold value (H_{nd}) is exceeded. The damage to the armour layer increases with increasing wave height until the armour layer is severely damaged and the core of the breakwater is exposed. This occurs at an actual wave height $H = 1.45\ H_{nd}$. It is assumed that damage is then so extensive that repair is impossible and that the entire structure must be rebuilt. For intermediate wave heights, a gradual increase in damage is assumed, which is expressed as a percentage of the number of armour units to be replaced (Table A7-2).

The breakwater is designed for a number of design wave heights, where a higher design wave requires a heavier and more costly armour layer, whereas the core remains unchanged. The cost of construction is I. The cost of rebuilding the breakwater is assumed to be equal to the estimated construction cost, while the cost of repairing damage is assumed to be double the unit price of the armour units. It is then possible to list the construction cost and the anticipated cost of repair, still split over the three categories of damage (4%, 8% and collapse). Adding together the three categories of damage for a particular design-wave height yields the average annual risk anticipated for that design if all damage is repaired in the year the damage took place. If it is decided not to repair the breakwater except in case of collapse, the risk is only the risk caused by the category collapse.

Since the risk is still expressed in terms of a value per annum, it is necessary to assess what amount of money should be reserved at the time of construction to allow for payment of the average annual repair cost during the lifetime of the structure. Although money is regularly spent from this repair fund, the balance still accrues interest at a rate of δ % per annum. At the end of the calculated life time, the balance of the fund can become zero.

If the annual expense is s, the interest rate δ%, and the lifetime of the structure T, it can easily be derived that the fund to be reserved (S) is:

$$S = \int_0^T e^{-\frac{\delta}{100}t} dt = s\frac{100}{\delta}\left(1 - e^{-\frac{\delta T}{100}}\right) \tag{15.1}$$

for $T = 100$ years $S = s \cdot 100/\delta$, and
for $T = 10$ years $S = 0.35s \cdot 100/\delta$.

An interest rate in the order of 3.5% is usually set.

By adding the initial construction cost (I) and the capitalized risk (S), one arrives at the total cost of the structure. When this total cost ($I + S$) is plotted as a function of the design wave height, it appears that there is an optimum design wave height or an optimum strength for the structure.

16 FLOW DEVELOPMENT IN CLOSURE GAPS

In Chapter 5, it was indicated that in the case of the closure of a river branch or an estuary, it is very important to know the velocities in the closure gap. This is because velocities in the gaps determine the stability of the material used and the choice of the closing strategy. There are fundamentally two ways to determine the velocities in the gap, both of which are useful at different stages in the design process and depending on circumstances.

At an early stage, it can be very useful to get a rough idea of the situation by making a simple manual calculation using the theory of Chezy. At a more definitive stage or in a more complex situation a mathematical model is often used. The most simple model is a 1-D model like DUFLOW. In this chapter, examples of these two options are presented.

16.1 Calculation of flow in a river channel

The example that is calculated in this chapter deals with a closure in a river system in which two main rivers flow more-or-less parallel in the same direction, each having its own catchment area and discharge characteristics. A connecting branch provides a link between the two. It is assumed that tidal influence is negligible. Such situations occur in various river deltas. This example is taken from the rivers Waal and Meuse (see Figure 4-9), with the connection near Heusden. (In the present situation the connection is closed near Andel, which is halfway along its length, by a dam and a ship lock).

In the calculation example, the river branch connecting the two points A and B of the main river system (Figure 16-1) has a fall of 2 m over a distance of 13 km. For the river flow, Chezy's formula for open channel flow applies:

$$u = C\sqrt{Ri} \tag{16.1}$$

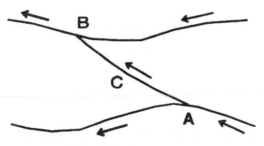

Figure 16-1 Channel view

The flow velocity will be in the order of 0.80 m/s. It is assumed that the water levels at A and B do not change. A closure is planned halfway, at point C. As the closure progresses, three river sections have to be considered:

At C, in the closure gap, the profile diminishes while the flow velocity increases. However, the resulting discharge also diminishes and finally reaches zero.

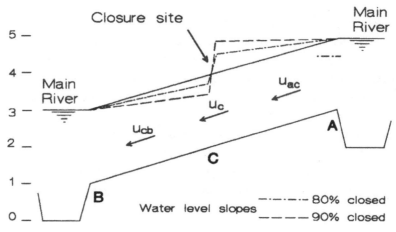

Figure 16-2 Closing a river channel

Section AC is a river section in which the diminished discharge causes the flow velocity to drop while the water level is pushed up by backwater from C. CB, however, is a river section in which the level is drawn down (see Figure 16-2).

In the two river sections, the flow and water level calculation is a backwater curve calculation approximated by Chezy's formula. In the gap, the weir formulae detailed in Section 5.2.3 must be used. The known parameters are the water levels at A and B, and the fact that the discharge quantity in all three section is equal (quasi static, neglecting water storage in the sections of the river during the process). The unknown variables are the water levels on both sides of C and the magnitude of the discharge.

The sum of the reductions in head loss over AC and CB is equal to the fall over the gap at C, which is responsible for the high flow at C. In the example, a 90% blockage results in flow velocities of 0.40 m/s in AC, 0.55 m/s in BC and of 4.50 m/s in the gap. During the final stage, the water levels in the sections AC and CB are nearly horizontal, while the head over the gap is nearly the full difference in level between A and B, in this case about 2 metres. Applying the formula (5.5), indicates that the flow velocity in the final gap would thus rise to about 6 m/s.

In deltas, rivers may bifurcate repeatedly to form a complex branching system. Closing one of these branches is comparable to the above situation, with the assumption that the water level at either end of the channel is fully determined by the regime of the delta. The length of the river section to be closed may be considerable and likewise the fall in head during the final stage of closure may be quite large.

The calculations for the intermediate phases of closure may sometimes be rather complex in which case a mathematical model like DUFLOW can be used to make the calculation. For example, this model was used during the Jamuna bridge construction in Bangladesh for the closure of two secondary channels in the braided river system of the Jamuna River.

16.2 Calculation of flow in the entrance of a tidal basin

As an example of the flow velocities occurring during the closure of a tidal basin, a calculation has been made to illustrate the change in the tidal characteristics. The calculation has been made for two different basins, each having the same water surface area, 50 km^2. This area is kept constant for all tidal levels, which implies that no parts of the basins dry out during low water periods.

One basin is relatively short and rectangular, viz. 5 km wide and 10 km long. The bottom elevation is taken as constant all over the basin at a level of 6 m below mean sea level (MSL). For this basin, the water level will be nearly horizontal at all tidal heights over the full area. Thus, a calculation based only on conservation of mass may be appropriate. This can be ascertained by calculating the length of the tidal wave, for which the celerity is:

$$c = \sqrt{gd} \tag{16.2}$$

which is 7.7 m/s, and the period is in the order of 44700 seconds. Thus, the tidal wave has a length of about 350 km. As the length of the basin is well under 1/20 of this wavelength, the simplification is acceptable.

The other basin is a long, narrow, funnel-shaped estuary, in which the tidal wave propagates. The total length is 41.750 km and the width narrows from 2000 m to 800 m at the end. The bottom level rises from MSL.–6 m at the entrance to MSL.-2 m at the end.

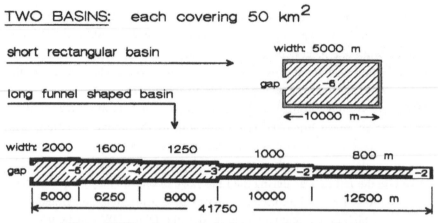

Figure 16-3 Two basins on which the calculations are based

On average, the celerity of the tidal wave is about 5 m/s and thus the tide at the end of the basin will lag more than 2 hours behind. This is confirmed by the length relation, which is much more than 1/20. The calculation needs to consider the conservation of mass as well as of momentum.

For comparison, both basins are closed by horizontal closure as well as by vertical closure. The results show the influences of the shape of the basin and the closure method. The four cases are illustrated by four sets of graphs showing curves for:
a. water level just outside the basin near the gap
b. water level just inside the basin near the gap
c. water level at the end of the basin
d. flow velocity in the closure gap
(Note: the water level curves for b and c are identical for the short basin)

The cases thus are characterized by:
case 1, horizontal closure of short basin
case 2, horizontal closure of long basin
case 3, vertical closure of short basin
case 4, vertical closure of long basin

All four cases have been calculated for stepwise reduction of the gap size in 5 stages (initial gap taken as 100%, 50%, 25%, 10% and 3%). For a vertical closure it is more difficult to define the gap size than it is for a horizontal closure, since for high sills the gap size for ebb differs a lot from the one for flood. Expressed in m² relative to MSL, a negative size is even possible. The sill levels have been selected such that comparison of the horizontal and vertical characteristics is possible. The different stages are shown in Figure 16-4.

5 closing stages

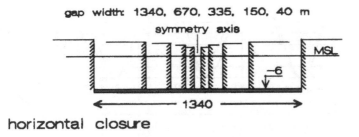

gap width: 1340, 670, 335, 150, 40 m

horizontal closure

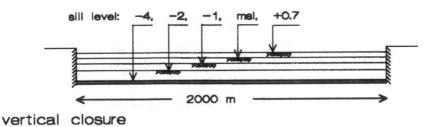

vertical closure

Figure 16-4 Closure gap dimensions and stages

For reasons of fair comparison, in the calculations various parameters have been taken as identical and constant (which is in fact not true). These are:

- all discharge coefficients are taken as 1.0 for all gap dimensions, in both, horizontal and vertical closures
- all constructed parts of the closure dam are considered impermeable for water
- the tidal wave at the entrance is a single sine wave with a range of 3 m, which does not change as consequence of the progressing closure
- the Chezy-values of all sections in the calculation network are taken as 50 for all water depths
- the calculation is made using the mathematical model DUFLOW

(Note: the DUFLOW model has also been used for calculating the short basin. This enables confirmation of the former remark that curves b and c are identical.)

It should be realized that in a practical case the model has to be calibrated by reproducing an actually measured situation. The results of this will give the required data for Chezy-values and dimensions of the sections (depth-width relation to represent an irregular gully profile).

Water levels in m above MSL (upper graph) and Flow velocities in m/s (lower graph) in relation to Closure Stage.

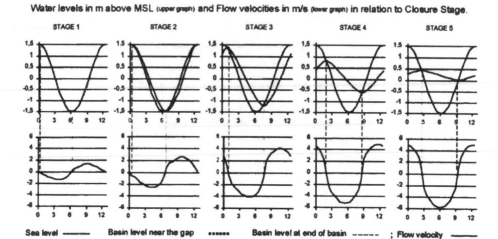

Figure 16-5 Horizontal closure of the short basin.

Water levels in m above MSL (upper graph) and Flow velocities in m/s (lower graph) in relation to Closure Stage.

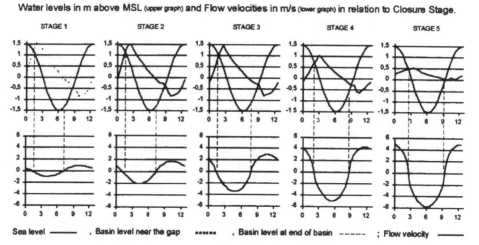

Figure 16-6 Horizontal closure of the long basin.

Water levels in m above MSL (upper graph) and Flow velocities in m/s (lower graph) in relation to Closure Stage.

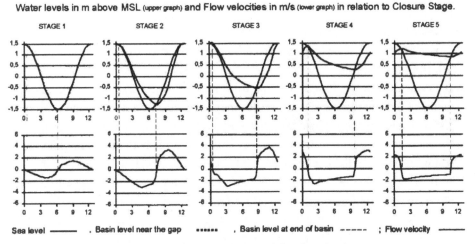

Sea level ——— , Basin level near the gap ••••••• , Basin level at end of basin ––––– ; Flow velocity ———

Figure 16-7 Vertical closure of the short basin.

Water levels in m above MSL (upper graph) and Flow velocities in m/s (lower graph) in relation to Closure Stage.

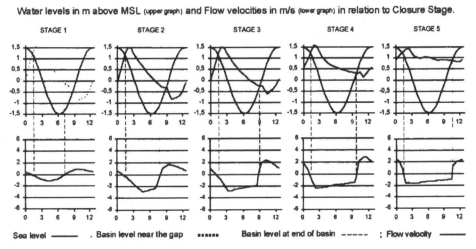

Sea level ——— , Basin level near the gap ••••••• Basin level at end of basin ––––– ; Flow velocity ———

Figure 16-8 Vertical closure of the long basin.

A few typical characteristics can be observed in the results. For the calculations with the short basin (cases 1 and 3), the water levels at the end of the basin do not show clearly as they are identical with the basin level near the gap. For the long basin (cases 2 and 4), the water level at the far end (dotted line) is lagging behind, as expected. Moreover, at the end of the basin, the high water reaches about the same level as it does near the gap but low water is much higher. Consequently, the mean level in the estuary rises towards the end by about 0.25 m.

For all cases, the water levels in the basin near the gap show a diminishing range for each succeeding stage and the moments of high and low water occur later. These are identical for the horizontal closures (cases 1 and 2) and for the vertical closures (cases 3 and 4). They differ in the rise in mean level, however. For the horizontal closures in stage 5 the rise is about 0.20 m. This is caused by the fact that during ebb the water level in the gap is slightly lower than during flood. For the vertical closures this rise is about 1 m, which is caused by the high level of the sill, preventing the basin from discharging.

Very typically for the long basin, the difference between water level curves at the gap and at the end diminishes as the closure progresses. In stage 5 there is hardly any difference and the basin apparently behaves as if it were a short basin with an almost horizontal water level all over the basin. This is true for both horizontal and vertical closure. In addition, the rise in mean level is identical to that in the short basin.

The flow velocities for the horizontal closures increase with every stage in a regular pattern. In the first few stages, the values for the short basin are slightly higher than for the travelling wave in the long basin, but in the last stage, they are identical. There is completely different flow behaviour in the vertical closures, however. In stage 2, during low water at sea, the critical-flow situation already occurs. This shows in the straight cut-off line of the curve. Although in later stages the duration of the critical flow is greater, the maximum value of the ebb flow is diminishing. (In these calculations the critical-flow does not show in the flood period. (This may also happen in other cases.)

Although the flood does not reach critical-flow conditions, the maximum flood value also diminishes. This is caused by the fact that the mean level in the basin rises and thus reduces the head loss during flood. Again, the curves for short and for long basins are nearly identical. For the vertical closures, during low water at sea there is a very large drop of the water level over the sill in stages 4 and 5. As the sill level is MSL and +0.7 respectively in these stages, it can be concluded that the sill does not fall dry. The basin level always remains slightly higher than the sill, which is logical. (In actual cases, when large quarry stone is used sometimes the dam is quite permeable. In such cases the basin level may go down further.)

Comparing the flow in the short and long basins for vertical closures leads to the same conclusion as with the horizontal closures. In the first stages, it is slightly smaller in the long basin, in later stages, it is identical.

Keeping flow velocities low by closing vertically is clearly demonstrated. This can be further illustrated in a graph showing the flow maxima against the profile of the gap, as is shown here for the short basin and the long basin. Both show identical trends. The results of the five stages calculated are presented, going from about 8000 m^2 down to nearly closed. For the gap profile the sectional area below MSL is taken. For the vertical closure in the last stage this would give a negative value. For that stage the data has been put in line with a 3% open gap. In the initial stages, the velocities during vertical closure are slightly higher than during horizontal closure. However, as vertical closure reaches critical-flow, the first ones diminish while the others still increase.

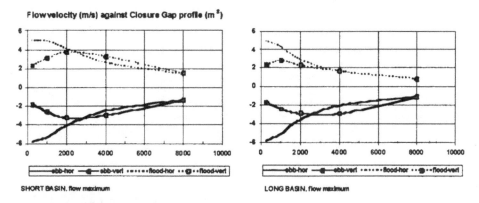

Figure 16-9 Flow velocities depending on closure method and gap-size

17 REVIEW

Chapter 17 reviews the result of the design process as a whole. It repeats the main choices to be made during design. To pass the examination it is essential for the student to master and to understand the contents of this chapter.

17.1 Breakwaters

17.1.1 Rubble or monolithic

The main choice facing the designer of a breakwater is the choice between a structure of the rubble mound type and one of the monolithic type. The advantages and disadvantages of each are therefore repeated here. Some of these are site specific and some are valid for the present time only. The designer must therefore carefully assess in which direction he should move.

Advantages of the rubble mounds are:
- Simple construction
- Withstands unequal settlements
- Large ratio between initial damage and collapse
- Many guidelines available for the designer

Disadvantages of the rubble mound are:
- Dependence on the availability of adequate quarry
- Large quantity of material required in deeper water
- Large space requirement
- Difficult to use as a quay wall

Advantages of the monolithic breakwater are:
- Short construction time on location
- Can function as quay wall
- Economical use of material in deeper water

Disadvantages of the monolithic breakwater are:

- Sensitivity to poor foundation conditions (settlements and liquefaction)
- Uncertainty about wave loads in breaking waves
- Complete and sudden failure when overloaded
- Reflection against vertical wall
- Limited support for the designer from guidelines and literature

In practice, this means that the choice of the basic type of breakwater will largely be made on the basis of wave-climate and soil conditions.

17.1.2 Quarry stone or concrete armour units

Generally, the use of quarry stone will be cheaper than the use of concrete blocks, even if the availability of quarry stone is limited. A problem of using quarry stone is the fact that it is a natural material, so its quality and properties are governed by nature. This means that neither density nor maximum size can be selected freely by the designer.

The main tool for the designer who is facing problems with the stability of quarry stone armour is reduction of the slope. The decision to reduce the slope is accompanied by a big increase in the volume of material required. At a certain point, the step towards concrete armour units becomes inevitable. In that case, the question that arises is whether to use simple blocks, cubes (or similar shapes) or more complicated shapes that rely on their interlocking capabilities. In the circumstances prevailing in the Netherlands and Belgium, preference is given to the simpler blocks, mainly because of the ease of construction and handling. Nevertheless, if a decision to use the more complicated units is made, the utmost care must be taken to avoid breakage.

17.1.3 Which design formula?

There are many design formulae available to the designer of both rubble mound and monolithic breakwaters.

For rubble mound breakwaters, the approach of Van der Meer has gained worldwide support, although the structure of the formula set is not very satisfactory as it lacks a direct link with physical understanding. Therefore it is still recommended that a final design be checked in a physical model. Irregular waves must certainly be used in the model study. It is also recommended that the behaviour of the structure under overloading should be checked to establish the condition where it fails. If the ratio between no damage and failure is small, this must have repercussions on the choice of the Ultimate Limit State.

For the time being it is recommended that the Goda formula should be used for monolithic breakwaters. The disadvantage is that, like the Van de Meer formulae, it has an inadequate theoretical base. Moreover, the Goda formula has acquired far less experimental support from elsewhere in the world. In this case also, physical model tests are strongly recommended. Proper attention must be paid to dynamic loads and their effect on the structure and the foundation.

17.1.4 Service limit state

The design and the cost estimates are very sensitive to the level at which the breakwater must exercise its functions. It is therefore essential that the functional requirements are analyzed carefully and that a clear distinction is made between ULS (survival) and SLS (functioning). A frequent mistake is to overestimate the functional conditions, which results in structures with too high a crest level. Since the volume in the cross-section is proportional to the square of the height, this has serious cost consequences.

17.2 Closure dams

For closure dams, there are a few main directions the designer can follow. The first one is the choice between basic methods, the second one is the optimal use of the natural conditions and boundary conditions and the third concerns the selection of materials and equipment.

17.2.1 Decisive circumstances

There is no single prescription suitable for all closures because there are too many variables and boundary conditions. The unequivocal case is that of a well-defined tidal basin with a single closure gap of uniform dimensions. In practice, the situation is frequently more complex. Sometimes special conditions may so strongly determine a case that they either restrict or offer possibilities. Five typical examples of such criteria are detailed below:

1. The area of the basin can be easily subdivided into separate compartments

In essence, this is a matter of cost. Subdividing the area diminishes the storage capacity of the individual areas. Each closure can therefore be a lot easier, probably permitting the use of locally available materials so the total cost of these small-sized closures may be less than the cost of a single closure of the total area. However, additional costs may be incurred in the construction of the embankments separating the compartments.

Because of the later use of the area, the embankments may have to be removed. Sometimes, re-use of materials is a possibility, but some of the material is certainly lost. Depending on the layout of the area, subdivision can be designed in two ways. An elongated basin with a single channel can be taken in successive sections, while a complex channel pattern may require the successive closure of adjacent sections.

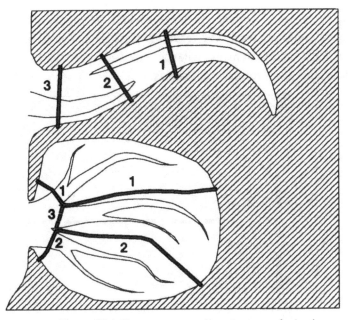

Figure 17-1 Two ways to reduce the area of a basin

2. The basin is penetrated by tide via two separate entrances

Closure of the basin means closure of both entrances, with the option to close either one first or to close both simultaneously by any combination of methods, materials and phasing. All actions in one entrance will certainly affect the conditions in the other entrance and the balance between the two may be quite sensitive. In the case of a major imbalance, the tide conditions in the entire basin will also be affected and this will lead to changes in flow and subsequent erosion at several locations.

In such a case, a mathematical hydraulic model may be complex and difficult to calibrate. The problem is that somewhere in the basin the tidal waves will meet. Since these waves have a different history, their shapes, phases and amplitudes are not exactly the same. Nevertheless, generally the tidal meeting (in Dutch called "wantij") is characterized by low flow velocities and an unusual relation between water level and flow.

The difficulty is how to estimate the correct Chezy-value for the gully system in this meeting area. For the existing situation, a wide range of values used in the mathematical model may give acceptable results and thus calibration gives no clue. However, as soon as the tide changes owing to the progress of the closure works and the meeting area moves, the unchecked value may be very important. Calculations with various assumed values will at least show the possible impact on the conditions.

For simultaneous closure, the impact of every combination of construction phases on the tide penetration has to be determined. As for a single closure, this is done by calculating with weir formulae. In these several schematic simplifications and practical coefficients are used. The resulting deviation has little impact on a single closure but for a dual system the balance may soon become unstable. Therefore, closure plans must allow for these deviations.

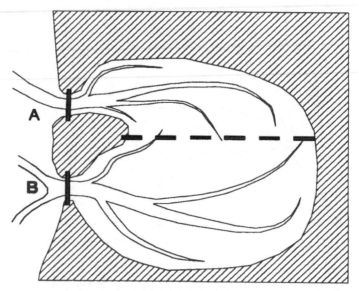

Figure 17-2 A basin with two entrances

Even in the case of a very well thought-out plan of concurrent closure methods, a setback in the execution of one, also affects the other. Moreover, a major failure in one closure may lead to a complete disaster, as the other one has to be dismantled to maintain the balance of the basin.

The easiest way to overcome the problem is to make a temporary or permanent closure dam across the meeting point, which separates the two tidal systems and divides the basin into two compartments. Then, the two primary closures of the basin area are fully independent. The mathematics is more simple and reliable. The closure design for each one is independent of the other, as is the execution. Constructing the separating dam is a partial closure and frequently an obvious solution. However, in some cases this may not be allowed, for instance because it blocks a navigation route.

Another method is to plan to close the two entrances one after the other. The order of activities then is:

- fix the bottom topography of entrance "A" by protecting bottom and shores against future scour.
- close the entrance "B" by any closure method and accept the change in tide and conditions in the basin as well as in entrance "A".
- next, close the entrance "A".

The advantage is that the closures are independent in design, method and execution. The uncertainty about the response of the basin to the imbalance after the first closure has to be covered. This can be done by assuming that deep gullies scour across the meeting area and by taking the whole basin as a storage area for the tide calculations for the second closure. Compared with the closures for separated basins, the closure "B" may be easier because "A" is still an open entrance. However, "A" with the full basin behind the gap will be much more comprehensive.

It might have been possible to stabilize the meeting area and prevent the erosion of deep gullies. Then, the flow velocities would have increased but the topography would have remained intact. However, the cost involved in such erosion prevention will generally be higher than that of providing a temporary partial-closure dam to fully separate the systems.

3. The closure profile consists of two (or more) main gullies and shallows

Between the main gullies there will be an area of tidal flats. These more or less separate the gullies during the low water periods. During rising and falling tides, on the one hand they are storage areas and on the other hand they ensure balance between the gullies. Although not considered a tidal meeting, this has a lot in common with such a meeting. For this case too, the first problem for the designer is to prepare the mathematical model. Tide penetration is calculated by adopting a gully network, while tidal flats are assumed to provide a storage area only. However, imbalance creates flow, which results in erosion and the development of a gully across the shallow area.

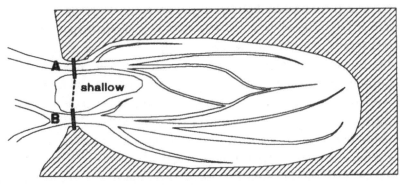

Figure 17-3 An estuary with two channels

How quickly will that occur, how deep will this gully be and what will be the Chezy roughness? Separating the systems by dividing the basin is not logical as both gullies run into the same main storage area. Therefore, after construction of the dam-section across the shallow, the only possibilities remaining are:

- to close the two gullies simultaneously in a very balanced way.
- to close one gully first, accept or prevent erosion across the tidal flat and then close the other gap, taking into account a fully adapted situation.

In the second case, it is most likely that the dam section across the tidal flat will be built before the closure of the first gully. Since the main gullies are relatively close to this, the erosion of the short cut across the tidal flat will most likely develop along this section of the dam. Thus, the toe of this dam has to be heavily protected. In addition, flow conditions in the remaining gap will be very adversely affected. A better solution may be to create a short cut by dredging at an appropriate location to guide the tide towards the last gap.

4. The gap to be closed is not in an equilibrium state

For a detailed example, see Section 2.5.5. This situation occurs in the case of a calamitous breach of a dam or dike. It may also happen when a construction phase goes wrong and creates unexpected conditions at the site of the gap. In these cases, time is a very important factor. Every day natural processes will try to achieve the equilibrium state and change the existing situation.

A first consideration is to analyze how quickly definite measures can be taken. Over this period, the situation will adapt and the magnitude of the change has to be estimated in order to plan the right measures. If this change is undesirable, temporary measures to halt or retard the deterioration can be considered. Such temporary measures include:

- stabilizing the attacked bottom of the gap by dropping coarse material. Stabilizing the sides of the gap is easier but may induce deeper scour. Generally, deeper scouring is worse than wider scouring.
- trying to avoid the erosion of gullies in the storage area, for instance by protecting critical spots with mats or quarry stone. A more developed gully network will result in easier penetration of the tide and increase the tidal volume.

In the meantime, data on the existing conditions can be measured and recorded, while a definite closure strategy is being drafted.

Usually, the existing situation has to be determined and secured before any construction phase can start. In some cases, such as calamities where life is endangered, a direct counter-attack is justified. The risk in such a case is that if the action fails, the situation is usually much worse than it was before the action. If an emergency closure cannot be obtained in a few days, certainty is more important than speed.

An example of a successful emergency closure was the blocking of the dike breach near Ouderkerk aan den IJssel (Holland) during the major storm flood in February

1953. A relatively small breach cut the dike, which secured a vast, densely populated area of Holland, north of Rotterdam. Several hours later on the same night, a small vessel was taken and put onto the remainder of the outer slope of the dike, with neither any erosion protection nor any re-profiling of the gap to fit the vessel's shape. Piping under the vessel and around stem and stern could easily have scoured another gap. Then, the vessel would have broken and been pushed away, leaving a very large gap. However, the piping was blocked by using tarpaulins ballasted with sandbags (see Figure 17-4). The closure was a success and this central area of Holland remained dry.

Figure 17-4 Closure dike breach Ouderkerk aan den IJssel 1953

5. Various alignments with different longitudinal profiles can be selected

These occur for instance, in a river branch with variable bottom topography. In a river bend there may be a deep triangular channel while in the cross-sections between bends a shallow box-profile may be available. (As in the alignments 1 or 2 in the upper half of the Figure 17-1, but this time as alternative locations.) Which of the two alignments is preferable?

Another example gives the situation which occurs after a dike located in a shallow area breaches. The breach will erode a deep scour hole very close to the original alignment of the dike. Owing to spreading of the flow, the surrounding shallow area will remain intact for some time, although erosion will gradually create gullies. The option is to restore the original dike or to build around the scour hole, either along the river (or sea) side or via the inside. Various considerations determine which option is the most attractive.

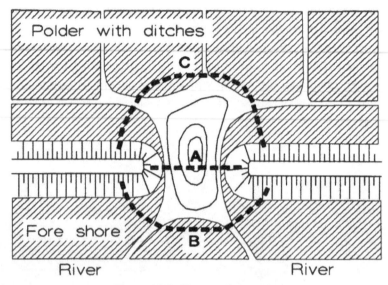

Figure 17-5 Closure alignments

For many dike breaches in the past, closing around the scour hole was preferred. The old method involved sinking mattresses vertically. In order not to lose area, where possible, the alignment at the river side was taken, so that the scour hole became situated within the enclosed polder. Nowadays, in the Netherlands, these former scour holes still can be seen in the landscape as small circular ponds just at the inner toe of the dike where the alignment of the dike winds around them in a semi-circle. In Dutch such a pond is called a "wiel".

An important parameter is the amount of material needed to block the gap. The flow is determined by the nett cross-sectional profile of the gap in m^2, while the gap has to be blocked by m^3 of material. For instance, a dam with slopes 1 in 1 with height "s" along a gap length "l", used to block a profile "$l \times s$", has a volume of "$l \times s^2$". An identical dam, of half the height along twice the length, blocks the same profile but requires only half the volume. On the other hand, the bottom protection (if needed) is twice as wide but may be more than half as long (in the flow direction). Other parameters relate to the equipment, the materials and the closure method used. A shallow gap may be difficult for large operating vessels to approach. It is preferable to use caissons in deep gullies. For a vertical closure, however, a long gap is advantageous because of the resulting lower current velocities. This is demonstrated in the examples of Sections 12.2 and 12.3.

APPENDIX 1
Glossary

The glossary that follows was compiled and reviewed by the staff of the Coastal Engineering Research Center. Although the terms came from many sources, the following publications were of particular value:

American Geological Institute (1957), "Glossary of Geology and Related Sciences with Supplement", 2nd Edition

American Geological Institute (1960), "Dictionary of Geological Terms", 2nd Edition

American Meteorological Society (1959), "Glossary of Meteorology"

Johnson, D.W. (1919), "Shore Process and Shoreline Development", John Wiley and Sons, Inc., New York

U.S. Army Coastal Engineering Research Center (1966), "Shore Protection, Planning and Design", Technical Report No. 4, 3d Edition

U.S. Coast and Geodetic Survey (1949), "Tide and Current Glossary", Special Publication No. 228, Revised (1949) Edition

U.S. Navy Oceanographic Office (1966), "Glossary of Oceanographic Terms", Special Publication (SP-35), 2d Edition

Wiegel, R.L. (1953), "Waves, Tides, Currents and Beaches: Glossary of Terms and List of Standard Symbols". Council on Wave Research, The Engineering Foundation, University of California.

Subsequently a selection from the glossary was made for the purpose of this book.

Terms	
ALONGSHORE	Parallel to and near the shoreline; LONGSHORE.
AMPLITUDE, WAVE	(1) The magnitude of the displacement of a wave from a mean value. An ocean wave has an amplitude equal to the vertical distance from still-water level to wave crest. For a sinusoidal wave, the amplitude is one-half the wave height. (2) The semirange of a constituent tide.
ANTINODE	See LOOP.
ARMOUR UNIT	A relatively large quarrystone or concrete shape that is selected to fit specified geometric characteristics and density. It is usually of nearly uniform size and usually large enough to require individual placement. In normal cases it is used as primary wave protection and is placed in layers that are at least two units thick.
BACKRUSH	The seaward return of the water following the uprush of the waves. For any given tide stage the point of farthest return seaward of the backrush is known as the LIMIT of BACKRUSH or LIMIT BACKWASH.
BACKWASH	(1) See BACKRUSH. (2) Water or waves thrown back by an obstruction such as a ship, breakwater, or cliff.

BATHYMETRY	The measurement of depths of water in oceans, seas, and lakes; also information derived from such measurements.
BENCH MARK	A permanently fixed point of known elevation. A primary benchmark is one close to a tide station to which the tide staff and tidal datum originally are referenced.
BOTTOM	The ground or bed under any body of water; the bottom of the sea. (See Figure A-1.)
BOTTOM (nature of)	The composition or character of the bed of an ocean or other body of water (e.g., clay, coral, gravel, mud, ooze, pebbles, rock, shell, shingle, hard, or soft).
BOULDER	A rounded rock more than 10 inches in diameter. A stone larger than a cobblestone. See SOIL CLASSIFICATION.
BREAKER	A wave breaking on a shore, over a reef, etc. Breakers may be classified into four types.
	SPILLING--bubbles and turbulent water spill down front face of wave. The upper 25 percent of the front face may become vertical before breaking. Breaking generally occurs over quite a distance.
	PLUNGING--crest curls over air pocket; breaking is usually with a crash. Smooth splash up usually follows.
	COLLAPSING--breaking occurs over lower half of wave, with minimal air pocket and usually no splash up. Bubbles and foam present.
	SURGING--wave peaks up, but bottom rushes forward from under wave, and wave slides up beach face with little or no bubble production. Water surface remains almost plane except where ripples may be produced on the beach face during runback.
BREAKER DEPTH	The still-water depth at the point where a wave breaks. Also called **BREAKING DEPTH**.
BREAKWATER	A structure protecting a shore area, harbor, anchorage, or basin from waves.
BUOYANCY	The resultant of upward forces, exerted by the water on a submerged or floating body, equal to the weight of the water displaced by this body.
BYPASSING, SAND	Hydraulic or mechanical movement of sand from the accreting updrift side to the eroding downdrift side of an inlet or harbor entrance. The hydraulic movement may include natural movement as well as movement caused by man.
CANAL	An artificial watercourse cut through a land area for such uses as navigation and irrigation.
CELERITY	Wave speed.
CHANNEL	(1) A natural or artificial waterway of perceptible extent which either periodically or continuously contains moving water, or which forms a connecting link between two bodies of water. (2) The part of a body of water deep enough to be used for navigation through an area otherwise too shallow for navigation. (3) A large strait, as the English Channel. (4) The deepest part of a stream, bay, or strait through which the main volume or current of water flows.
CHART DATUM	The plane or level to which soundings (or elevations) or tide heights are referenced (usually LOW WATER DATUM). The surface is called a tidal datum when referred to a certain phase of tide. To provide a safety factor for navigation, some level lower than MEAN SEA LEVEL is generally selected for hydrographic charts, such as MEAN LOW WATER or MEAN LOWER LOW WATER. See DATUM PLANE.
CHOP	The short-crested waves that may spring up quickly in a moderate breeze, and which break easily at the crest. Also WIND CHOP.

CLAPOTIS	The French equivalent for a type of STANDING WAVE. In American usage it is usually associated with the standing wave phenomenon caused by the reflection of a non-breaking wave train from a structure with a face that is vertical or nearly vertical. Full clapotis is one with 100 percent reflection of the incident wave; partial clapotis is one with less than 100 percent reflection.
CLAY	See SOIL CLASSIFICATION.
CNOIDAL WAVE	A type of wave in shallow water (i.e., where the depth of water is less than 1/8 to 1/10 the wavelength). The surface profile is expressed in terms of the Jacobian elliptic function *cn u;* hence the term cnoidal.
COASTLINE	(1) Technically, the line that forms the boundary between the COAST and the SHORE. (2) Commonly) the line that forms the boundary between the land and the water.
COBBLE (COBBLESTONE)	See SOIL CLASSIFICATION.
CONTOUR	A line on a map or chart representing points of equal elevation with relation to a DATUM. It is called an ISOBATH when connecting points of equal depth below a datum. Also called DEPTH CONTOUR.
CONVERGENCE	(1) In refraction phenomena, the decreasing of the distance between orthogonals in the direction of wave travel. Denotes an area of increasing wave height and energy concentration. (2) In wind-setup phenomena, the observed increase in setup (above the increase that would occur in an equivalent rectangular basin of uniform depth), that is caused by changes in plan form or depth. Also the decrease in basin width or depth causing such an increase in setup.
CORAL	(1) (Biology) Marine coelenterates (Madreporaria), solitary or colonial, which form a hard external covering of calcium compounds or other materials. The corals that form large reefs are limited to warm, shallow waters, while that forming solitary, minute growths may be found in colder waters to great depths. (2) (Geology) The concretion of coral polyps composed almost wholly of calcium carbonate, forming reefs and tree-like and globular masses. May also include calcareous algae and other organisms producing calcareous secretions, such as bryozoans and hydrozoans.
CORE	A vertical cylindrical sample of the bottom sediments from which the nature and stratification of the bottom may be determined.
CREST LENGTH, WAVE	The length of a wave *along* its crest. Sometimes called CREST WIDTH.
CREST OF WAVE	(1) The highest part of a wave. (2) That part of the wave above still-water level.
CREST WIDTH, WAVE	See CREST LENGTH, WAVE.
CURRENT	A flow of water.
CURRENT, COASTAL	One of the offshore currents flowing generally parallel to the shoreline in the deeper water beyond and near the surf zone. These are not genetically related to waves and the resulting surf, but may be related to tides, winds, or distribution of mass.
CURRENT, EBB	The tidal current away from shore or down a tidal stream. Usually associated with the decrease in the height of the tide.
CURRENT, FLOOD	The tidal current toward shore or up a tidal stream. Usually associated with the increase in the height of the tide.
CURRENT, LITTORAL	Any current in the littoral zone caused primarily by wave action; e.g., LONGSHORE CURRENT, RIP CURRENT. See also CURRENT, NEAR-SHORE.

CURRENT, LONGSHORE	The littoral current in the breaker zone moving essentially parallel to the shore that is usually generated by waves breaking at an angle to the shoreline.
CURRENT, TIDAL	The alternating horizontal movement of water associated with the rise and fall of the tide caused by the astronomical tide-producing forces. Also CURRENT, PERIODIC. See also CURRENT, FLOOD and CURRENT, EBB.
DATUM, CHART	See CHART DATUM.
DATUM, PLANE	The horizontal plane to which soundings, ground elevations, or water surface elevations are referred. Also REFERENCE PLANE. The plane is called a TIDAL DATUM when defined by a certain phase of the tide. The following datums are ordinarily used on hydrographic charts
	MEAN LOW WATER--Atlantic coast (US), Argentina, Sweden, and Norway
	MEAN LOWER LOW WATER--Pacific coast (US)
	MEAN LOW WATER SPRINGS--United Kingdom) Germany, Italy, Brazil, and Chile.
	LOW WATER DATUM--Great Lakes (U. S. and Canada).
	LOWEST LOW WATER SPRINGS--Portugal.
	LOW WATER INDIAN SPRINGS--India and Japan (See INDIAN TIDE PLANE).
	LOWEST LOW WATER--France, Spain, and Greece.
	A common datum used on topographic maps is based on MEAN SEA LEVEL. See also BENCHMARK.
DECAY DISTANCE	The distance waves travel after leaving the generating area (FETCH).
DECAY OF WAVES	The change waves undergo after they leave a generating area (FETCH) and pass through a calm, or region of lighter winds. In the process of decay, the significant wave height decreases and the significant wavelength increases.
DEEP WATER	Water so deep that surface waves are little affected by the ocean bottom. Generally, water deeper than one-half the surface wavelength is considered deep water. Compare SHALLOW WATER.
DEPTH	The vertical distance from a specified tidal datum to the sea floor.
DEPTH OF BREAKING	The still-water depth at the point where the wave breaks. Also BREAKER DEPTH. (See Figure A-2.)
DEPTH CONTOUR	See CONTOUR.
DIFFRACTION (of water waves)	The phenomenon by which energy is transmitted laterally along a wave crest. When a part of a train of waves is interrupted by a barrier) such as a breakwater) the effect of diffraction is manifested by propagation of waves into the sheltered region within the barrier's geometric shadow.
DIURNAL	Having a period or cycle of approximately one TIDAL DAY.
DIURNAL TIDE	A tide with one high water and one low water in a tidal day.
DIVERGENCE	(1) In refraction phenomena) the increasing of distance between orthogonals in the direction of wave travel. Denotes an area of decreasing wave height and energy concentration. (2) In wind-setup phenomena) the observed decrease in setup under that which would occur in an equivalent rectangular basin of uniform depth) that is caused by changes in plan form or depth. Also the increase in basin width or depth causing such a decrease in setup.
DURATION	In wave forecasting, the length of time the wind blows in nearly the same direction over the FETCH (generating area).
DURATION, MINIMUM	The time necessary for steady-state wave conditions to develop for a given wind velocity over a given fetch length.
EBB CURRENT	The tidal current away from shore or down a tidal stream and usually associated with the decrease in height of the tide.

EBB TIDE	The period of tide between high water and the succeeding low water; a falling tide.
ECHO SOUNDER	An electronic instrument used to determine the depth of water by measuring the time interval between the emission of a sonic or ultrasonic signal and the return of its echo from the bottom.
EDDY	A circular movement of water formed on the side of a main current. Eddies may be created at points where the main stream passes projecting obstructions or where two adjacent currents flow counter to each other. Also EDDY CURRENT.
EDGE WAVE	An ocean wave parallel to a coast, with crests normal to the shoreline. An edge wave may be STANDING or PROGRESSIVE. Its height diminishes rapidly seaward and is negligible at a distance of one wavelength offshore.
EMBANKMENT	An artificial bank such as a mound or dike generally built to retain (hold back) water or to carry a roadway.
ENERGY COEFFICIENT	The ratio of the energy in a wave per unit crest length transmitted forward with the wave at a point in shallow water to the energy in a wave per unit crest length transmitted forward with the wave in deep water. On refraction diagrams this is equal to the ratio of the distance between a pair of orthogonals at a selected shallow-water point to the distance between the same pair of orthogonals in deep water. Also the square of the REFRACTION COEFFICIENT.
EROSION	The wearing away of land by the action of natural forces. On a beach, the carrying away of beach material by wave action, tidal currents, littoral currents, or by deflation.
FAIRWAY	The parts of a waterway that are open and unobstructed for navigation. The main travelled part of a waterway; a marine thoroughfare.
FATHOM	A unit of measurement used for soundings equal to 1.83 meters (6 feet).
FATHOMETER	The copyrighted trademark for a type of echo sounder.
FEELING BOTTOM	The initial action of a deepwater wave, in response to the bottom, upon running into shoal water.
FETCH	The area in which SEAS are generated by a wind having a fairly constant direction and speed. Sometimes used synonymously with FETCH LENGTH. Also GENERATING AREA.
FETCH LENGTH	The horizontal distance (in the direction of the wind) over which a wind generates SEAS or creates a WIND SETUP.
FLOOD CURRENT	The tidal current toward shore or up a tidal stream, which is usually associated with the increase in the height of the tide.
FLOOD TIDE	The period of tide between low water and the succeeding high water; a rising tide.
FORESHORE	The part of the shore, lying between the crest of the seaward berm (or upper limit of wave wash at high tide) and the ordinary low-water mark, that is ordinarily traversed by the uprush and backrush of the waves as the tides rise and fall. See BEACH FACE.
FREEBOARD	The additional height of a structure above design high water level that is added to prevent overflow. Also, the vertical distance between the water level and the top of the structure at a given time. On a ship, the distance from the waterline to main deck or gunwale.
FROUDE NUMBER	The dimensionless ratio of the inertial force to the force of gravity for a given fluid flow. It may be given as $Fr = V/Lg$ where V is a characteristic velocity, L is a characteristic length, and g the acceleration of gravity--or as the square root of this number.

GENERATION OF WAVES	(1) The creation of waves by natural or mechanical means. (2) The creation and growth of waves caused by a wind blowing over a water surface for a certain period of time. The area involved is called the GENERATING AREA or FETCH.
GRADIENT (GRADE)	See SLOPE. With reference to winds or currents, the rate of increase or decrease in speed, usually in the vertical; or the curve that represents this rate.
GRAVEL	See SOIL CLASSIFICATION.
GRAVITY WAVE	A wave whose velocity of propagation is controlled primarily by gravity. Water waves more than 2 inches long are considered gravity waves. Waves longer than 1 inch and shorter than 2 inches are in an indeterminate zone between CAPILLARY and GRAVITY WAVES. See RIPPLE.
GROIN (British, GROYNE)	A shore protection structure built (usually perpendicular to the shoreline) to trap littoral drift or retard erosion of the shore.
GROIN SYSTEM	A series of groins acting together to protect a section of beach. Commonly called a groin field.
GROUP VELOCITY	The velocity of a wave group. In deep water, it is equal to one-half the velocity of the individual waves within the group.
HALF-TIDE LEVEL	see MEAN TIDE LEVEL.
HARBOR	(British, HARBOUR). Any protected water area affording a place of safety for vessels. See also PORT.
HEIGHT OF WAVE	See WAVE HEIGHT.
HIGH TIDE, HIGH WATER (HW)	The maximum elevation reached by each rising tide. See TIDE.
HIGH WATER	See HIGH TIDE.
HIGH WATER LINE	Strictly speaking, the intersection of the plane of mean high water with the shore. The shoreline delineated on the nautical charts of the National Ocean Service approximates the high water line. For specific occurrences, the highest elevation on the shore reached during a storm or rising tide, including meteorological effects.
HIGH WATER OF ORDINARY SPRING TIDES (HWOST)	A tidal datum appearing in some British publications that is based on high water of ordinary spring tides.
HIGHER HIGH WATER (HHW)	The higher of the two high waters of any tidal day. The single high water occurring daily during periods when the tide is diurnal is considered to be a higher high water.
HIGHER LOW WATER (HLW)	The higher of two low waters of any tidal day.
HINDCASTING, WAVE	The use of historic synoptic wind charts to calculate characteristics of waves that probably occurred at some past time.
HURRICANE	An intense tropical cyclone in which winds tend to spiral inward toward a core of low pressure. Maximum surface wind velocities equal or exceed 33.5 meters per second (75 mph or 65 knots) for several minutes or longer at some points. TROPICAL STORM is the term applied if maximum winds are less than 33.5 meters per second.
HURRICANE PATH or TRACK	Line of movement (propagation) of the eye through an area.
HYDROGRAPHY	(1) The configuration of an underwater surface including its relief, bottom materials, coastal structures, etc. (2) The description and study of seas, lakes, rivers, and other waters.

HYPOTHETICAL HURRICANE ("HYPO-HURRICANE")	A representation of a hurricane, with specified characteristics, that is assumed to occur in a particular study area, following a specified path and timing sequence.
INLET.	(1) A short, narrow waterway connecting a bay) lagoon) or similar body of water with a large parent body of water. (2) An arm of the sea (or other body of water) that is long compared to its width and may extend a considerable distance inland. See also TIDAL INLET.
IRROTATIONAL WAVE	A wave with fluid particles that do not revolve around an axis through their centers, although the particles themselves may travel in circular or nearly circular orbits. Irrotational waves may be PROGRESSIVE, STANDING, OSCILLATORY, or TRANSLATORY. For example, the Airy, Stokes, cnoidal, and solitary wave theories describe irrotational waves. Compare TROCHOIDAL WAVE.
JETTY	(1) (United States usage) On open seacoasts, a structure extending into a body of water, which is designed to prevent shoaling of a channel by littoral materials and to direct and confine the stream or tidal flow. Jetties are built at the mouths of rivers or tidal inlets to help deepen and stabilize a channel. (2) (British usage) WHARF or PIER. See TRAINING WALL.
KINETIC ENERGY (OF WAVES)	In a progressive oscillatory wave, a summation of the energy of motion of the particles within the wave.
KNOT	The unit of speed used in navigation equal to 1 nautical mile (6,076.115 feet or 1,852 meters) per hour. (approximately equal to 0.5 m/s)
LEE	(1) Shelter, or the part or side sheltered or turned away from the wind or waves. (2) (Chiefly nautical) The quarter or region toward which the wind blows.
LEEWARD	The direction *toward* which the wind is blowing; the direction toward –which waves are traveling.
LENGTH OF WAVE	The horizontal distance between similar points on two successive waves measured perpendicularly to the crest.
LEVEE	A dike or embankment to protect land from inundation.
LITTORAL TRANSPORT	The *movement* of littoral drift in the littoral zone by waves and currents. Includes movement parallel (longshore transport) and perpendicular (on-offshore transport) to the shore.
LITTORAL TRANSPORT RATE	Rate of transport of sedimentary material parallel or perpendicular to the shore in the littoral zone. Usually expressed in cubic meters (cubic yards) per year. Commonly synonymous with LONGSHORE TRANSPORT RATE.
LOAD	The quantity of sediment transported by a current. It includes the suspended load of small particles and the bed load of large particles that move along the bottom.
LONGSHORE	Parallel to and near the shoreline; ALONGSHORE.
LOOP OR ANTINODE	That part of a STANDING WAVE where the vertical motion is greatest and the horizontal velocities are least. Loops (sometimes called ANTINODES) are associated with CLAPOTIS and with SEICHE action resulting from wave reflections. Compare NODE.
LOW TIDE (LOW WATER, LW)	The minimum elevation reached by each falling tide. See TIDE.
LOW WATER DATUM	An approximation to the plane of mean low water that has been adopted as a standard reference plane. See also DATUM, PLANE and CHART DATUM.
LOW WATER LINE	The intersection of any standard low tide datum plane with the shore.

LOW WATER OF ORDINARY SPRING TIDES (LWOST)	A tidal datum appearing in some British publications that is based on low water of ordinary spring tides.
LOWER HIGH WATER (LHW)	The lower of the two high waters of any tidal day.
LOWER LOW WATER (LLW)	The lower of the two low waters of any tidal day. The single low water occurring daily during periods when the tide is diurnal is considered to be a lower low water.
MANGROVE	A tropical tree with interlacing prop roots that is confined to low-lying brackish areas.
MARSH, SALT	A marsh periodically flooded by salt water.
MASS TRANSPORT	The net transfer of water by wave action in the direction of wave travel. See also ORBIT.
MEAN HIGH WATER (MHW)	The average height of the high waters over a 19-year period. For shorter periods of observations, corrections are applied to eliminate known variations and reduce the results to the equivalent of a mean 19-year value. All high water heights are included in the average where the type of tide is either semidiurnal or mixed. Only the higher high water heights are included in the average where the type of tide is diurnal. So determined, mean high water in the latter case is the same as mean higher high water.
MEAN HIGH WATER SPRINGS	The average height of the high waters occurring at the time of spring tide. Frequently abbreviated to HIGH WATER SPRINGS.
MEAN HIGHER HIGH WATER (MHHW)	The average height of the higher high waters over a 19-year period. For shorter periods of observation, corrections are applied to eliminate known variations and reduce the result to the equivalent of a mean 19-year value.
MEAN LOW WATER (MLW)	The average height of the low waters over a 19-year period. For shorter periods of observations, corrections are applied to eliminate known variations and reduce the results to the equivalent of a mean 19-year value. All low water heights are included in the average where the type of tide is either semidiurnal or mixed. Only lower low water heights are included in the average where the type of tide is diurnal. So determined, mean low water in the latter case is the same as mean lower low water.
MEAN LOW WATER SPRINGS	The average height of low waters occurring at the time of the spring tides. It is usually derived by taking a plane depressed below the half-tide level by an amount equal to one-half the spring range of tide, necessary corrections being applied to reduce the result to a mean value. This plane is used to a considerable extent for hydrographic work outside of the United States and is the plane of reference for the Pacific approaches to the Panama Canal. Frequently abbreviated to LOW WATER SPRINGS.
MEAN LOWER LOW WATER (MLLW)	The average height of the lower low-waters, over a 19-year period. For shorter periods of observations, corrections are applied to eliminate known variations and reduce the results to the equivalent of a mean 19-year value. Frequently abbreviated to LOWER LOW WATER.
MEAN SEA LEVEL	The average height of the surface of the sea for all stages of the tide over a 19-year period. This is usually determined from hourly height readings and is not necessarily equal to MEAN TIDE LEVEL.
MEAN TIDE LEVEL	A plane midway between MEAN HIGH WATER and MEAN LOW WATER. This is not necessarily equal to MEAN SEA LEVEL. Also HALF-TIDE LEVEL.
MEDIAN DIAMETER	The diameter which marks the division of a given sand sample into two equal parts by weight, one part containing all grains larger than that diameter and the other part containing all grains smaller.

MIDDLE GROUND SHOAL	A shoal formed by ebb and flood tides in the middle of the channel of the lagoon or estuary end of an inlet.
MIXED TIDE	A type of tide in which the presence of a diurnal wave is conspicuous by a large inequality in either the high or low water heights, with two high waters and two low waters usually occurring each tidal day. In strictness, all tides are mixed, but the name is usually applied without definite limits to the tide intermediate to those predominantly semidiurnal and those predominantly diurnal (See Figure A-9).
MONO-CHROMATIC WAVES	A series of waves generated in a laboratory; each wave has the same length and period.
MONOLITHIC	Like a single stone or block. In coastal structures, the type of construction in which the structure's component parts are bound together to act as one.
NAUTICAL MILE	The length of a minute of arc, 1/21,600 of an average great circle of the Earth. Generally, one minute of latitude is considered equal to one nautical mile. The accepted United States value as of 1 July 1959 is 1,852 meters (6,076.115 feet), approximately 1.15 times as long as the U.S. statute mile of 5,280 feet. Also geographical mile.
NEAP TIDE	A tide occurring near the time of quadrature of the moon with the sun. The neap tidal range is usually 10 to 30 percent less than the mean tidal range.
NODE.	That part of a STANDING WAVE where the vertical motion is least and the horizontal velocities are greatest. Nodes are associated with CLAPOTIS and with SEICHE action resulting from wave reflections. Compare LOOP.
OCEANOGRAPHY	The study of the sea, embracing and indicating all knowledge pertaining to the sea's physical boundaries, the chemistry and physics of seawater, and marine biology.
OFFSHORE	(1) In beach terminology, the comparatively flat zone of variable width, extending from the breaker zone to the seaward edge of the Continental Shelf. (2) A direction seaward from the shore.
ORBIT	In water waves, the path of a water particle affected by the wave motion. In deepwater waves, the orbit is nearly circular, and in shallow-water waves, the orbit is nearly elliptical. In general, the orbits are slightly open in the direction of wave motion, giving rise to MASS TRANSPORT.
ORBITAL CURRENT	The flow of water accompanying the orbital movement of the water particles in a wave. Not to be confused with wave-generated LITTORAL CURRENTS.
ORTHOGONAL	On a wave-refraction diagram, a line drawn perpendicularly to the wave crests. WAVE RAY.
OSCILLATION	(1) A periodic motion backward and forward. (2) Vibration or variance above and below a mean value.
OSCILLATORY WAVE	A wave in which each individual particle oscillates about a point with little or no permanent change in mean position. The term is commonly applied to progressive oscillatory waves in which only the form advances, the individual particles moving in closed or nearly closed orbits. Compare WAVE OF TRANSLATION. See also ORBIT.
OVERTOPPING	Passing of water over the top of a structure as a result of wave runup or surge action.
OVERWASH	That portion of the uprush that carries over the crest of a berm or of a structure.
PARAPET	A low wall built along the edge of a structure such as a seawall or quay.
PARTICLE VELOCITY	The velocity induced by wave motion with which a specific water particle moves within a wave.
PERMEABLE GROIN	A groin with openings large enough to permit passage of appreciable quantities of LITTORAL DRIFT.

PLUNGING **BREAKER**	See BREAKER.
PORT	A place where vessels may discharge or receive cargo. It may be the entire harbor, including its approaches and anchorages, or only the commercial part of a harbor, where the quays, wharves, facilities for transfer of cargo, docks, and repair shops are situated.
POTENTIAL ENERGY OF **WAVES**	In a progressive oscillatory wave, the energy resulting from the elevation or depression of the water surface from the undisturbed level.
PRISM	See TIDAL PRISM.
PROBABLE MAXIMUM WATER LEVEL	A hypothetical water level (exclusive of wave runup from normal wind-generated waves) that might result from the most severe combination of hydrometeorological, geoseismic, and other geophysical factors, and that is considered reasonably possible in the region involved, when each of these factors exerts a maximum effect on the locality.
	This level represents the physical response of a body of water to maximum applied phenomena such as hurricanes, moving squall lines, other cyclonic meteorological events, tsunamis, and astronomical tide, combined with maximum probable ambient hydrological conditions such as wave setup, rainfall, runoff, and river flow. There is virtually no risk of this water level being exceeded.
PROFILE, BEACH	The intersection of the ground surface with a vertical plane; may extend from the top of the dune line to the seaward limit of sand movement.
PROGRESSION (of a beach)	See ADVANCE.
PROGRESSIVE WAVE	A wave that moves relative to a fixed co-ordinate system in a fluid. The direction in which it moves is termed the direction of wave propagation.
PROPAGATION OF WAVES	The transmission of waves through water.
PROTOTYPE	In laboratory usage, the full-scale structure, concept, or phenomenon used as a basis for constructing a scale model or copy.
QUARRYSTONE	Any stone processed from a quarry.
QUAY (Pronounced KEY)	A stretch of paved bank or a solid artificial landing place parallel to the navigable waterway, for use in loading and unloading vessels.
QUICKSAND	Loose, yielding, wet sand which offers no support to heavy objects. The upward flow of the water has a velocity that eliminates contact pressures between the sand grains and causes the sand-water mass to behave like a fluid.
RECESSION (of a beach)	(1) A continuing landward movement of the shore line. (2) A net landward movement of the shoreline over a specified time. Also RETROGRESSION.
REFLECTED WAVE	That part of an incident wave that is returned seaward when a wave impinges on a steep beach, barrier, or other reflecting surface.
REFRACTION (of water waves)	(1) The process by which the direction of a wave moving in shallow water at an angle to the contours is changed. The part of the wave advancing in shallower water moves more slowly than that part still advancing in deeper water, causing the wave crest to bend toward alignment with the underwater contours. (2) The bending of wave crests by currents.
REFRACTION COEFFICIENT	The square root of the ratio of the distance between adjacent orthogonals in deep water to their distance apart in shallow water at a selected point. When multiplied by the SHOALING FACTOR and a factor for friction and percolation, this becomes the WAVE HEIGHT COEFFICIENT or the ratio of the refracted wave height at any point to the deepwater wave height. Also, the square root of the ENERGY COEFFICIENT.

REFRACTION DIAGRAM	A drawing showing positions of wave crests and/or orthogonals in a given area for a specific deepwater wave period and direction.
RESONANCE	The phenomenon of amplification of a free wave or oscillation of a system by a forced wave or oscillation of exactly equal period. The forced wave may arise from an impressed force upon the system or from a boundary condition.
REVETMENT	A facing of stone, concrete, etc., built to protect a scarp, embankment, or shore structure against erosion by wave action or currents.
REYNOLDS NUMBER	The dimensionless ratio of the inertial force to the viscous force in fluid motion, $R_e = LV/v$, where L is a characteristic length, v the kinematic viscosity, and V a characteristic velocity. The Reynolds number is important in the theory of hydrodynamic stability and the origin of turbulence.
RIP CURRENT	A strong surface current flowing seaward from the shore. It usually appears as a visible band of agitated water and is the return movement of water piled up on the shore by incoming waves and wind. With the seaward movement concentrated in a limited band its velocity is somewhat accentuated. A rip consists of three parts: the FEEDER CURRENTS flowing parallel to the shore inside the breakers; the NECK, where the feeder currents converge and flow through the breakers in a narrow band or rip"; and the HEAD, where the current widens and slackens outside the breaker line. A rip current is often miscalled a rip tide. Also RIP SURF. See NEARSHORE CURRENT SYSTEM.
RIPRAP	A protective layer or facing of quarrystone, usually well graded within wide size limit, randomly placed to prevent erosion, scour, or sloughing of an embankment of bluff; also the stone so used. The quarrystone is placed in a layer that is at least twice the thickness of the 50 percent size or 1.25 times the thickness of the largest size stone in the gradation.
RUBBLE	(1) Loose angular water worn stones along a beach. (2) Rough, irregular fragments of broken rock.
RUBBLE-MOUND STRUCTURE	A mound of random-shaped and random-placed stones protected with a cover layer of selected stones or specially shaped concrete armor units. (Armor units in a primary cover layer may be placed in an orderly manner or dumped at random.)
RUNUP	The rush of water up a structure or beach on the breaking of a wave. Also UPRUSH, SWASH. The amount of runup is the vertical height above still-water level to which the rush of water reaches.
SCOUR	Removal of underwater material by waves and currents, especially at the base or toe of a shore structure.
SEA LEVEL	See MEAN SEA LEVEL.
SEAS	Waves caused by wind at the place and time of observation.
SEA STATE	Description of the sea surface with regard to wave action. Also called state of sea.
SEAWALL	A structure separating land and water areas and primarily designed to prevent erosion and other damage due to wave action. See also BULKHEAD.
SEICHE	(1) A standing wave oscillation of an enclosed water body that continues, pendulum fashion, after the cessation of the originating force, which may have been either seismic or atmospheric. (2) An oscillation of a fluid body in response to a disturbing force that has the same frequency as the natural frequency of the fluid system. Tides are now considered to be seiches that are induced primarily by the periodic forces caused by the Sun and Moon. (3) In the Great Lakes area, any sudden rise in the water of a harbor or a lake whether or not it is oscillatory (although inaccurate in a strict sense, this usage is well established in the Great Lakes area).

SEISMIC SEA WAVE	See TSUNAMI.
SEMIDIURNAL TIDE	A tide with two high waters and two low waters in a tidal day with comparatively little diurnal inequality.
SETUP, WAVE	Super-elevation of the water surface over normal surge elevation due to onshore mass transport of the water by wave action alone.
SETUP, WIND	See WIND SETUP.
SHALLOW WATER	(1) Commonly, water of such a depth that surface waves are noticeably affected by bottom topography. It is customary to consider water of depths less than one-half the surface wavelength as shallow water. See TRANSITIONAL ZONE and DEEP WATER. (2) More strictly, in hydrodynamics with regard to progressive gravity waves, water in which the depth is less than 1/25 the wavelength; also called VERY SHALLOW WATER.
SHINGLE	(1) Loosely and commonly, any beach material coarser than ordinary gravel, especially any having flat or almost flat pebbles. (2) Strictly and accurately, beach material of smooth, well-rounded pebbles that are roughly the same size (fine 20 mm -- coarse 60 mm). The spaces between pebbles are not filled with finer materials. Shingle often gives out a musical sound when stepped on.
SHOAL (noun)	A detached elevation of the sea bottom comprised of any material except rock or coral, which may endanger surface navigation.
SHOAL (verb)	(1) To *become* shallow gradually. (2) To *cause* to become shallow. (3) To *proceed* from a greater to a lesser depth of water.
SHOALING COEFFICIENT	The ratio of the height of a wave in water of any depth to its height in deep water with the effects of refraction, friction, and percolation eliminated. Sometimes SHOALING FACTOR or DEPTH FACTOR. See also ENERGY COEFFICIENT and REFRACTION COEFFICIENT.
SHOALING FACTOR	See SHOALING COEFFICIENT.
SIGNIFICANT WAVE	A statistical term relating to the one-third highest waves of a given wave group and defined by the average of their heights and periods. The composition of the higher waves depends upon the extent to which the lower waves are considered. Experience indicates that a careful observer who attempts to establish the character of the higher waves will record values that approximately fit the definition of the significant wave.
SIGNIFICANT WAVE HEIGHT	The average height of the one-third highest waves of a given wave group. Note that the composition of the highest waves depends upon the extent to which the lower waves are considered. In wave record analysis, the average height of the highest one-third of a selected number of waves, this number being determined by dividing the time of record by the significant period. Also CHARACTERISTIC WAVE HEIGHT.
SIGNIFICANT WAVE PERIOD	An arbitrary period generally taken as the period of the one-third highest waves within a given group. Note that the composition of the highest waves depends upon the extent to which the lower waves are considered. In wave record analysis, this is defined as the average period of the most frequently recurring larger well-defined waves in the record under study.
SINUSOIDAL WAVE	An oscillatory wave having the form of a sinusoid.
SLACK TIDE (SLACK WATER)	The state of a tidal current when its velocity is near zero, especially the moment when a reversing current changes direction and its velocity is zero. Sometimes considered the intermediate period between ebb and flood currents during which the velocity of the currents is less than 0.05 meter per second (0.1 knot). See STAND OF TIDE.

SLOPE	The degree of inclination to the horizontal. Usually expressed as a ratio, such as 1:25 or 1 on 25, indicating 1 unit vertical rise in 25 units of horizontal distance; or in a decimal fraction (0.04); degrees (20°18'); or percent (4 percent).
SOIL CLASSIFICATION (size)	An arbitrary division of a continuous scale of grain sizes such that each scale unit or grade may serve as a convenient class interval for conducting the analysis or for expressing the results of an analysis. There are many classifications used.
SOLITARY WAVE	A wave consisting of a single elevation (above the original water surface), whose height is not necessarily small compared to the depth and that is neither followed nor preceded by another elevation or depression of the water surfaces.
SOUNDING	A measured depth of water. On hydrographic charts, the soundings are adjusted to a specific plane of reference (SOUNDING DATUM).
SOUNDING DATUM	The plane to which soundings are referred. See also CHART DATUM.
SOUNDING LINE	A line, wire, or cord used in sounding, which is weighted at one end with a plummet (sounding lead). Also LEAD LINE.
SPIT	A small point of land or a narrow shoal projecting into a body of water from the shore (See Figure A-8)
SPRING TIDE	A tide that occurs at or near the time of new or full moon (SYZYGY) and which rises highest and falls lowest from the mean sea level.
STANDING WAVE	A type of wave in which the surface of the water oscillates vertically between fixed points, called nodes, without progression. The points of maximum vertical rise and fall are called antinodes or loops. At the nodes, the underlying water particles exhibit no vertical motion, but maximum horizontal motion. At the antinodes, the underlying water particles have no horizontal motion, but maximum vertical motion. They may be the result of two equal progressive wave trains travelling through each other in opposite directions. Sometimes called CLAPOTIS or STATIONARY WAVE.
STILL-WATER LEVEL	The elevation that the surface of the water would assume if all wave action were absent.
STOCKPILE	Sand piled on a beach foreshore to nourish downdrift beaches by natural littoral currents or forces. See FEEDER BEACH.
STORM SURGE	A rise above normal water level on the open coast due to the action of wind stress on the water surface. Storm surge resulting from a hurricane also includes that rise in level due to atmospheric pressure reduction as well as that due to wind stress. See WIND SETUP.
SURF	The wave activity in the area between the shoreline and the outermost limit of breakers.
SURF BEAT	Irregular oscillations of the nearshore water level with periods on the order of several minutes.
SURF ZONE	The area between the outermost breaker and the limit of wave uprush. (See Figures A-2 and A-S.)
SURGE	(1) The name applied to wave motion with a period intermediate between that of the ordinary wind wave and that of the tide, say from 1/2 to 60 minutes. It is low in height, usually less than 0.9 meter (0.3 foot). See also SEICHE. (2) In fluid flow, long interval variations in velocity and pressure, not necessarily periodic, perhaps even transient in nature. (3) see STORM SURGE.
SURGING BREAKER	See BREAKER.

SWELL	Wind-generated waves that have traveled out of their generating area. Swell characteristically exhibits a more regular and longer period and has flatter crests than waves within their fetch (SEAS).
SYNOPTIC CHART	A chart showing the distribution of meteorological conditions over a given area at a given time. Popularly called a weather map.
TIDAL DAY	The time of the rotation of the Earth with respect to the Moon, or the interval between two successive upper transits of the Moon over the meridian of a place, approximately 24.84 solar hours (24 hours and 50 minutes) or 1.035 times the mean solar day. Also called lunar day.
TIDAL FLATS	Marshy or muddy land areas which are covered and uncovered by the rise and fall of the tide.
TIDAL PRISM	The total amount of water that flows into a harbor or estuary or out again with movement of the tide, excluding any freshwater flow.
TIDAL RANGE	The difference in height between consecutive high and low (or higher high and lower low) waters.
TIDAL RISE	The height of tide as referred to the datum of a chart.
TIDAL WAVE	(1) The wave motion of the tides. (2) In popular usage, any unusually high and destructive water level along a shore. It usually refers to STORM SURGE or TSUNAMI.
TIDE	The periodic rising and falling of the water that results from gravitational attraction of the Moon and Sun and other astronomical bodies acting upon the rotating Earth. Although the accompanying horizontal movement of the water resulting from the same cause is also sometimes called the tide, it is preferable to designate the latter as TIDAL CURRENT, reserving the name TIDE for the vertical movement.
TIDE STATION	A place at which tide observations are being taken. It is called a primary tide station when continuous observations are to be taken over a number of years to obtain basic tidal data for the locality. A *secondary* tide station is one that is operated over a short period of time to obtain data for a specific purpose.
TIDE, STORM	See STORM SURGE.
TROCHOIDAL WAVE	A theoretical, progressive oscillatory wave first proposed by Gerstner in 1802 to describe the surface profile and particle orbits of finite amplitude, non-sinusoidal waves. The waveform is that of a prolate cycloid or trochoid, and the fluid particle motion is rotational as opposed to the usual irrotational particle motion for waves generated by normal forces. Compare IRROTATIONAL WAVE
TROUGH OF WAVE	The lowest part of a waveform between successive crests. Also, that part of a wave below still-water level.
TSUNAMI	A long-period wave caused by an underwater disturbance such as a -volcanic eruption or earthquake. Also SEISMIC SEA WAVE. Commonly miscalled "tidal wave".
UPLIFT	The upward water pressure on the base of a structure or pavement.
UPRUSH	The rush of water up onto the beach following the breaking of a wave. Also SWASH, RUNUP.
VARIABILITY OF WAVES.	(1) The variation of heights and periods between individual waves within a WAVE TRAIN. (Wave trains are not composed of waves of equal height and period, but rather of heights and periods which vary in a statistical manner.) (2) The variation in direction of propagation of waves leaving the generating area. (3) The variation in height along the crest, usually called "variation along the wave."
VELOCITY OF WAVES	The speed at which an individual wave advances. See WAVE CELERITY.

VISCOSITY (or internal friction)	That molecular property of a fluid that enables it to support tangential stresses for a finite time and thus to resist deformation.
WAVE FORECASTING	The theoretical determination of future wave characteristics, usually from observed or predicted meteorological phenomena.
WAVE GROUP	A series of waves in which the wave direction, wavelength, and wave height vary only slightly. See also GROUP VELOCITY.
WAVE HEIGHT	The vertical distance between a crest and the preceding trough. See also SIGNIFICANT WAVE HEIGHT.
WAVE HEIGHT COEFFICIENT	The ratio of the wave height at a selected point to the deepwater wave height. The REFRACTION COEFFICIENT multiplied by the shoaling factor.
WAVE PERIOD	The time for a wave crest to traverse a distance equal to one wavelength. The time needed for two successive wave crests to pass a fixed point. See also SIGNIFICANT WAVE PERIOD.
WAVE PROPAGATION	The transmission of waves through water.
WAVE SPECTRUM	In ocean wave studies, a graph, table, or mathematical equation showing the distribution of wave energy as a function of wave frequency. The spectrum may be based on observations or theoretical considerations. Several forms of graphical display are widely used.
WAVE OF TRANSLATION	A wave in which the water particles are permanently displaced to a significant degree in the direction of wave travel. Distinguished from an OSCILLATORY WAVE.
WAVE TROUGH	The lowest part of a wave form between successive crests. Also that part of a wave below still-water level.
WAVE VELOCITY	The speed at which an individual wave advances.
WAVELENGTH	The horizontal distance between similar points on two successive waves measured perpendicular to the crest.
WIND SETUP	On reservoirs and smaller bodies of water (I) the vertical rise in the still-water level on the leeward side of a body of water caused by wind stresses on the surface of the water; (2) the difference in still-water levels on the windward and the leeward sides of a body of water caused by wind stresses on the surface of the water. STORM SURGE (usually reserved for use on the ocean and large bodies of water).
WIND WAVES	(1) Waves formed and built up by the wind. (2) Loosely, any wave generated by wind.
WINDWARD	The direction from which the wind is blowing.

APPENDIX 2
Quarry operations

Reconnaissance

Basically, two types of quarries can be discerned:

1. Producing aggregates for concrete etc. A fine fragmentation is required. It is achieved by special drilling and blasting techniques. Classification is done by sieving.

2. Producing blockstones. The aim of the quarry operation is here to produce the largest possible blocks by sawing and cutting or by drilling and blasting. Classification takes place by picking up individual blocks.

For the construction of rubble mound breakwaters, quarries of the b-type are indispensable.

The size of the blocks obtained from the quarry is limited by the geological properties of the stone massif. Whatever is the origin of the geological formation, there will be discontinuities restricting the block size. To a certain extent, the size of the blocks can be influenced by the drilling and blasting pattern, but the size of a block will never exceed the distance between the natural cracks in the material.

When assessing suitable locations for a quarry a geological survey should be carried out, paying attention to the following points (Simons [1981]):

Joints (see Figure A2-1 and Figure A2-2)

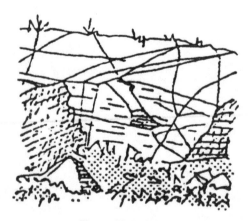

Figure A2-1 Joints

A break of geological origin in the continuity of a body of rock along which there has been no visible displacement. A group of parallel joints. is called a set and joint sets intersect to form a joint system. Joints can be open, filled or healed.

Joints frequently form parallel to the bedding-planes, foliation and cleavage and may be termed bidding-joints, foliation joints and cleavage-joints accordingly.

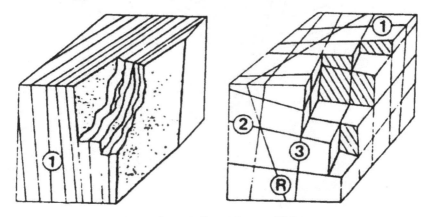

Figure A2-2 Different forms of joints

Fault *(see Figure A2-3 and Figure A2-4)*

A fracture or fracture zone along which there has been recognizable displacement from a few centimeters to a few kilometers in scale. The walls are often striated and polished (slickensided) resulting from the shear- displacement.

Frequently rock on both sides of a fault is shattered and altered or weathered, resulting in fillings such as breccia and gouge. Fault width may vary from millimeters to hundreds of meters.

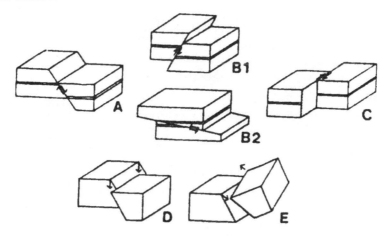

Figure A2-3 Types of fracture

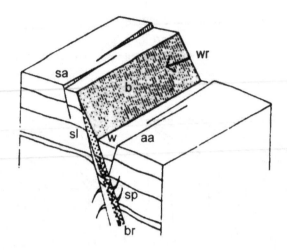

Figure A2-4 Types of fracture

Discontinuities

The general term for any mechanical discontinuity in a rock mass having zero or low tensile strength.

It is the collective term for most types of joints, weak bedding planes, weak schistocity-planes, weakness zones and faults.

The ten parameters selected to describe discontinuities and rockmasses are as follows:

1) Orientation:

Attitude of discontinuity in space (see Figure A2-5)

2) Spacing:

Perpendicular distance between adjacent discontinuities (see Figure A2-6 and Figure A2-7).

3) Persistence:

Discontinuity trace length as observed in an exposure.

4) Roughness:

Inherant surface roughness and waviness relative to the mean plane of a discontinuity.

5) Wall strength:

Equivalent compression strength of the adjacent rockwalls of a discontinuity. Maybe lower than rock block strength due to weathering or alteration of the walls.

6) Aperture:

Perpendicular distance between rock-walls of a discontinuity in the intervening space is air or waterfilled.

7) Filling:

Material that separates the adjacent rock-walls of a discontinuity and that is usually eeaker than the parent-rock.

8) Seepage:

Water-flow and free-moisture visible in individual discontinuities or in the rock mass as a whole.

9) Number of sets:

The number of joint sets comprising the intersecting joint system the rock mass may be divided by individual discontinuities.

10) Block-size (see Figure A2-8):

Rock-block dimensions resulting from the mutual orientation of intersecting joint sets and resulting from the spacing of the individual sets. Individual discontinuities may further influence the block and the shape. Block-size can be described either by means of the average dimension of typical blocks (block-size index l_b) or by the total number of joints intersecting a unit volume of the rockmass (Volumetric Joint Count J_v) (see Table A2-1).

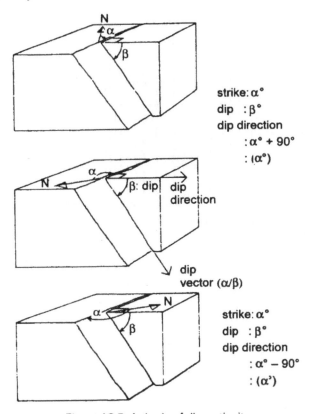

strike: $\alpha°$
dip : $\beta°$
dip direction
    : $\alpha° + 90°$
    : $(\alpha°)$

β: dip| dip direction

dip vector (α/β)

strike: $\alpha°$
dip : $\beta°$
dip direction
    : $\alpha° - 90°$
    : (α')

Figure A2-5 Attitude of discontinuity

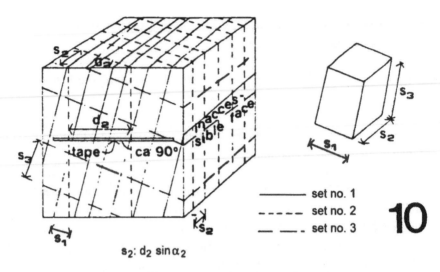

$s_2: d_2 \sin \alpha_2$

Figure A2-6 Distance between adjacent discontinuities

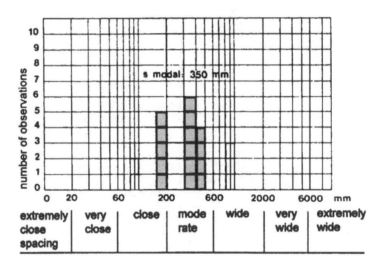

Figure A2-7 Distance between adjacent discontinuities

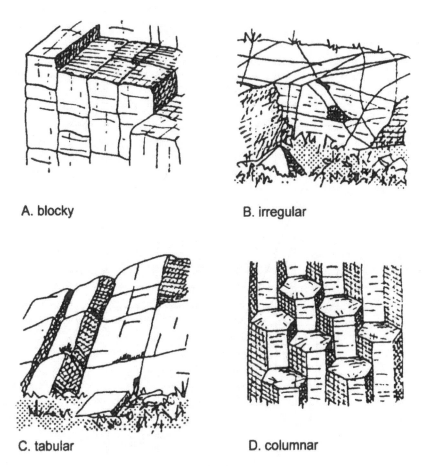

A. blocky   B. irregular

C. tabular   D. columnar

Figure A2-8 Block-size

The following descriptive terms give an impression of the corresponding block size:

Description	J_v (Joints/m^3)
Very large blocks	< 1.0
Large blocks	1-3
Medium-sized blocks	3 – 10
Small blocks	10 – 30
Very small blocks	> 30

Table A2-1 Block size

Values of J_v > 60 would represent crushed rock, typical of a clay-free crushed zone. On the basis of this information an experienced geologist is able to provide an expected fragmentation curve. See Figure A2-9 as example.

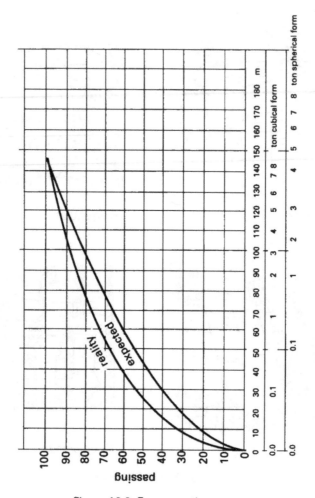

Figure A2-9 Fragmentation curve

Apart from these data, information must be obtained on the density, the mechanical strength, the abrasive resistance and the chemical durability (in sea water!).

Before a prospective location can be selected to establish the quarry, it should be ascertained that the following requirements are met:

- easy accessibility
- volume of the formation must be enough to serve the whole job
- blasting must be possible without excessive damage to human life or the environment in general
- concessions must be made available
- in the near vicinity of the quarry sufficient space should be available to open work yards, depots, etc.

Operation of the quarry

The planning of the quarry operation is mainly based on the expected fragmentation curve. According to Figure A2-9, 10% of each blast will be in blocks of 5 ton ($5 \cdot 10^4$ N) and larger. Consequently it is necessary to blast 10 X ton of material to obtain X ton of blocks of 5 ton and larger. The other 90% of the material must, however, also be classified, transported, stored and eventually be disposed of. In view of the cost involved it is often necessary to search for productive use of this finer material.

Figure A2-9 shows at the same time the dramatic consequence of a slight deviation from the expected curve. This would double the quantity of material to be blasted in order to deliver the required X tonnes of stone larger than 5 ton. Therefore a test blast of up to 100.000 ton of material is a necessary investment in the pretender stage.

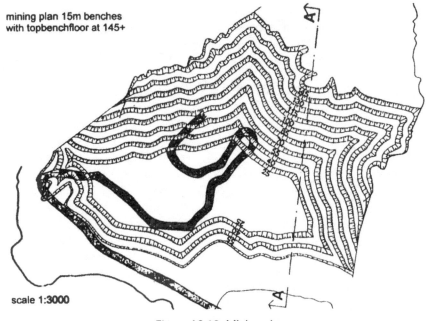

mining plan 15m benches
with topbenchfloor at 145+

scale 1:3000

Figure A2-10 Mining plan

The mining operation itself should be done in a systematic way, following a pre-designed mining plan. During the blasting, bench floors are created. The sequence of blasting depends on the overall pit slope (soil mechanical stability!). The width on each bench floor should be sufficient to create working space for classification, loading and transport (Figure A2-10 and Figure A2-11). See also Escher [1940].

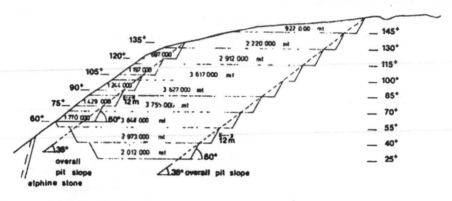

mining plan 15 m benches
with topbenchfloor at 120+
total quantity approx. 6.337.000 mt

with topbenchfloor at 145+
total quantity aprox. 25.886.000 mt

scale 1:1000

section A-A

Figure A2-11 Mining plan

APPENDIX 3
Concrete armour units

A3.1 Introduction

As compared with quarry stone, concrete armour units have the advantage that shape; the designer or contractor can choose size and density at liberty. These aspects of armour units will be treated briefly in the following. When adapting the size, concrete quality and fabrication methods become important. Therefore, this subject will also be discussed.

A3.2 Shape

As indicated in Chapter 7, a large number of different shapes has been developed by consultants, researchers and contractors.

The simplest shapes are the *Cube and the Parallelepid*. The advantage of these shapes is that moulds are simple and that handling and storage is also quite simple. Some modifications of the shape are used to improve the behaviour of the blocks under wave attack. Amongst others, one may discern sleeves in the side plains, ears at the corners or holes in the centre.

It has been common practice to use special shapes as well, providing interlocking properties that lead to an enhanced stability. Most units in this category are slender units like *Tetrapod, Akmon, Dolos, etc.* The slender shape creates a much better interlocking than present in quarry stone and the simple blocks. Also the porosity of these units is slightly larger than the porosity of the more traditional shapes, which probably adds to the stability as well. This has resulted in (Hudson) stability numbers K_D that are ranging from 8 upwards. (see Chapter 7). Drawings of these units are given in Figure A3-1, Figure A3-2, Figure A3-3 and Figure A3-4, along with the relation between size and volume. Because of the slender shapes, the characteristic dimensions may be much larger than the nominal diameter D_n, defined as the cubic root of the volume.

$$D_n = 3\sqrt{\frac{M}{\rho_r}}$$

The latest developments turn away from the very slender units because of structural problems encountered, and discussed elsewhere in this Annex. These latest units are Accropode and Core lock, developed respectively by SOGREAH and the Waterways Experiment Station. Dimensions of these blocks are also given in the Figures.

Because SOGREAH strongly recommends to place the Accropodes in a particular pattern, these recommendations are also added to this appendix.

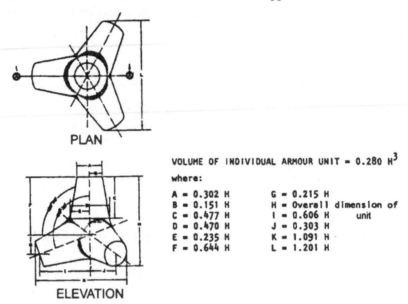

PLAN

ELEVATION

VOLUME OF INDIVIDUAL ARMOUR UNIT = 0.280 H^3

where:

A = 0.302 H		G = 0.215 H	
B = 0.151 H		H = Overall dimension of	
C = 0.477 H		I = 0.606 H	unit
D = 0.470 H		J = 0.303 H	
E = 0.235 H		K = 1.091 H	
F = 0.644 H		L = 1.201 H	

Figure A3-1 Tetrapod

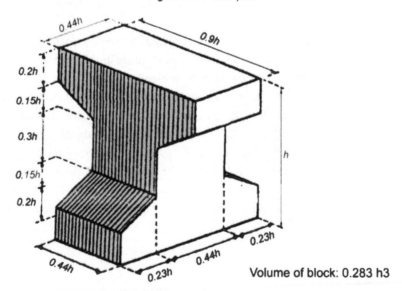

Volume of block: 0.283 h3

Figure A3-2 Akmon

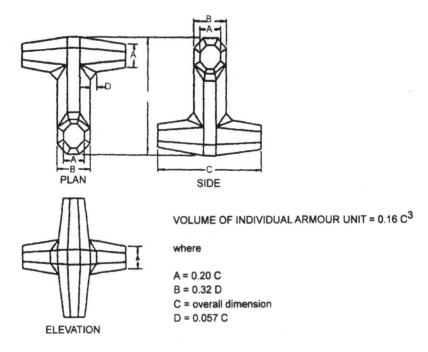

VOLUME OF INDIVIDUAL ARMOUR UNIT = 0.16 C^3

where

A = 0.20 C
B = 0.32 D
C = overall dimension
D = 0.057 C

Figure A3-3 Dolos

A3.3 Size

Although the size of the concrete armour units seems unlimited, there is a limitation in practice. Since the units are in principle not reinforced, to avoid corrosion problems, the structural integrity depends largely on the tensile strength of the concrete. Increasing the linear size of the units leads to an increase of mass and forces proportional to the third power of the size. ($V :: D^3$).

The cross-sectional area that provides the structural strength increases, however with the square of the size only. This means that the tension in any cross section increases basically linear with the dimension of the unit. Since the strength is constant, a larger block becomes more and more vulnerable to structural damage when the actual tension exceeds the available tensile strength. This is a failure mechanism that was certainly overlooked in the (small-scale) model investigations aimed at the hydraulic stability of the units.

A3.4 Density

The non-reinforced concrete as used in armour units will have a density ρ_r of 2200 to 2400 kg/m^3 if no special measures are taken. Care shall be taken to achieve a high density so as to avoid penetration of chloride and chemical damage. Since the volume of the units is rather large, proper attention shall be paid to the granulometry of the aggregates forming the skeleton (sand, gravel).

Higher densities can be achieved by selecting heavier aggregates like basalt, slacks or ore. In that case, the durability of the end product shall be ascertained by an adequate test programme. In this respect, the use of sulphate resisting cement is always recommended.

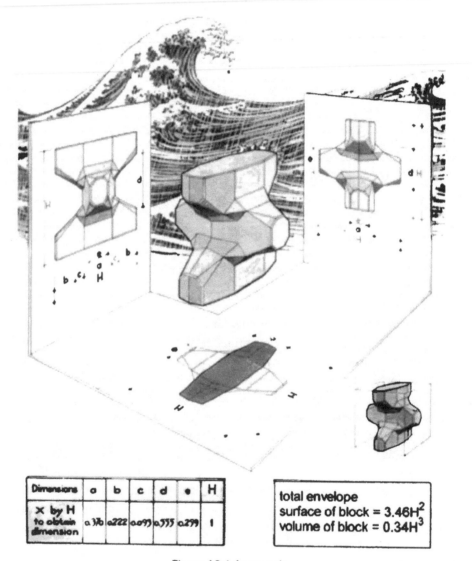

Dimensions	a	b	c	d	e	H
x by H to obtain dimension	0.37b	0.222	0.099	0.555	0.299	1

total envelope
surface of block = $3.46H^2$
volume of block = $0.34H^3$

Figure A3-4 Accropode

A3-5 Fabrication

For the fabrication of armour units, a special concrete mixing plant is almost a requirement to achieve a good and constant quality.

For economic reasons, the contractor wants a mixture that enables him to achieve a quick turn around time for the moulds. This reduces the number of moulds required. Since the moulds of the special shaped blocks are costly, there is a pressure to use a quick hardening mixture, so that the mould can be removed within say 24 hours after casting.

Because of the large mass of the units, attention is required for the heat generated during the hardening process. This is concentrated in the heart of the units. Specifically when the mould is removed, the surface of the units can cool rapidly, which leads to large tensions in the fresh concrete. In many cases these temperature gradients initiate cracks. It can easily be demonstrated that such cracks have a large influence on the eventual strength of the unit, and thus on the breakage of units during handling or during exposure to high (wave) loads. The problem of temperature gradients plays a very dominant role in places with a strong wind and a low humidity (cooling of the surface) and in regions with a large range between daily maximum and minimum temperatures. The latter occurs in tropical areas and in areas with a desert climate.

This problem can be tackled by the following measures:

- Reduction of the cement content
- Use of low heat cement
- Use of slower hardening cement
- Insulation of the units after removal of the moulds
- Spraying curing compound after removal of the moulds.

A3.6 Placement

Concrete units are lifted using slings or clamps. The use of steel hooks is not recommended because they will initiate spalling due to corrosion.

Most concrete blocks are placed at random. Special placement is difficult when working in deeper water and under exposed conditions where no diver assistance is available on a regular basis. Another disadvantage of special placing is the difficulty of repairing damage.

SOGREAH recommends a placing method for its Accropodes, which avoids these disadvantages to a large extent. This method prescribes a grid, but leaves the orientation of the block free (see Figures A3-5, A3-6 and A3-7).

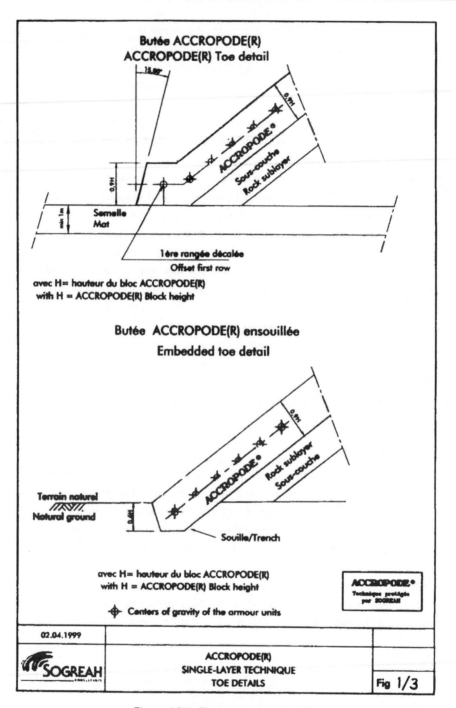

Butée ACCROPODE(R)
ACCROPODE(R) Toe detail

avec H= hauteur du bloc ACCROPODE(R)
with H = ACCROPODE(R) Block height

Butée ACCROPODE(R) ensouillée
Embedded toe detail

avec H= hauteur du bloc ACCROPODE(R)
with H = ACCROPODE(R) Block height

Centers of gravity of the armour units

02.04.1999

SOGREAH

ACCROPODE(R)
SINGLE-LAYER TECHNIQUE
TOE DETAILS

Fig 1/3

Figure A3-5 Placing accropodes (1)

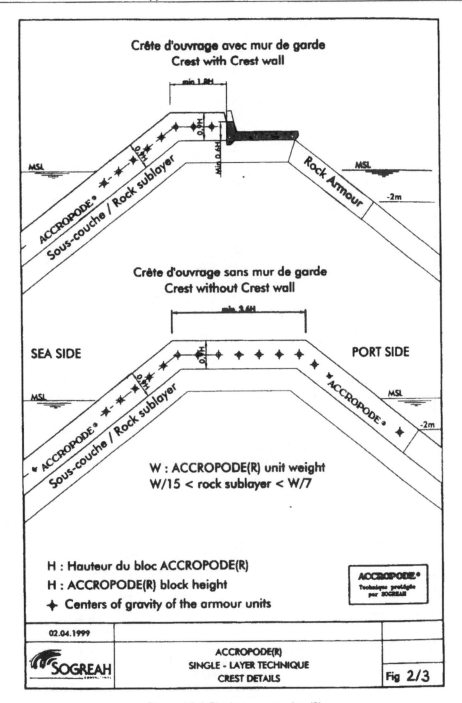

Figure A3-6 Placing accropodes (2)

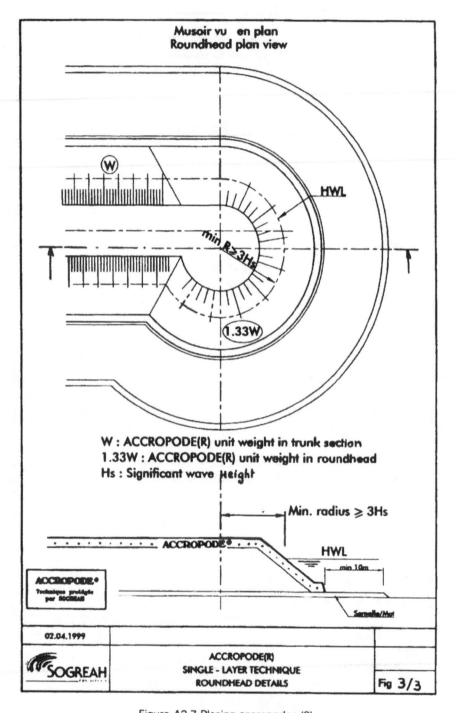

Figure A3-7 Placing accropodes (3)

APPENDIX 4
Goda's principles for breakwater design

From: the 1992 short course for the ICCE '92: "Design and Reliability of Coastal Structures", published by Instituto di Idraulica, Universita di Bologna, Italy.

THE DESIGN OF UPRIGHT BREAKWATERS

Yoshimi Goda
Department of Civil Engineering Yokohama National University, Japan

ABSTRACT
The historical development of upright breakwaters in Japan is briefly reviewed as an introduction. Various wave pressure formulas for vertical walls are discussed, and then the design formulas currently employed in Japan are presented with an example of calculation. Several design factors are also discussed.

TABLE OF CONTENTS

1   Introduction

An upright breakwater is defined here as a structure having an upright section rested upon a foundation It is often called a vertical breakwater or composite breakwater. The former is sometimes referred to a structure directly built on the rock foundation without layers of rubble stones. The latter on the other hand means a breakwater functioning as a sloping-type structure when the tide level is low but as a vertical-wall structure when the tide level is high. Because the terminology may vary from person to person, the definition above is given here in order to avoid further confusion.

Upright breakwaters are of quite old structural type. Old ports in the Roman Empire or ports in even older periods had been provided with breakwaters with upright structures. The upright breakwaters of recent construction have the origin in the 19th century. Italian ports have many upright breakwaters as discussed in the following lecture by Dr. L. Franco. British ports also have a tradition of upright breakwater construction as exemplified in Dover Port. The British tradition can be observed in old breakwaters of Indian ports such as Karachi, Bombay, and Madras. Japanese ports owes this tradition of upright breakwaters to British ports, because the modern breakwater construction began at Yokohama Port in 1890 under supervision of British army engineer, retired Major General H. S. Palmer. Since then Japan has built a large number of upright breakwaters along her long coastline extending over 34,000 km. The total length of upright breakwaters in Japan would exceed several hundred kilometers, as the total extension of breakwaters is more than 1,000 km.

The present note is intended to introduce the engineering practice of upright breakwater design to coastal and harbor engineers in the world, based on the experience of Japanese engineers.

2   Historical development of upright breakwaters in Japan

2.1   Examples of upright breakwaters in modern history of Japanese ports

Figure 1 illustrates typical cross sections of upright breakwaters in Japan in time sequences, which is taken from Goda [1985]. The east breakwater of Yokohama Port in Fig. 1 (a) utilized the local material of soft clayey stones for rubble foundation and minimized the use of concrete blocks in the upright section. The stone-filled middle section was replaced by concrete blocks during reconstruction after the storm damage in 1902. The wave condition in Yokohama was not severe with the design height of 3 m.

The structural type of upright breakwaters was adopted at a more exposed location of Otaru Port as shown in Fig. 1 (b) by 1. Hiroi in 1897, who was the chief engineer of regional government, later became a professor of the Tokyo Imperial University, and established the framework of Japanese harbor engineering. The first reinforced concrete caisson breakwater in Japan was built at Kobe in 1911, based on the successful construction of caisson-type quaywall at Rotterdam in 1905. Then Hiroi, immediately seeing the bright future of caisson breakwaters, employed the concept to an island breakwater of Otaru Port shown in Fig. 1 (c), where the design wave was 6 m high. He carried out various field measurements, including wave pressures on a vertical wall, for his finalization of breakwater design. Through these efforts, he came to propose the wave pressure formula for breakwater design, which is to be discussed in the next section.

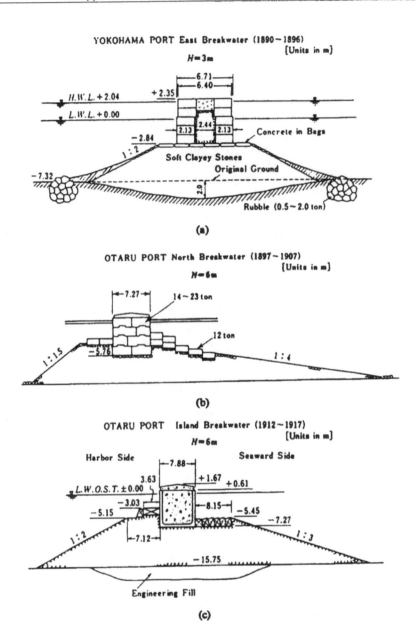

YOKOHAMA PORT East Breakwater (1890–1896)

[Units in m]

$H = 3$m

OTARU PORT North Breakwater (1897–1907)

[Units in m]

$H = 6$m

(b)

OTARU PORT   Island Breakwater (1912–1917)

[Units in m]

$H = 6$m

(c)

Fig. 1 (a-c) Historical development of upright breakwater in Japan after Goda [1985].

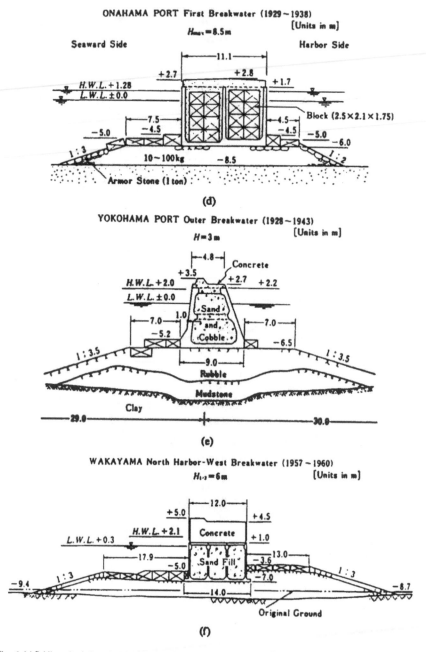

Fig. 1 (d-f) Historical development of upright breakwater in Japan (continued) after Goda [1985].

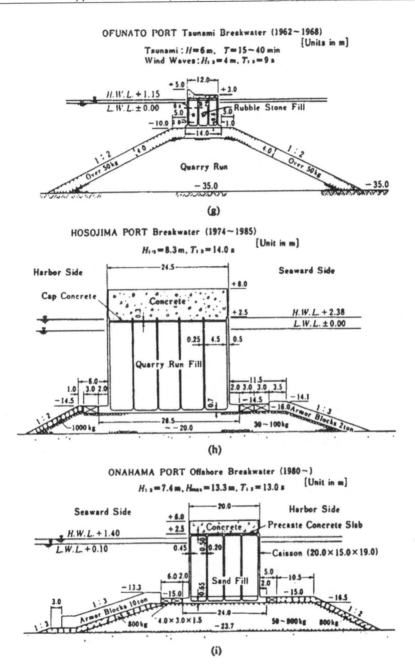

Fig. 1 (g-i) Historical development of upright breakwater in Japan (continued) after Goda [1985].

Hiroi's breakwater caissons were filled with concrete for durability and stability. The work time for concrete placement was sometimes saved by the use of precast blocks as in the example of Onahama Port in Fig. 1 (d). Concrete filling of breakwater caisson had been a tradition before the end of World War II, but a pioneering construction of reinforced concrete caisson breakwater with sand filling was carried out in Yokohama Port during the period of 1928 to 1943: Fig. 1 (e) shows its cross section. After World War II the use of sand as the filler material of caisson cells gradually became a common practice in Japan.

The breakwater of Wakayama Port shown in Fig. 1 (f) was built upon a quite soft ground so that it was provided with a wide foundation for the purpose of counter-balancing the weight of upright section. The breakwater of Ofunato Port in Fig. 1 (g) was built to reduce the inflow of tsunami waves into the bay. The water depth of 35 m below the datum level was the deepest one at the time of construction in 1962, but the present record of the deepest breakwater in Japan is held at Kamaishi Port with the depth of 60 m. Some design features and wave pressures on this breakwater have been discussed by Tanimoto and Goda [1991b]. One of the widest breakwaters is that of Hosojima Port shown in Fig. 1 (h): the widest at present is found at Hedono Port in a remote island with the width 38m (see Tanimoto and Goda 1991a). The breakwater of Onahama Port shown in Fig. 1 (i) is of recent design using Goda's wave pressure formulas to be discussed later.

2.2 Some features of Japanese upright breakwaters

As seen in these examples, Japanese breakwaters of upright type have a few common features. One is the relatively low crest elevation above the high water level. Presently, the recommendation for ordinary breakwaters is the crest height of $0.6 H_{1/3}$ above the high water level for the design condition.

For the design storm condition, this elevation is certainly insufficient to prevent wave agitations by the overtopped waves. However, it is a way of thinking of harbor engineers in Japan that the design waves are accompanied by strong gale and storm winds in any case and safe mooring of large vessels within a limited area of harbor basin cannot be guaranteed even if wave agitations are reduced minimum. As the storm waves with the return period of one year or less are much lower than the design wave, the above crest elevation is thought to be sufficient for maintaining a harbor basin calm at the ordinary stormy conditions.

Another feature of Japanese upright breakwaters is a relatively wide berm of rubble foundation and provision of two to three rows of large foot (toe) protection blocks. There is no fixed rule for selection of the berm width and engineers always consult with the examples of existing breakwaters in the neighborhood or those at the location of similar wave conditions. It is somewhat proportional to the size of concrete caisson itself, but the final decision must await good judgment of the engineer in charge. The foot protection concrete blocks have the size ranging from 2 to 4 m in one direction and the height of 1.5 to 2 m, weighing 15 to 50 tf. Though these blocks used to be solid ones, recent blocks are provided with several vertical holes to reduce the uplift force and thus to increase the stability against wave action.

A new development in upright breakwaters of Japan is the employment of various modifications to the shape of concrete caissons, such as perforated walls, vertical slits, curved slits with circular arc members, dual cylindrical walls and others (see Tanimoto and Goda 1991a). These new caisson shapes have been developed to actively dissipate wave energy and thus to reduce wave reflection and wave pressures. A number of these breakwaters have been built and functioning as expected.

3   Review of wave pressure formulae for vertical wall

3.1   Hiroi's formula

Prof. Hiroi published the wave pressure formula for breakwater design in 1919. It is a quite simple formula with the uniform pressure distribution of the following intensity:

$$p = 1.5 \, w_0 H \tag{1}$$

where w_0 denotes the specific weight of sea water and H the incident wave height. This pressure distribution extends to the elevation of 1.25 H above the design water level or the crest of breakwater if the latter is lower, as shown in Fig. 2.

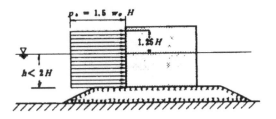

Fig. 2 Wave pressure distribution by Hiroi's formula.

Prof. Hiroi explained the phenomenon of wave pressure exerted upon a vertical wall as the momentum force of impinging jet flow of breaking waves and gave the reasoning for its quantitative evaluation. However, he must have had some good judgment on the magnitude of wave pressure from his long experience of harbor construction and several efforts of pressure measurements in situ. He states that he obtained the records of wave pressure exceeding 50 tf/m^2 by the pressure gauges set at a concrete wall in water of several meters deep. Nevertheless, he did not incorporate such high pressures into the formula of breakwater design, by saying that the high wave pressure must have lasted for only a short duration and are ineffective to cause appreciable damage to breakwaters.

Hiroi's wave pressure formula was intended for use in relatively shallow water where breaking waves are the governing factor. He also recommended to assume the wave height being 90% of water depth if no reliable information is available on the design wave condition. Hiroi's wave pressure formula was soon accepted by harbor engineers in Japan, and almost all breakwaters in Japan had been designed by this formula till the mid 1980s.

The reliability of Hiroi's formula had been challenged thrice at least. The first challenge was the introduction of Sainflou's formula in 1928 for standing wave pressures. Differentiation of two formulas was made, by referring to the recommendation of PIANC in 1935, in such a way that Hiroi's formula was for the case of the water depth above the rubble foundation being less than twice the incident wave height, while Sainflou's formula was for the water depth equal to or greater than twice the wave height. The second challenge was raised when the concept of significant wave was introduced in early 1950s. Which one of H_{max}, $H_{1/10}$, or $H_{1/3}$ is to be used in Hiroi's formula was the question. A consensus was soon formed as the recommendation for the use of $H_{1/3}$ based on the examination of existing breakwater designs and wave conditions. The third challenge was made by Goda [1973] against the insensitivity of the estimated pressure intensity to the variations in wave period and other factors. Hiroi's formula could not meet this challenge and is not used presently for the design of major breakwaters.

Though the pressure formula by Hiroi was so simple, the total wave force thus estimated was quite reliable on the average. Thanks to this characteristic, Japanese breakwaters had rarely experienced catastrophic damage despite the very long extension around the country.

3.2 Sainflou's formula

As well known, Saiflou published a theory of trochoidal waves in front of a vertical wall in 1928 and presented a simplified formula for pressure estimation. The pressure distribution is sketched as in Fig. 3, and the pressure intensities and the quantity of water level rise δ_0 are given as

$$
\left.
\begin{aligned}
p_1 &= (p_2 + w_0 h)(H + \delta_0)/(h + H + \delta_0) \\[2mm]
p_2 &= w_0 H/\cosh kh \\[2mm]
\delta_0 &= (\pi H^2/L) \coth kh
\end{aligned}
\right\}
\tag{2}
$$

where L is the wavelength and k is the wavenumber of $2\pi/L$.

Sainflou [1928] presented the above formula for standing wave pressures of nonbreaking type and the formula has been so utilized. The formula was derived for the purpose of practical application from the standpoint of a civil engineer and it has served its objective quite well. Just like the case of Hiroi's formula, it was born when the concept of wave irregularity was unknown. There seems to exist no established rule for the choice of representative wave height to be used with Sainflou's formula. Some advocates the use of $H_{1/3}$, some favors $H_{1/10}$, and the other prefers the selection of $H_{1\%}$.

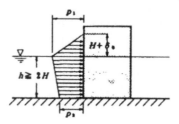

Fig. 3 Wave pressure distribution by Sainflou's formula.

It was customarily in Japan to use $H_{1/3}$ with Sainflou's formula but in a modified form. Through examinations of several minor damage of breakwaters, it had been revealed that a simple application of Sainflou's formula had yielded underestimation of wave pressures under storm conditions. For the zone extending $\pm H/2$ around the design water level, the wave pressure by Sainflou's foumula was replaced with that by Hiroi's formula. The modified formula was sometimes called the partial breaking wave pressure formula in Japan, because it was aimed to introduce the effect of partial wave breaking in relatively deep water. The dual system of Hiroi's wave pressure formula for breaking waves and of modified Sainflou's formula for standing waves had been the recommended engineering practice of breakwater design in Japan for the period from around 1940 to the early 1980s.

3.3   Minikin's formula and others

Although Hiroi's formula had been regarded as the most dependable formula for breaking wave pressures in Japan, it remained unknown in Europe and America. As the field measurement at Dieppe revealed the existence of very high pressures caused by impinging breaking waves and the phenomenon was confirmed by laboratory experiments by Bagnold [1939], harbor engineers in western countries began to worry about the impact breaking wave pressures. Then in 1950, Minikin proposed the following formula for breaking wave pressures, which consisted of the dynamic pressure p_m and the hydrostatic pressure p_s as sketched in Fig. 4:

Dynamic pressure:

$$\left.\begin{array}{l} p_n = p_{max}(1 - 2|z|/H)^2 : |z| \leq H/2 \\ p_{max} = 101 w_0 d(1 + d/h)H/L \end{array}\right\}\qquad(3)$$

Hydrostatic pressure:

$$p_s = \begin{cases} 0.5 w_0 H(1 - 2z/H) & : 0 \leq z < H/2 \\ 0.5 w_0 H & : z < 0 \end{cases}\qquad(4)$$

Because it was the first descriptive formula for breaking wave pressures, it was immediately accredited as the design formula and listed in many textbook and engineering manuals. Even in present days, technical papers based on Minikin's formula are published in professional journals from time to time.

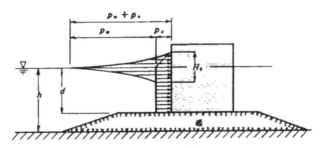

Fig. 4 Wave pressure distribution by Minikin's formula.

Minikin [1950] did not give any explanation how he derived the above formulation except for citing the experiments of Bagnold. In the light of present knowledge on the nature of impact breaking wave pressures, the formula has several contradictory characteristics. First, the maximum intensity of wave pressure increases as the wave steepness increases, but the laboratory data indicates that waves with long periodicity tends to generate well developed plunging breakers and produce the impact pressure of high intensity. In fact, Bagnold carried out his experiments using a solitary wave.

Second, Eq. 3 yields the highest p_{max} when d is equal to h or when no rubble foundation is present. It is harbor engineers' experience that a breakwater with a high rubble mound has a larger possibility of being hitten by strong breaking wave pressures than a breakwater with a low rubble mound.

Third, Minikin's formula yields excessively large wave force against which no rational upright breakwater could be designed. To the author's knowledge, no prototype breakwater has ever been constructed with the wave pressures estimated by Minikin's formula. Reanalysis of the stability of prototype breakwaters in Japan which experienced storm waves of high intensity, some undamaged and others having been displaced over a few meters, has shown that the safety factor against sliding widely varies in the range between 0.09 and 0.63 [Goda 1973b and 1974]. The safety factors of undamaged and displaced breakwaters were totally mixed together and no separation was possible. Thus the applicability of Minikin's formula on prototype breakwater design has been denied definitely.

There have been several proposals of wave pressure formulas for breakwater design. Among them, those by Nagai [1968, 1969] and Nagai and Otusbo [1968] are most exhaustive. Nagai classified the various patterns of wave pressures according to the wave conditions and the geometry of breakwater, and presented several sets of design formulas based on many laboratory data. However, his system of wave pressure formulas was quite complicated and these formulas gave different prediction of wave pressures at the boundaries between the zones of their applications. Another problem in the use of Nagai's method was the lack of specification for representative wave height for irregular waves. There was only a few cases of verification of the applicability of his method for breakwater design using the performance data of prototype breakwaters. Because of these reasons, the method is not used in Japan presently.

The Miche-Rundgren formula for standing wave pressure [CERC 1984] represents an effort to improve the accuracy of Sainflou's formula for engineering application. Certainly, the formula would give better agreement with the laboratory data than Sainflou's one. However, it has not been verified with any field data and its applicability for breakwater design is not confirmed yet.

4   Design formulae of wave pressures for upright breakwaters

4.1   Proposal of universal wave pressure formulae

It is a traditional approach in wave pressure calculation to treat the phenomena of the standing wave pressures and those by breaking waves separately. Casual observations of wave forms in front of a vertical wall could lead to a belief that breaking wave pressures are much more intensive than nonbreaking wave pressures and they should be calculated differently. The previous practice of wave pressure calculation with the dual formulas of Hiroi's and Sainflou's in Japan was based on such belief. The popularity of Minikin's formula prevailing in western countries seems to be owing to the concept of separation of breaking and nonbreaking wave pressures.

The difference between the magnitudes of breaking and nonbreaking wave pressures is a misleading one. The absolute magnitude of breaking wave pressures is certainly much larger than that of nonbreaking one. The height of waves which break in front of a vertical wall, however, is also greater than that of nonbreaking waves. The dimensionless pressure intensity, p/w_0H, therefore, increases only gradually with the increase of incident wave height beyond the wave breaking limit, as demonstrated in the extensive laboratory data by Goda [1972].

A practical inconvenience in breakwater design with the dual pressure formula system is evident when a breakwater is extended offshoreward over a long distance from the shoreline. While the site of construction is in shallow water, the wave pressures are evaluated with the breaking wave pressure formula. In the deeper portion, the breakwater would be subject to nonbreaking waves. Somewhere in between, the wave pressure formula must be switched from that of breaking to nonbreaking one. At the switching section, the estimated wave pressures jump from one level to another. With the Japanese system of the combined formulas of Hiroi's and modified Sainflou's, the jump was about 30%. To be exact with the pressure calculation, the width of upright section must be changed also. However, it is against the intuition of harbor engineers who believe in smooth variation of the design section. The location of switching section is also variable, dependent on the design wave height. If the design wave height is modified by a review of storm wave conditions after an experience of some damage on the breakwater, then an appreciable length of breakwater section would have to be redesigned and reconstructed.

The first proposal of universal wave pressure formula for upright breakwater was made by Ito et al. [1966] based on the sliding test of a model section of breakwaters under irregular wave actions. Then Goda [1973b, 1974] presented another set of formulas based on extensive laboratory data and being supported by verification with 21 cases of breakwater displacement and 13 cases of no damage under severe storm conditions. The proposed formulas were critically reviewed by the corps of engineers in charge of port and harbor construction in Japan, and they were finally adopted as the recommended formulas for upright breakwater design in Japan in 1980, instead of the previous dual formulas of Hiroi's and modified Sainflou's.

4.2 Design wave

The upright breakwater should be designed against the greatest force of single wave expected during its service life. The greatest force would be exerted by the highest wave among a train of random waves corresponding to the design condition on the average. Thus the wave pressure formulas presented herein are to be used together with the highest wave to be discussed below.

(1) *Wave height*

$$H_{max} = \begin{cases} 1.8H_{1/3} & : h/L_0 \geq 0.2 \\ \min\{(\beta_0{}^**H_0' + \beta_1{}^**h), \beta_{max}{}^**H_0', 1.8H_{1/3}\} & : h/L_0 < 0.2 \end{cases} \tag{5}$$

$$H_{1/3} = \begin{cases} K_s H_0' & : h/L_0 \geq 0.2 \\ \min\{(\beta_0 H_0' + \beta_1 h), \beta_{max}{}^**H_0', K_s H_0'\} & : h/L_0 < 0.2 \end{cases} \tag{6}$$

in which the symbol min$\{a,b,c\}$ stands for the minimum value among a, b and c, and H_0' denotes the equivalent deepwater significant height. The coefficients β_0 and others have empirically been formulated from the numerical calculation data of random wave breaking in shallow water as follows, after Goda [1975]:

$$\beta_0 = 0.028 \, (H_0'/L_0)^{-0.38} \exp[20 \tan^{1.5}\theta]$$

$$\beta_1 = 0.52 \exp[4.2 \tan\theta] \qquad\qquad (7)$$

$$\beta_{max} = \max \{0.92, 0.32 \, (H_0'/L_0)^{-0.29} \exp[2.4 \tan\theta]$$

$$\beta_0^* = 0.052 \, (H_0'/L_0)^{-0.38} \exp[20 \tan^{1.5}\theta]$$

$$\beta_1^* = 0.63 \exp[3.8 \tan\theta] \qquad\qquad (8)$$

$$\beta_{max}^* = \max \{1.65, 0.53 \, (H_0'/L_0)^{-0.29} \exp[2.4 \tan\theta]$$

in which the symbol $\max\{a,b\}$ stands for the larger of a or b, and $\tan\theta$ denotes the inclination of sea bottom.

The shoaling coefficient K_s is evaluated by taking the finite amplitude effect into consideration. Figure 5 has been prepared for this purpose based on the theory of Shuto [1974].

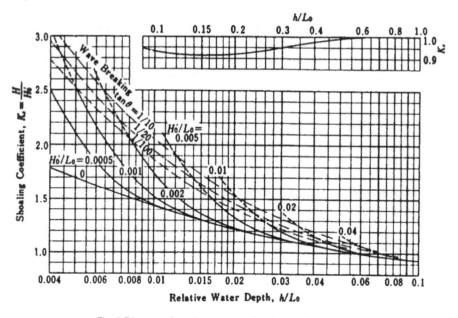

Fig. 5 Diagram of nonlinear wave shoaling coefficient K_s.

The selection of the fixed relation $H_{max} = 1.8 \, H_{1/3}$ outside the surf zone was based on three factors of reasoning. First, the fixed ratio was preferred to an introduction of duration dependent relation based on the Rayleigh distribution of wave heights, because such variability in the design wave height would cause some confusion in design procedures. Second, the examination of prototype breakwater performance under severe storm wave actions yielded reasonable results of safety factor against sliding by using the above fixed relation. Third, a possible deviation of the ratio $H_{max}/H_{1/3}$ from 1.8 to 2.0, say, corresponds to an increase of 11% and it can be covered within the margin of safety factor which is customarily taken at 1.2. However, it is a recommendation and an engineer in charge of breakwater design can use other criterion by his own judgment.

For evaluation of H_{max} by the second part of Eq. 7 or within the surf zone, the water depth at a distance 5 $H_{1/3}$ seaward of the breakwater should be employed. This adjustment of water depth has been introduced to simulate the nature of breaking wave force which becomes the greatest at some distance shoreward of the breaking point. For a breakwater to be built at the site of steep sea bottom, the location shift for wave height evaluation by the distance 5 $H_{1/3}$ produces an appreciable increase in the magnitude of wave force and the resultant widening of upright section.

(2) *Wave Period*

The period of the highest wave is taken as the same with the significant wave period of design wave, i. e.,

$$T_{max} = T_{1/3} \tag{9}$$

The relation of Eq. 9 is valid as the ensemble mean of irregular waves. Though individual wave records exhibit quite large deviations from this relation, the use of Eq. 9 is recommended for breakwater design for the sake of simplicity.

(3) *Angle of Wave Incidence to Breakwater*

Waves of oblique incidence to a breakwater exert the wave pressure smaller than that by waves of normal incidence, especially when waves are breaking. The incidence angle β is measured as that between the direction of wave approach and a line normal to the breakwater. It is recommended to rotate the wave direction by an amount of up to 15° toward the line normal to the breakwater from the principal wave direction. The recommendation was originally given by Prof. Hiroi together with his wave pressure formula, in consideration of the uncertainty in the estimation of wave direction, which is essentially based on the 16 points-bearing of wind direction.

4.3 Wave pressure, buoyancy and uplift pressure

(1) *Elevation to which the the wave pressure is exerted*

The exact elevation of wave crest along a vertical wall is difficult to assess because it varies considerably from $1.0H$ to more than $2.0H$, depending on the wave steepness and the relative water depth. In order to provide a consistency in wave pressure calculation, however, it was set as in the following simple formula:

$$\eta^* = 0.75(1 + \cos\beta)H_{max} \tag{10}$$

For waves of normal incidence, Eq. 10 gives the elevation of $\eta^* = 1.5\ H_{max}$.

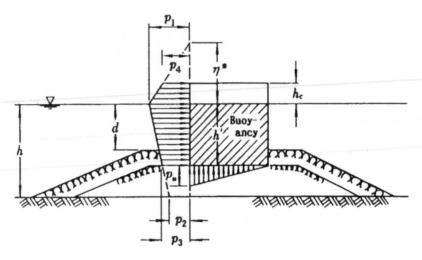

Fig. 6 Wave pressure distribution by Goda's formulas.

(2) *Wave pressure exerted upon the front face of a vertical wall*

The distribution of wave pressure on an upright section is sketched in Fig. 6. The wave pressure takes the largest intensity p_1, at the design water level and decreases linearly towards the elevation η^* and the sea bottom, at which the wave pressure intensity is designated as p_2.

The intensities of wave pressures are calculated by the following:

$$\left.\begin{array}{l} p_1 = 0.5\,(1 + \cos\beta)\,(\alpha_1 + \alpha_2 + \cos^2\beta)w_0\,H_{max} \\[2mm] p_2 = p_1/\cosh kh \\[2mm] p_3 = \alpha_3 p_1 \end{array}\right\} \qquad (11)$$

in which

$$\left.\begin{array}{l} \alpha_1 = 0.6 + 0.5\,[2kh/\sinh 2kh]^2 \\[2mm] \alpha_2 = \min\,\{[(h_b - d)/3h_b](H_{max}/d)^2,\ 2d/H_{max}\} \\[2mm] \alpha_3 = 1 - (h'/h)[1 - 1/\cosh kh] \end{array}\right\} \qquad (12)$$

where h_b denotes the water depth at the location at a distance $5H_{1/3}$ seaward of the breakwater.

The coefficient α_1 takes the minimum value 0.6 for deepwater waves and the maximum value 1.1 for waves in very shallow water. It represents the effect of wave period on wave pressure intensities. The coefficient α_2 is introduced to express an increase of wave pressure intensities by the presence of rubble mound foundation. Both coefficients α_1 and α_2 have empirically been formulated, based on the data of laboratory experiments on wave pressures. The coefficient α_3 is derived by the relation of linear pressure distribution. The above pressure intensities are assumed to remain the same even if wave overtopping takes place.

The effect of the incident wave angle on wave pressures is incorporated in η^* and p_1 with the factor of $0.5\,(1 + \cos\beta)$ and a modification to the term of α_2 with the factor of $\cos^2\beta$.

(3) Buoyancy and uplift pressure

The upright section is subject to the buoyancy corresponding to its displacement volume in still water below the design water level. The uplift pressure acts at the bottom of the upright section, and its distribution is assumed to have a triangular distribution with the toe pressure p_u given by Eq. 13.

$$p_u = 0.5\,(1 + \cos\beta)\,\alpha_1\,\alpha_3 w_0 H_{max} \tag{13}$$

The toe pressure p_u is set smaller than the wave pressure p_3 at the lowest point of the front wall. This artifice has been introduced to improve the accuracy of the prediction of breakwater stability, because the verification with the data of prototype breakwater performance indicated some overestimation of wave force if p_u were taken the same with p_3. When the crest elevation of breakwater h_c is lower than η^*, waves are regarded to overtop the breakwater. Both the buoyancy and the uplift pressure, however, are assumed to be unaffected by wave overtopping.

4.4 Stability analysis

The stability of an upright breakwater against wave action is examined for the three modes of failure: i.e., sliding, overturning, and collapse of foundation. For the first two modes, the calculation of safety factor is a common practice of examination. The safety factors against sliding and overturning are defined by the following:

Against sliding: $S.F. = \mu(W - U)/P$ (14)

Against overturning: $S.F. = (Wt - M_U)/M_P$ (15)

The notations in the above equations are defined as follows:
M_P: moment of total wave pressure around the heel of upright section
M_U: moment of total uplift pressure around the heel of upright section
P: total thrust of wave pressure per unit extension of upright section
t: horizontal distance between the center of gravity and the heel of upright section
U: total uplift pressure per unit extension of upright section
W: weight of upright section per unit extension in still water
μ: coefficient of friction between the upright section and the rubble mound

The safety factors against sliding and overturning are dictated to be equal to or greater than 1.2 in Japan. The friction coefficient between concrete and rubble stones is usually taken as 0.6. The coefficient seems to have a smaller value in the initial phase of breakwater installment, but it gradually rises to the value around 0.6 through consolidation of the rubble mound by the oscillations of the upright section under wave actions. The fact that most of breakwater displacements by storm waves occur during the construction period or within a few years after construction supports the above conjecture.

The bearing capacity of the rubble mound and the sea bottom foundation was used to be examined with the bearing pressures at the heel of upright section and at the interface between the rubble mound and the foundation. However, a recent practice in Japan is to make analysis of circular slips passing through the rubble mound and the foundation, by utilizing the simplified Bishop method (see Kobayashi et al. 1987). For the rubble mound, the apparent cohesion of $c = 2$ tf/m^2 and the angle of internal friction of $\phi =35°$ are recommended.

4.5   Example of wave pressure calculation

An example of calculation is given here in order to facilitate the understanding of the breakwater design procedure. The design wave and site conditions are set as in the following:

Waves: $H_0' = 7.0$ m, $T_{1/3} = 11$ s, $\beta = 10°$
Depth etc.: $h = 18$ m, $d = 10$ m, $h' = 11.5$ m, $h_c = 4.5$ m
Bottom slope: tan $\theta = 1/50$

The incident wave angle is the value after rotation by the amount up to 15°. The geometry of upright breakwater is illustrated in Fig. 7.

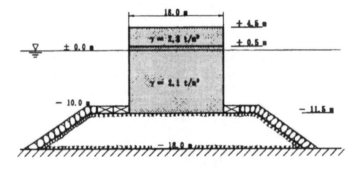

Fig.7 Sketch of upright breakwater for stability analysis.

i) *Design wave height* H_{max} *and the maximum elevation of wave pressure* $\eta*$

The coefficients for wave height calculation are evaluated as

$L_0 = 188.8$ m, $H_0'/L_0 = 0.0371$, $h/L_0 = 0.0953$, $K_s = 0.94$
$\beta_0 = 0.1036$, $\beta_1 = 0.566$, $\beta_{max} = $ min $\{0.92, 0.84\} = 0.92$
$\beta_0* = 0.1924$, $\beta_1* = 0.680$, $\beta_{max}* = $ min $\{1.65, 1.39\} = 1.65$

Then, the wave heights and the maximum elevation are obtained as

$H_{1/3} = $ min $\{10.91, 6.44, 6.58\} = 6.44$ m
$h_b = 18.0 + 5×6.44/50 = 18.64$ m
$H_{max} = $ min $\{14.02, 11.55, 11.84\} = 11.55$ m
$\eta* = 0.75× (1 + \cos 10°) × 11.55 = 17.19$ m

ii) *Pressure components*

The wavelength at the depth 18 m is $L = 131.5$ m. The coefficients for wave pressure are evaluated as

$kh = 2\pi \times 18/131.5 = 0.860$

$\alpha_1 = 0.6 + 0.5\times [2 \times 0.860/ \sinh(2 \times 0.860)]^2 = 0.802$

$\alpha_2 = \min \{ [(18.64 - 10.0)/ (3 \times 18.64)] \times (11.55/10)^2,$

$\qquad\qquad 2 \times 10/11.55 \}$

$\qquad = \min \{0.206, 1.732\} = 0.206$

$\alpha_3 = 1 - 11.5/18.0 \times [1 - 1/\cosh(0.860)] = 0.820$

Then, the intensities of wave pressure and uplift pressure are calculated as

p_1 = $0.5 \times (1+0.9848) \times [0.802+0.206 \times (0.9848)^2] \times 1.03 \times 11.55$

   = 11.83 tf/m^2

p_2 = $11.83/\cosh(0.860) = 8.49$ tf/m^2

p_3 = $0.820 \times 11.83 = 9.70$ tf/m^2

p_4 = $11.83 \times (1 - 4.5/17.19) = 8.73$ tf/m^2

p_u = $0.5 \times (1+0.9848) \times 0.802 \times 0.820 \times 1.03 \times 11.55 = 7.76$ tf/m^2

The symbol p_4 denotes the pressure intensity at the top of upright section.

iii) *Total pressure and uplift, and their moments*

$P = 0.5 \times (11.83+9.70) \times 11.5+0.5 \times (11.83+7.76) \times 4.5 = 167.9$ tf/m

$M_P = 1366.2$ tf-m/m

$U = 0.5 \times 18.0 \times 7.76 = 69.8$ tf/m

$M_U = (2/3) \times 69.8 \times 18 = 837.6$ tf-m/m

iv) *Stability of upright section against wave action*

The specific weight of upright section is assumed as in the following:

The portion above the elevation + 0.5 m : $\gamma_c = 2.3$ tf/m^3

The portion below the elevation +0.5 m: $\gamma_c' = 2.1$ tf/m^3

The difference in the specific weight reflects a current practice of sand filling in the cells of concrete caisson. The weight of upright section is calculated for the dry and in situ conditions, respectively, as

$W_a = 2.1 \times (11.5 + 0.5) \times 18.0+2.3 \times (4.5 - 0.5) \times 18.0 = 619.2$ tf/m

$W = 619.2 - 1.03 \times 11.5 \times 18.0 = 406.0$ tf/m

The safety factors against sliding and overturning of the upright section are calculated as in the following:

Against sliding: $S.F. = 0.6 \times (406.0 - 69.8)/167.9 = 1.20$

Against overturning: $S.F. = (406.0 \times 9.0 - 837.6)/1366.2 = 2.06$

Therefore, the upright breakwater with the uniform width of $B = 18.0$ m sketched in Fig. 7 is considered stable against the design wave of $H_0' = 7.0$ m and $T_{1/3} = 11.0$ s.

5   Discussion of several design factors

5.1   Precautions against impulsive breaking wave pressure

The universal wave pressure formulas described hereinbefore do not address to the problem of impulsive breaking wave pressure in a direct manner. The coefficient α_2, however, has the characteristic of rapid increase with the decrease of the ratio d/H_{max}. This increase roughly reflects the generation of impulsive breaking wave pressure.

Though the impact pressure of breaking waves exerted upon a vertical wall is much feared by coastal and harbor engineers, it occurs under the limited conditions only. If waves are obliquely incident to a breakwater, the possibility of impact pressure generation is slim. If a rubble mound is low, the sea bottom should be steep and waves be of swell type for the impact pressure to be generated. A most probable situation under which the impact pressure is exerted upon an upright breakwater is the case with a high rubble mound with an appreciable berm width (see Tanimoto et al. 1987). Most of breakwater failures attributed to the action of the impulsive breaking wave pressure are due to the wave forces of normal magnitude, which could be estimated by the universal wave pressure formulas described in the present lecture note.

The impact pressure of breaking waves last for a very short time duration, which is inversely proportional to the peak pressure intensity. In other words, the impulse of impact pressure is finite and equal to the forward momentum of advancing wave crest which is lost by the contact with the vertical wall. The author has given an estimate of the average value of the impact pressure effective in causing sliding of an upright section, by taking into account the elastic nature of a rubble mound and foundation [Goda 1973a]. Because the major part of impact is absorbed by the horizontal oscillations and rotational motion of the upright section, the impact pressure effective for sliding is evaluated as $(2 \sim 3)\, w_0 H_{max}$.

Nevertheless, the pressure intensity of the above order is too great to be taken into the design of upright breakwaters: the mean intensity of wave pressure employed for the stability analysis of the breakwater sketched in Fig. 7 is only $0.91\, w_0 H_{max}$. Engineers in charge of breakwater design should arrange the layout and the cross section of breakwater in such way to avoid the danger of impact pressure generation. If the exertion of impulsive breaking wave pressure on the upright section seems inevitable, a change in the type of breakwater structure, such as a sloping type breakwater or a vertical breakwater protected by a mound of concrete blocks, should be considered.

5.2   Structural aspects of reinforced concrete caisson

The upright section of vertical breakwater is nowadays made by reinforced concrete caisson. The width is determined by the stability condition against wave action. The height of caisson or the base elevation is so chosen to yield the minimum sum of the construction cost of rubble mound and upright section.

The length of caisson is governed by the capacity of manufacturing yard. In March 1992, Kochi Port facing the Pacific in Shikoku, Japan, set a breakwater caisson with the length 100 m in position. It is of hybrid structure with steel frames and prestressed concrete.

A concrete caisson is divided into a number of inner cells. The size of inner cells is limited to 5 m or less in ordinary design. The outer wall is 40 to 50 cm thick, the partition wall 20 to 25 cm thick, and the bottom slab 50 to 70 cm thick. These dimensions are subject to the stress analysis of reinforced concrete. As the upright breakwater withstands the wave force mainly with its own weight, the use of prestressed concrete for breakwater caisson is not advantageous in the ordinary situations. For the caisson of special shapes for enhancing wave dissipation such as the caisson with circular arc members, prestressed concrete is utilized.

5.3 Armor units for rubble mound

The berm and slope of a rubble mound needs to be protected with armor units against the scouring by wave action. Foot-protection blocks weighing from 15 to 50 tf are placed in front of an upright section. The rest of the berm and slope are covered by heavy stones and/or specially shaped concrete blocks. The selection of armor units is left to the judgment of engineers, with the aid of hydraulic model tests if necessary.

A formula for the weight of armor stones on the berm of rubble mound has been proposed by Tanimoto et al. [1982] as the results of systematic model tests with irregular waves. The minimum weight of armor stones can be calculated by a formula of the Hudson type:

$$W = \gamma_r H_{1/3}{}^3 / [N_s{}^3 (S_r - 1)^3] \tag{16}$$

in which W is the weight of armor stones, γ_r the specific weight of armor stones, S_r the ratio of γ_r to the specific weight of seawater, and N_s, the stability number, the value of which depends on the wave conditions and mound dimensions.

For waves of normal incidence, Tanimoto et al. [1982] gave the following function for armor stones:

$$N_s = \max \{1.8, [1.3 \; \frac{1-\kappa}{\kappa^{1/3}} \frac{h'}{H_{1/3}} + 1.8 \exp [-1.5 \; \frac{(1-\kappa)^2}{\kappa^{1/3}} \frac{h'}{H_{1/3}}]]\} \tag{17}$$

in which the parameter κ is calculated by

$$\kappa = [2kh' / \sinh 2kh'] \sin (2\pi B_M / L') \tag{18}$$

and where h' denotes the water depth at which armor stones are placed, L' the wavelength at the depth h', and B_M the berm width.

Though the stability number for concrete blocks has not been formulated, a similar approach to the data of hydraulic model tests on concrete blochs will enable the formulation of the stability number for respective types of concrete blocks.

6   Concluding remarks

The design and construction of upright breakwaters is a well established, engineering practice, at least in Japan, Korea, and Taiwan. A large number of these breakwaters have been built and will be built to protect ports and harbors. In these countries, the problem of impulsive breaking wave pressure is rather lightly dealt with. The tradition owes to Prof. Hiroi, who established the most reliable wave pressure formula in shallow water and showed the upright breakwaters could be successfully constructed against breaking waves.

This is not to say that no breakwaters have failed by the attack of storm waves. Whenever a big storm hits the coastal area, several reports of breakwater damage are heard. However, the number of damaged caissons is very small compared with the total number of breakwater caissons installed along the whole coastline. Probably the average rate per year would be less than 1%, though no exact statistic is available. Most cases of breakwater damage are attributed to the underestimation of the storm wave condition when they were designed.

In the past, the majority of breakwaters were constructed in relatively shallow water with the depth up to 15 m, for example, because the vessels calling ports were relatively small. In such shallow water, the storm wave height is controlled by the breaking limit of the water depth. One reason for the low rate of breakwater failure in the past could be this wave height limitation at the locations of breakwaters.

The site of breakwater construction is moving into the deeper water in these days. Reliable evaluation of the extreme wave condition is becoming the most important task in harbor engineering, probably much more than the improvement of the accuracy of wave pressure prediction.

References

Bagnold, R.A. [1939]: Interim report on wave-pressure research, J. Inst. Civil Engrs., Vol.12, pp.202-226.

CERC (Coastal Engineering Research Center, U.S. Army Corps of Engrs.) [1984]: Shore Protection Hanual, U.S. Government Printing Office, pp.7-161~173. Goda, Y. [1972]: Experiments on the transition from nonbreaking to postbreaking wave pressures, Coastal Engineering in Japan, Vol.15, pp.81-90.

Goda, Y. [1973a]: Motion of composite breakwater on elastic foundation under the action of impulsive breaking wave pressure, Rept. Port and Harbor Pes. Inst., Vol.12, No.3, pp.3-29 (in Japanese).

Goda, Y. [1973b]: A new method of wave pressure calculation for the design of composite breakwater, Rept. Port and Harbor Res. Inst., Vol. 12, No. 3, pp.31 70 (in Japanese).

Goda, Y. [1974]: New wave pressure formulae for composite breakwater, Proc. 14th Int. Conf. Coastal Eng., pp.1702-1720.

Goda, Y. [1975]: Irregular wave deformation in the surf zone, Coastal Engineering in Japan, Vol.18, pp.13-26.

Goda, Y. [1985]: Random Seas and Design of Maritime Structures, University of Tokyo Press., pp.108-110.

Hiroi, I. [1919]: On a method of estimating the force of waves, Memoirs of Engg. Eaculty, Imperial University of Tokyo, Vol. X, No.1, p.19.

Ito, Y., Fujishima, M., and Kitatani, T. [1966]: On the stability of breakwaters, Rept. Port and Harbor Res. Inst., Vol. 5, No. 14, pp.1-134 (in Japanese).

Kobayashi, M., Terashi, M., and Takahashi, K. [1987]: Bearing capacity of a rubble mound supporting a gravity structure, Rept. Port and Harbor Res. Inst., Vol.26, No.5, pp.215-252.

Minikin, R. R. [1950]: Winds, waves and Maritime Structures, Griffin, London, pp.38-39.

Nagai, S. [1968]: Pressures of partial standing waves, J. Waterways and Harbors Div., Proc. ASCE, Vol.94, No.WW3, pp.273-284.

Nagai, S. [1969]: Pressures of standing waves on a vertical wall, J. Waterways and Harbors Div., Proc. ASCE, Vol.95, No.WW1, pp.53-76.

Nagai, S. and Otsubo, T. [1968]: Pressure by breaking waves on composite breakwaters, Proc. 11th Int. Conf. Coastal Engg, pp.920-933.

Sainflou, G. [1928]: Essai sur les digues maritimes, verticales, Annales Ponts et Chaussées, Vol.98, No.4.

Shuto, N. [1974]: Nonlinear long waves in a channel of variable section, Coastal Engineering in Japan, Vol.17, pp.1-14.

Tanimoto, K., Yagyu, T., and Goda, Y. [1982]: Irregular wave tests for composite breakwater foundation, Proc. 18th Int. Conf. Coastal Engg., pp.2144-2163.

Tanimoto, K., Takahashi, S., and Kimura, K. [1987]: Structures and hydraulic characteristics of breakwaters - The state of arts of breakwater design in Japan, Rept. Port and Harbor Res. Inst., Vol.26, No.5, pp.11-55..

Tanimoto, K. and Goda, Y. [1991a]: Historical development of breakwater structures in the world, Proc. Conf. on Coastal Structures and Breakwaters, Inst. Civil Engrs., pp.153-166.

Tanimoto, K. and Goda, Y. [1991b]: Stability of deepwater caisson breakwater against random waves, Proc. Conf. on Coastal Structures and Breakwaters, Inst. Civil Engrs., pp.181-193.

APPENDIX 5
Filter rules

Filter rules were developed initially for uniform flow conditions. Terzaghi defined a set of rules (as early as 1922) for a geometrically impermeable filter:

- For stability of the interface between base material (subscript B) and filter material (subscript F):

$$\frac{d_{15F}}{d_{85B}} < 5$$

1   For an adequate permeability of the filter layer to avoid its uplifting:

$$\frac{d_{15F}}{d_{15B}} > 5$$

2   For internal stability of either base or filter material:

$$\frac{d_{60}}{d_{10}} < 10$$

The filter works in such a way that the voids in the filter material are closed off by the larger grains from the base material. In this process, some smaller grains from the base material may be lost before an arch of the larger fractions of the base material forms an effective barrier. The process can be compared with the plastering of a riverbed with coarser material.

In breakwater design, stricter rules are used.

From the Shore Protection Manual, we know the rule that the weight ratio between armour and first underlayer should be between 10 and 15. In Chapter 11, we have seen that a ratio of 25 is still acceptable. This leads to a diameter ratio of 2 to 3, instead of a ratio of more than 5 as can be derived on the basis of the Terzaghi formulae.

The reasons for this difference are amongst others:

- In breakwaters, it is not acceptable that some of the finer grains would disappear from a layer that is only two units thick. The size of these grains is still so large that considerable settlements would occur
- Because of the constriction of the layer thickness to two units, arches can not be formed in intermediate layers
- Because of the cyclic character of the wave load, arches if formed at all, would break up at flow reversal

- Finer material in the underlayers have a negative influence on the stability of the armour (P – in the Van der Meer formula)

De Graauw et al [1983] who performed well known experiments on non ideal filters concluded that for cyclic loading a ratio of $d_{50F}/d_{50B} = 3$ would be the limit for penetration of base material into the filter. This is well in accordance with the practical weight ratio of 25.

Two Japanese researchers (Kawakami and Esashi, [1961]) established also a relation based on the absolute size of the grains:

$$\log\left(\frac{d_{10F}}{d_{10B}} - 2\right) < \frac{1.9}{\log(d_{10B} - 0.001) - 3}$$

Grain sizes in this formula expressed in mm!

This again would lead to a ratio between 3 and 4 for the very large units (> 1m) in the outer part of the breakwater.

From these considerations it will be clear that the filter rules for the armour layer, the first underlayer, the core and the toe should be followed in a strict way. Further in the interior of the breakwater at the interface between core and subsoil, and in the filters under the toe, the filter rules can be taken closer to the values of Terzaghi, thus allowing a larger step in the grain size between successive layers.

APPENDIX 6
Breakwater examples

A6.1 Alternatives for Hook of Holland breakwaters (Europoort)

A6.1.1 Caisson breakwaters

90° CAISSON

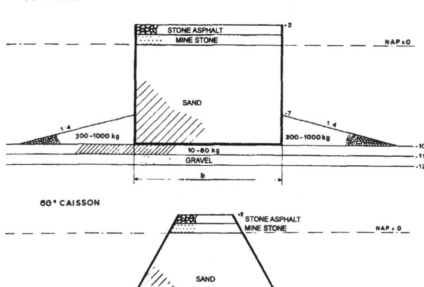

60° CAISSON

„HANSTHOLM" CAISSON

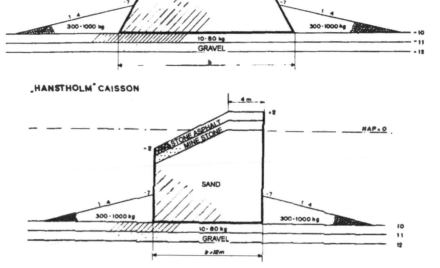

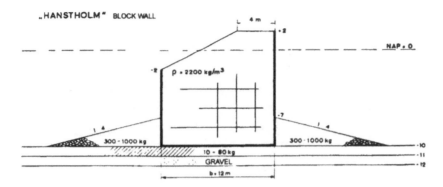

„HANSTHOLM" BLOCK WALL

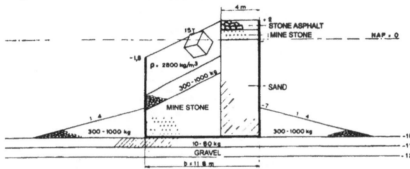

„HANSTHOLM" CAISSON WITH BLOCKS

A6.1.2 Stone breakwaters

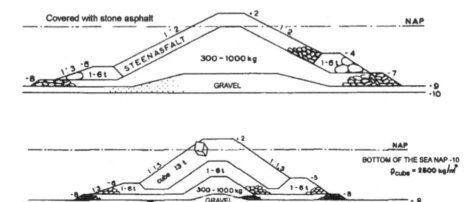

A6.1.3 Combinations

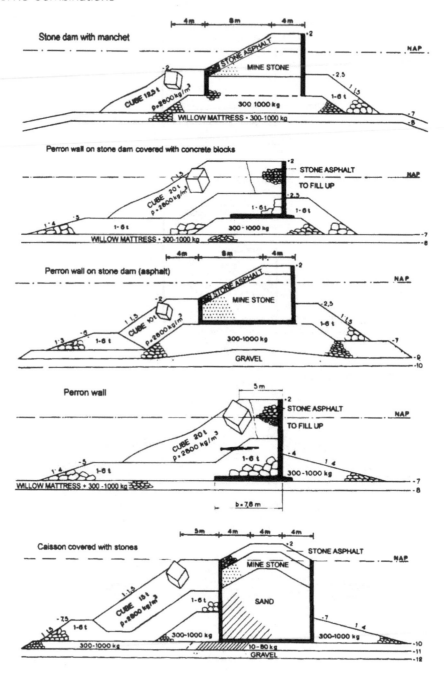

A6.2 IJmuiden

As another example the development - and history - of the design of the breakwater of IJmuiden will be described. The various cross sections which were considered are given in figures 1 through 9:

1. The original cross section. Failure occurs due to damage on the harbour side of the crest (inner slope).

2. In order to avoid this, these armour units have been removed. In order to protect the much lighter rock blocks of one to five tons under the first cover layer, these rock blocks have been penetrated with asphalt.

3. Due to the typical lay-out of the moles, waves will reach the inner slope of the mole almost parallel to the breakwater, with the result that the armour units on the inner slope just below the water surface are attacked.

4. In order to avoid the necessity of penetration, the cap construction of rock asphalt has been extended below water level. When the amour units on the inner slope move (due to the oblique waves) the stability is endangered.

5. In order to overcome these difficulties, the entire inner slope is made from rock asphalt. The disadvantage of this solution is, however, that the layer can be lifted due to pressure differences across this layer. This layer, therefore, has to be of sufficient thickness and weight.

   Note 1: When a decision was taken on the cross-section,the model technique had not yet progressed to the extent that the plastic mass of (impermeable) stone asphalt could be reproduced in the model. Therefore, the initial decisions on the design in stone asphalt were based upon calculations only.

6. For this reason the inner slope is not covered completely with rock asphalt, but only in spots. These spots increase the stability sufficiently without the danger of uplifting.

7. In order to avoid or to decrease the uplift forces, the cover of rock asphalt has been extended to the inner and outer slope of the breakwater.

8. Since this breakwater does not suffer from overtopping it can also be lowered.

   Note 2: The length of the scouring protection greatly influences the breakwater stability (Van Oorschot [1968])

9. A later stage of this design development, the crest has again been made higher in order to enable the transport of construction materials over this crest to the cranes standing at the construction area at both sides of the breakwater.

10. This figure shows the savings in material (by double hatching).

    Note 3: In the meantime, the design proved to be of insufficient stability, which was, afterwards, confirmed by model tests. To increase the stability, the asphalt cover layer was covered with concrete cubes, approximately according to the double hatched area of figure 9 (see also d'Angremond, Span, V.d. Weide and Woenstenk [1970]).

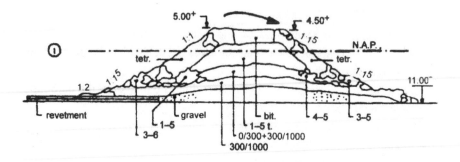

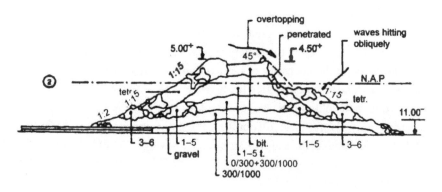

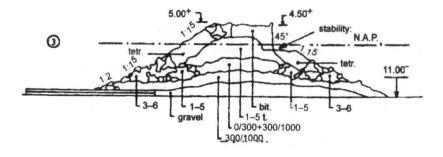

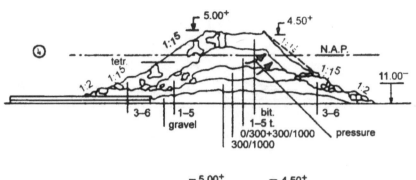

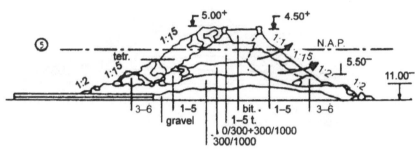

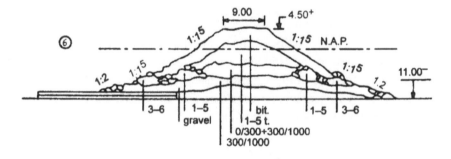

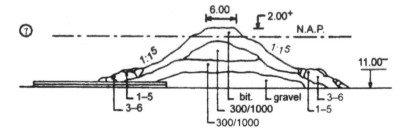

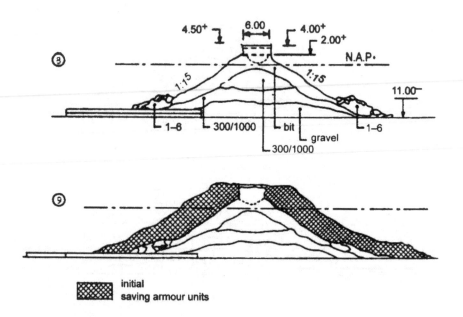

initial
saving armour units

A6.3 Scheveningen

Originally it was proposed to construct the breakwaters for the fishing harbour of Scheveningen following a modified IJmuiden design (Figures 1 through 5).

1. The disadvantages of the IJmuiden design, i.e. the lifting up of the slabs was solved by constructing a fully asphalt grouted superstructure above the level of -1 m., i.e. in the area where grouting could be done dry above water.

2. Model tests, however, showed that the toe at -1 m would not be stable. It was proposed to grout also the toe. This was rejected because grouting below the level of L.W. was not accepted

3. Therefore, the toe was then protected with cubes. This design proved to be relatively uneconomic because both the grouted superstructure and the cubes were designed to withstand the wave forces.

4. Finally this design was selected and constructed. For the deeper parts, the design was adapted to the more severe wave attack by heavier armour units and by a strong berm.

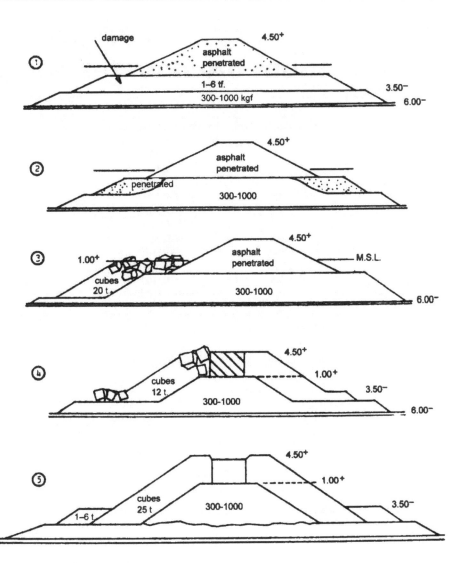

APPENDIX 7
Optimum breakwater design

Wave Height H (m)	Probability of Exceedance (times per annum)
4	1.11
5	$1.58*10^{-1}$
5.2	$8.4*10^{-2}$
5.5	$7.62*10^{-2}$
5.8	$3.8*10^{-2}$
6	$2.47*10^{-2}$
6.5	$7.35*10^{-3}$
7.15	$3.0*10^{-3}$
7.25	$2.63*10^{-3}$
7.8	$9.0*10^{-4}$
7.98	$8.0*10^{-4}$
8.7	$1.5*10^{-4}$

Table A7-1 Long-term wave climate

Actual Wave Height H	Damage in % of armour layer
$H < H_{nd}$	0
$H_{nd} < H < 1.3 H_{nd}$	4
$1.3 H_{nd} < H < 1.45 H_{nd}$	8
$H > 1.45 H_{nd}$	Collapse

Table A7-2 Development of damage

The initial construction cost I of the breakwater is estimated to be:
$ 8620 for the core and $ 1320.H_{nd} for the armour layer.
For design wave heights of 4, 5, 5.5 and 6 m this results in initial construction cost as per Table A7-3.

Design wave height H_{nd}	Initial cost breakwater "C"	Initial cost Amour Layer "A"
(m)	($) per running meter	($) per running meter
4	13900	5280
5	15220	6600
5.5	15900	7280
6	16540	7920

Table A7-3 Initial construction cost per running meter

H_{nd}	$1 < H < 1.3 H_{nd}$ n = 4% damage			$1.3 H_{nd} < H < 1.45 H_{nd}$ n = 8% damage			$H > 1.45 H_{nd}$ Collapse		
	Δp	Δw	$\Delta p.\Delta w$	Δp	Δw	$\Delta p.\Delta w$	Δp	Δw	$\Delta p.\Delta w$
(m)	(1/year)	($)	($/year)	(1/year)	($)	($/year)	(1/year)	($)	($/year)
4	1.02	420	430	$4.6\ 10^{-2}$	860	40	$3.8\ 10^{-2}$	13900	530
5	$1.5\ 10^{-1}$	530	80	$4.7\ 10^{-3}$	1060	5	$2.6\ 10^{-3}$	15220	40
5.5	$7.4\ 10^{-2}$	580	40	$2.2\ 10^{-3}$	1160	-	$8\ 10^{-4}$	15900	10
6	$2.4\ 10^{-2}$	630	15	$7.5\ 10^{-4}$	1260	-	$1.5\ 10^{-4}$	16540	3

Table A7-4 Annual risk for various values of H_{nd} per category of damage level

Note:

Δp $= p_l - p_{l+1}$ probability of occurrence of the wave height in the indicated interval

p_l = probability of exceedance of the wave height at the lower limit of the interval

p_{l+1} = probability of exceedance of the wave height at the upper limit of the interval

Δw = cost of repair of the armour layer (2*n*A) respectively cost of replacement (C)

This leads to the values of average annual risk $s = \Sigma(\Delta p \cdot \Delta w)$ as shown in Table A7-5.

H_{nd}	$s = \Sigma(\Delta p.\Delta w)$		
	Full repair of partial damage	Only repair of serious damage(>8%)	No repair of partial damage
(m)	($ per year)	($ per year)	($ per year)
4	1000	570	530
5	125	45	40
5.5	50	10	10
6	18	3	3

Table A7-5 Average annual maintenance cost for various maintenance strategies

For a lifetime of 100 years, which is a reasonable assumption for a breakwater, capitalisation on an interest rate of 3.33% leads to the figures as given in Table A7-6.

H_{nd}	Capitalised risk S		
	Full repair of partial damage	Only repair of serious damage(>8%)	No repair of partial damage
(m)	($)	($)	($)
4	30000	17100	15900
5	3750	1350	1200
5.5	1500	300	300
6	540	90	90

Table A7-6 Capitalised maintenance cost for various maintenance strategies

It is now a simple exercise to add the initial cost I and the capitalised maintenance cost S as in Table A7-7.

H_{nd}	Total cost $I + S$		
	Full repair of partial damage	Only repair of serious damage(>8%)	No repair of partial damage
(m)	($)	($)	($)
4	43900	31000	29800
5	18970	16570	16420
5.5	17400	**16200**	**16200**
6	**17080**	16630	16630
6.5	17300		

Table 7-7 Total cost for various maintenance strategies

The optimum values are printed boldly.

APPENDIX 8
Construction equipment

From: CUR/RWS-publication no. 169

A8.1 Land-based equipment

A8.1.1 Material in bulk

The land-based equipment must be split in highway and off-highway equipment. The off-highway equipment is designed to work on rough, uneven surfaces, where traditional vehicles can hardly move. In most countries, use of this equipment on public roads is not permitted since it causes excessive damage to the pavement.

If, nevertheless, material must be transported over long distances, there are a few options left:

- Construct special roads or tracks
- Use high capacity highway trucks
- Use existing or special rail connections

The most widely used off-highway equipment is shown in Figure A8-1. It shows a mix of tyre and track based vehicles. Use of tyre-fitted equipment in quarry stone operations leads to excessive wear and tear, although it can not always be avoided. It is often necessary to spread finer material over the larger size stone (with a bulldozer) to create an accessible surface for tyre mounted equipment.

type		capacity (m³)	weight (ton)	wheel load (ton) ground pressure	width (m)
(off higway) dump truck	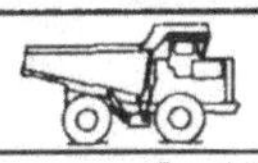	20 - 90	empty: 30 - 110 loaded: 60 - 270	front/rear (ton) empty: 15/15 - 50/60 loaded: 20/40 - 90/180	wheel base 3.7 - 5.7
articulated dump truck		12 - 27	empty: 20 - 40 loaded: 40 - 90	front/rear (ton) empty: 10/10 - 20/20 loaded: 14/26 - 30/60	wheel base 5.7 - 6 8
wheel loader		2.5 - 9	15 - 86		bucket width 2.7 - 4.7
track loader		2.5 - 3	25	60 - 90 kPa	bucket width 2.7
backhoe crane		0.5 - 15	15 - 200	40 - 150 kPa	track gauge 2 - 5
front shovel	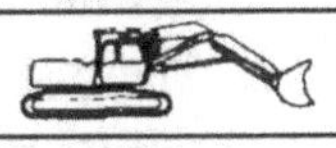	2 - 15	40 - 200	70 - 190 kPa	track gauge 2 - 5
bulldozer		blade width 2.5 - 5 m	10 - 80	50 - 100 kPa	track gauge 2 - 3

Figure A8-1 Review of land-based equipment

A comparison between a highway and an off-highway dumper is given in Figure A8-2.

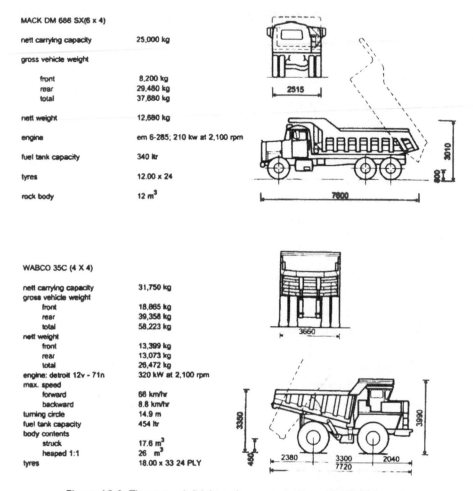

MACK DM 686 SX(6 x 4)

nett carrying capacity	25,000 kg
gross vehicle weight	
front	8,200 kg
rear	29,480 kg
total	37,680 kg
nett weight	12,680 kg
engine	em 6-285; 210 kw at 2,100 rpm
fuel tank capacity	340 ltr
tyres	12.00 x 24
rock body	12 m^3

WABCO 35C (4 X 4)

nett carrying capacity	31,750 kg
gross vehicle weight	
front	18,865 kg
rear	39,358 kg
total	58,223 kg
nett weight	
front	13,399 kg
rear	13,073 kg
total	26,472 kg
engine: detroit 12v - 71n	320 kW at 2,100 rpm
max. speed	
forward	66 km/hr
backward	8.8 km/hr
turning circle	14.9 m
fuel tank capacity	454 ltr
body contents	
struck	17.6 m^3
heaped 1:1	26 m^3
tyres	18.00 x 33 24 PLY

Figure A8-2 Tipper truck (highway) versus dump truck (off-highway)

Use of this heavy equipment requires a considerable space in the quarry, on the road, or on the crest of a breakwater (see Figure A8-3). If space is insufficient to provide two lanes, passing places must be created at a practical distance. Since backing up reduces the speed considerably, also turning places or even turntables must be provided sometimes.

If the construction material can not be placed by direct dumping methods, bulk handling is still possible by using skips or containers. These skips can be loaded at the stockpile and transported to the work front on trucks or trailers. They can also be filled by dump trucks at the work front. At the work front, they are handled by crane and emptied at the spot that was not accessible for the direct dumping procedure.

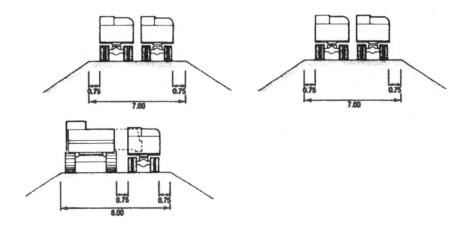

Figure A8-3 Space requirements for heavy vehicles

A8.1.2 Special placement

The larger size quarry stones and the concrete armour units are not placed in bulk but individually by crane. For this purpose, heavy cranes are used as indicated in Figure A8-4. These cranes can either be wheel mounted or tyre mounted. The lifting capacity decreases with the distance. It means that placing armour units near the toe of the structure is the most critical load condition for the crane. It is possible, however, to make use of the buoyancy of the elements by keeping the load just submerged when the crane is reaching out.

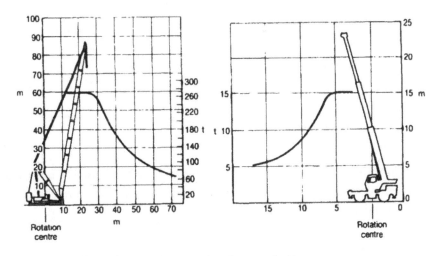

Figure A8-4 Lifting capacity of two typical heavy cranes

The cranes handle individual units with the aid of a clamshell grab or an orange peel grab (see Figure A8-5). A crane as indicated in Figure A8-4 obstructs the work front, for instance at the tipping end of a breakwater under construction. Although the crane is necessary to place armour units and heavier stone, it prevents   direct dumping of core material by dump trucks. It has been indicated that providing a skip or container can solve this, so that the crane can do the bulk handling as well. This method, however reduces the construction speed considerably. Therefore, sometimes the heavy crane is placed on a gantry, so that trucks can pass under the frame of the crane.

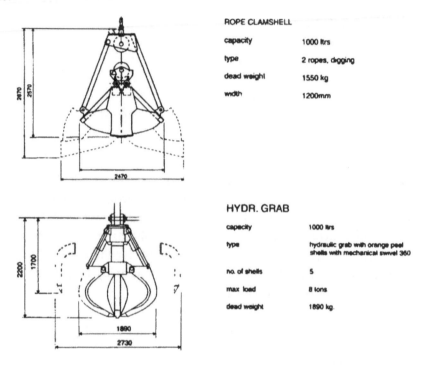

ROPE CLAMSHELL

capacity	1000 ltrs
type	2 ropes, digging
dead weight	1550 kg
width	1200mm

HYDR. GRAB

capacity	1000 ltrs
type	hydraulic grab with orange peel shells with mechanical swivel 360
no. of shells	5
max load	8 tons
dead weight	1890 kg.

Figure A8-5 Grab types

A8.2 Waterborne equipment

A8.2.1 Material in bulk

For the handling of material in bulk by waterborne equipment, we must make a distinction between very fine-grained material that can be handled by dredging equipment and the coarser material.

Pipeline transport
Fine-grained material like sand and gravel can still be handled by dredging equipment using hydraulic transport modes (pumps and pipelines). For transport over limited distances, pipeline transport is very common. In most cases, the pipeline is laid over already reclaimed land, and extended as the material pumped into the water

reaches the required level. Under water natural slopes are formed, generally with a rather gentle slope, so that large volumes of material are required per running meter of dam. The coarser the material, the steeper become the slopes. This leads eventually to much smaller quantities per running meter and thus to a faster forward movement of the work front.

Pipelines can also be laid over water by using floats or pontoons to carry the weight. In this way it is possible to apply sand or gravel in layers over the seabed. In order to prevent uncontrolled spreading of material, the end of the pipeline is often submerged and fitted with a diffuser.

Pipeline transport is not feasible in very rough seas, and not economic over large distances. Pipeline transport is than replaced by transport in barges or seagoing vessels like trailing suction hopper dredges. These vessels discharge the material either through openings in the bottom, or in the case of hopper dredges by pumping the material overboard via the suction pipe. In this way, a similar effect is achieved as in the case of a fixed pipeline with a diffuser.

Floating transport

By floating transport, we mean transport by barge or vessel. Some of these vessels can be used for both, the finer material like sand and gravel and the medium sized material like quarry stone to weights of say, 1 ton.

Amongst the barges and vessels we can distinguish the following types:

- Flat deck barges
- Bottom door barges
- Split barges
- Tilt barges
- Side unloading vessels

All barges can be either push or pull barges or self-propelled vessels, with a varying sophistication of propulsion. The more sophisticated the propulsion system, the better is the accuracy of working in waves and currents.

- Flat deck barges exist in a wide range of carrying capacity, the largest meant for overseas transport of quarry stone in batches up to 30,000 tonnes! Loading and unloading is mostly done by crane or by wheelloader. Smaller size barges have a draught up to 2.5 m, the large seaworthy barges may have a larger draught up to 5 m.
- Bottom door barges are used for fine-grained material and quarry stone as well. Their loaded draught is about 2.5 m. The load is dumped by opening bottom doors. Since the doors are opening all at once, the dumping is very uncontrolled. Care must be taken that the doors are not damaged when they are opened and

may hit the mass of fresh dumped stone. In some cases, the barges are constructed such that the bottom doors do not stick out below the keel of the vessel, so as to prevent damage to the doors. The vessels can not dump to a higher level than MSL –3m. The capacity is in the order of 600 tonnes.

- Split barges are again used for both fine-grained material and medium sized quarry stone. The barges unload by splitting the two halves of the hull in a longitudinal direction. Some barges can maintain a relatively small cleft. By moving the vessel sideways it is possible to dump a curtain of material that covers the seabed like a carpet. The capacity is in the order of 600 to 1000 tonnes.

- Tilt barges are in fact flat top barges that are unloaded by flooding a ballast compartment along the side of the vessel. In this way the vessel is listing so much that the load slides down over the side into the water. The dumping is rather uncontrolled. Tilt barges are gradually replaced by more modern equipment. The capacity is in the order of 600 tonnes.

Side unloading vessels are again flat deck vessels. Unloading is realised by pushing the material overboard by mechanical means. In this way accurate dumping is possible, specifically if the vessel is equipped with a sophisticated propulsion system. The capacity ranges from 600 to 2000 tonnes. The various types of vessels are shown in Figure A8-6.

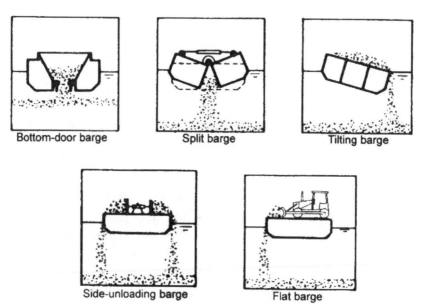

Figure A8-6 Barges for dumping material

A8.2.2 Special placement

Special placement of material with waterborne equipment is defined here as placing with a crane. This is a narrow definition because it has been indicated in § A8.2.1 that pipelines and some barges can also be used for (more or less) accurate placement.

Using a crane to places sometimes heavy material at sea is complicated because of the disturbance by waves and currents.

If the crane is fixed on a pontoon or vessel, proper care must be taken of the anchoring, so that the position can be maintained in spite of the prevailing currents. The action of the waves, however, is more difficult to compensate. There are always relative movements between the crane and the structure, and between the crane and a transport barge if the material is supplied from a vessel different from the crane pontoon. These relative movements make the handling (especially of heavy) material complicated and sensitive for weather delays. Therefore, it is often tried to reduce the relative movements between two vessels by combining the transport barge with the crane pontoon (Figure A8-7).

Figure A8-7 Example of combined transport and crane vessel

Another solution is to make the crane platform completely independent of the water motion. This can be done by methods developed in the offshore industry: the self-elevating platform. Such platforms have been used during the construction of the IJmuiden breakwater. They are floating pontoons during transport. The spuds or legs are lowered when the pontoon arrives at the right position. By hydraulic jacks, the pontoon is then lifted out of the water along the spuds and forms eventually a stable working platform for the crane.

A8.3 Tolerances

When considering construction equipment, it is impossible to neglect the accuracy of placing material under varying conditions. It makes a tremendous difference whether material is placed in the dry, where it is possible to visually inspect and control the operations, or the material is placed under water, where one must rely on instruments for observation and control. Fortunately, position fixing and other measuring instruments have improved over the last years, so that the difference between working above and under water has been reduced. Still, however, there is a considerable difference. As a rough indication, one is referred to Table A8-1.

Type of Material	Place of application			
	Above water	Less than 5 m below LW	5 to 15 m below LW	More than 15 m below LW
Gravel (on or from land)	0.05m	0.10 to 0.15m	0.10 to 0.15m	0.10 to 0.15m
Gravel waterborne (standard)	n.a.	0.3 to 0.5m	0.3 to 0.5m	0.3 to 0.5m
Gravel Waterborne (special)	n.a.	0.1 to 0.2m	0.1 to 0.2m	0.1 to 0.2m
Quarry stone (bulk) $W < 300kg$	0.25 to 0.5 D_{50} absolute ±0.2m	+0.5 to – 0.3m	+0.5 to – 0.3m	+0.5 to – 0.3m
Quarry stone (bulk) $W > 300$ kg	+0.4 to–0.2m	+0.8 to – 0.3m	+0.8 to – 0.3m	+0.8 to – 0.3m
Quarry stone (individual) $W > 300$ kg	± 0.3 D_{50}	± 0.5 D_{50}	± 0.5 D_{50}	± 0.5 D_{50}
Armour layer Design profile versus actual profile	+0.35 to–0.25m	+0.6m to –0.4m	+0.6m to –0.4m	+0.6m to –0.4m

Table A8-1   Vertical placing tolerances

A8.4 Moving on impassable sites

In closure design, acces on the worksite for the heavy equipment is sometimes difficult because of the poor conditions of the road. Bearing capacity is not the only determining factor that makes a site passable. This is clear for everyone who tried to drive with a normal car on a sandy beach. On the dry beach a four-wheel driven car with very low gear is needed. In non-granular soil the difference shows for instance in heavy consolidated laterite clay, which is a good base for driving when dry but absolutely impassable after some rain. Since damming activities generally take place in deltaic areas in which sand, sandy clay, clay, silt and peat frequently occur, moving across the site with all sorts of equipment and vehicles under various weather conditions may be a problem. Besides, driving on top of a quarry stone dam will be a problem if the stone is coarser than about 0.5 m in diameter.

A distinction has to be made between the heavy operational equipment like dump trucks, cranes, hydraulic excavators, bulldozers and the exploration-equipment used for soil-investigation, positioning finding and measuring. This last group generally consists of lightweight equipment, which is used frequently however, in original terrain conditions. Very special vehicles have been developed for these specific circumstances, although usually they are not readily available and quite expensive sometimes. Hovercraft or amphirol are examples of these vehicles. The hovercraft supports itself by an air-bell confined within rubber skirts, maintained underneath the vehicle by continuous pumping. It can be used on land, on mud flats and on water. The amphirol is supported by two horizontal cylinders provided with opposite winded archimedean screws. Motion is obtained by rotating the cylinders. Steering is realised by the horizontal orientation of the cylinders and the rotation direction. The amphirol moves well on soft grounds, mud flats and on water. These vehicles are very useful for small equipment and special assignments.

For the heavy equipment the right system of movement has to be selected. If moving speed is important or if driving on site and on public roads has to be combined, the only solution is to use pneumatic tires. Variables with tires are the dimension of the tire, the pressure inside, the number of tires and how many are driven and can be steered.

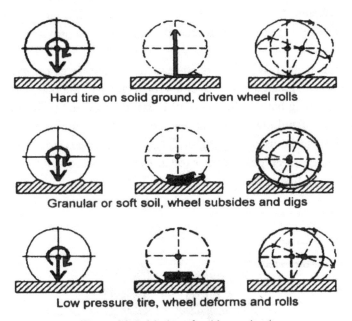

Hard tire on solid ground, driven wheel rolls

Granular or soft soil, wheel subsides and digs

Low pressure tire, wheel deforms and rolls

Figure A8-8 Motion of a driven wheel

A wheel has to transfer the forces onto the ground and its motion depends on the reaction forces of the ground. A non-driven wheel transfers a vertical load and a lateral force, while a driven wheel has the load and a rotational momentum. Usually the pressure in a truck's tire is quite high (300 to 500 kN/m^2) and the tire will hardly deform. The reaction force of the road surface or the ground is also high. The maximum lateral force is the friction coefficient times the support force. Deep profiles on the outside of the tire may give it sufficient grip in loose ground and then the shear force in the soil determines the friction coefficient. If the friction is exceeded, the wheel slips. This is the case in the above-mentioned example of wet laterite clay.

If the soil cannot stand the point load of the wheel, the tire will sink into the ground until the load is spread over sufficient bearing area. Consequently, when turning, the wheel has to move up against a slope, by which the support area shifts to the front side of the wheel. A much larger momentum is required to achieve driving and quite soon the friction reaches its maximum. Then the wheel starts turning without lateral movement and it digs itself further down into the soil. This is the case in the dry-beach sand example. A momentum exerted on all wheels (four-wheel drive) and very slow turning will improve the situation.

With a low air pressure in the tire, it is able to deform which gives it a larger support area without subsiding into the soil. Therefore, vehicles with special low-pressure tires (e.g. wheel-loaders) can move much easier but their motion is very springy. For heavy transport that is generally not allowed and then increasing the number of tires is the only solution. For exceptional heavy transport a large number of axles, each with a set of wheels is used. In that case all the wheels are provided with integrated steering capacity.

The next step is to take crawlers instead of wheels (e.g. crawler cranes and bulldozers). They spread the load over a very wide area. Support during driving is not very determining but the shifting of centre of the load and sufficient lateral force during pushing or pulling are important. For different purposes, wide or narrow crawler-tracks can be used. Driving on large quarry stones, for instance, requires narrow tracks. The high point loads exerted by the spanned stones have to be kept close to the centre-line of the track in order not to damage it.

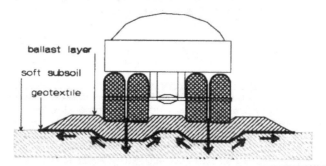

Figure A8-9 Temporary road on soft subsoil

Another solution is to prepare a number of roadways across the area along which surfacing is made to allow vehicles to move along. Those roads are of a temporary nature and should be removable and relatively inexpensive. Two systems are frequently used. One is to pave with large, steel-framed wooden slabs or steel planking. Draglines and cranes can position these themselves and then drive on them. The other system is to provide a road-base direct on the existing soft ground. A geotextile sheet is unrolled and ballasted by a layer of sand or gravel. As soon as the wheels of a truck move on to the ballast layer, the road-base underneath is pressed down into the soil. Due to the sag-bend in the sheet part of the vertical load is transferred sideways into tension in the textile. This horizontal force is taken by friction in the soil for which sufficient length and ballast has to be available. If the trail of the wheel into the ballast bed is too deep, direct sheer of the tire on the sheet will occur and the sheet will tear. Therefore the ballast layer needs to be quite thick and trucks have to prevent deep ruts by regularly shifting tracks. The geotextile has three functions. It separates the subsoil and the ballast material; it transfers the surcharge to the subsoil and spreads the load via tension and elongation to the sides. Besides, due to the separation between ballast and subgrade, the removal of the temporary road is simpler and the ballast material can be re-used easily. A typical event occurs with bulldozers driving on a hydraulic fill. The freshly settled sand still has a large pore volume and is saturated with water. The vibrations of the dozer and its weight (although spread over a large area), will fluidise the top layer of the sand. Generally, this remains within such limits that the bulldozer can continue driving while the sand resettles in a denser grain-structure. Bulldozing leads to densification. However, sometimes the liquefaction covers a too large area and the bulldozer may sink down into the fluidised sand. A fill of fine silty sand (smaller than about 120 micron) is very sensitive for this and hardly passable. Then a long time is needed to await evaporation of the pore water before driving is possible.

APPENDIX 9
Closures using ancient willow mattresses

In the old days, thick willow mattresses were the only suitable structures for this purpose and sand tightness was obtained by the thick layer of willow branches. In specific circumstances, like the sinking of a thick closure sill, the classical structure is still an option. In the classical bottom protection, the most important material used was branches cut from willow trees. These branches (osiers) were bound together in two ways.

Figure A9-1 Components of an ancient mattress

First, branches were bound together as a bundle making a sheaf-shape (called "rijsbos"), with a diameter of about 0.3 metres. Secondly, long strings of 0.1 to 0.2 m diameter and lengths of 50 m or more were composed. These strings (called "wiepen") were used to make a rectangular grating with a grid size of 3 feet (0.9 m).

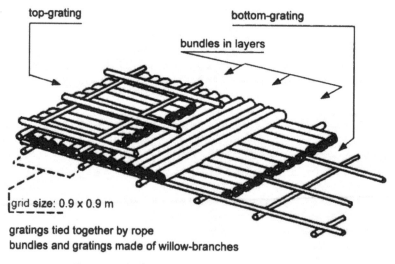

Figure A9-2 Composition of an ancient mattress

These strings and sheaves were used for several purposes but the main tool for closure works, made with these components, was a mattress (called "zinkstuk"). Such a mattress consisted of an upper and a lower grating with several layers of sheaves in between. It was made on a strip of beach (called "zate") during the low water period, came afloat during high water and was then launched and towed to the closure site.

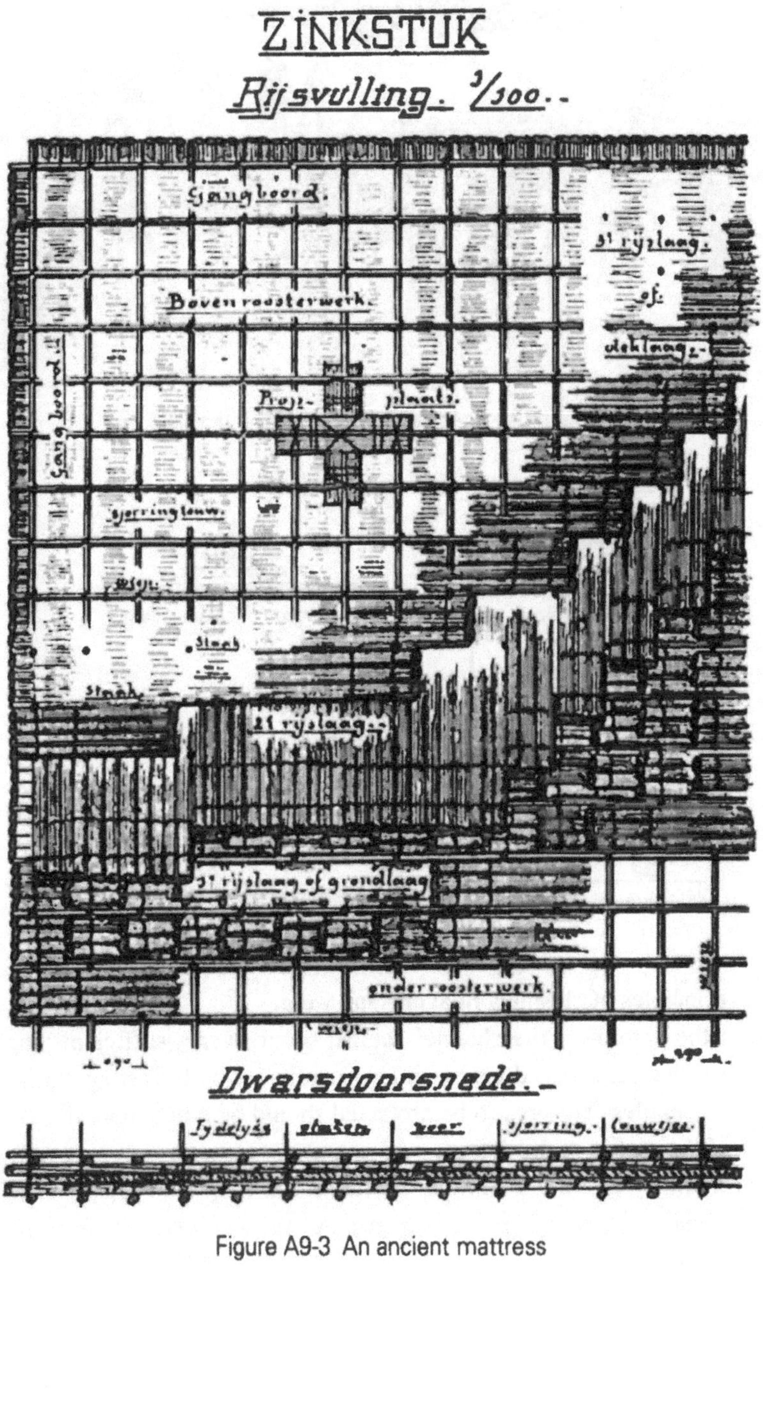

Figure A9-3 An ancient mattress

Figure A9-4 An ancient mattress in the field

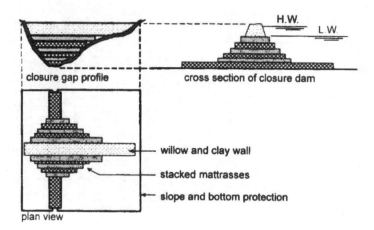

Figure A9-5 Closing by mattress sinking

The basic principles of closing a tidal channel were:
1. Protect the bottom of the channel against scour over a sufficient length in the flow direction. This was done by sinking mattresses by ballasting them with clay-lumps or cobbles. The area to be protected should be much more than the bottom width of the profile of the dam.
2. Raise a sill by sinking a number of thick mattresses on top of each other, a process called "sinking up" (opzinken). As the mattresses had an interwoven

structure with much hollow space, they were stable but rather permeable and very compressible. Every mattress added weight onto the sill and compressed it.

3. The sinking was executed during slack/still water and consequently could not be done to a higher level than the tide allowed. This was usually slightly above low-water level. At the seaward side, the water level would then fall below sill level for small periods.

4. Next, a dam had to be constructed over the sill, progressing from both ends towards the centre. Depending on the prevailing circumstances, this was done by depositing clay into a sort of clay wall, or by piling up these willow sheaves, knitting and fixing them together in a wall shape, ballasted with clay. (This willow-and-clay wall can be seen as an ancient way of "terre armee" and in Dutch is called "rijspakwerk".

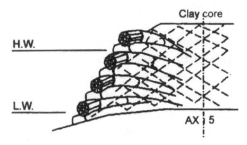

Figure A9-6 "Rijspakwerk"

When the closure had been completed, the actual dam had still to be made. The initial profile was too weak, still very compressible and risk of piping via the willow branches was high. Generally, the initial profile was made part of the definite profile by adding a clay profile against it.

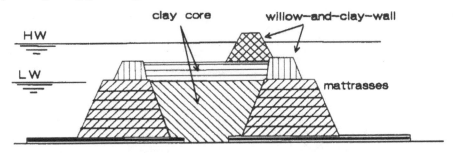

Figure A9-7 Mattress-sinking and clay core.

Sometimes, the risk of piping was considered too high during the construction of the closure profile, because of slow progress or of high head loss over the dam. If so, a different design was used.

The bottom and slope protection, was then executed in two parts leaving, an unprotected strip in the centre line of the profile. Then sills (as in 2) were made on each part leaving a dip in between (see Figure A9-7). This dip was filled by clay only, which, since there are no willow branches running from one side to the other, made an impermeable core in the centre of the dam. This method was much safer, but required much more material and took a longer time to construct. Longer time meant longer exposure to high flow conditions and more risk of adverse weather.

In addition to operation management and logistics, the skills required for these operations were the craftsmanship needed to make strong willow mattresses, manoeuvre them to the site under high flow conditions by rowing or sailing, position them on the exact location by using wires and anchors, and to sink and ballast them in the limited time of slack water. The higher the sill, the more difficult this was. However, a higher sill made the last step, the construction of the willow and clay cross-wall, easier.

A river with mainly permanent flow does not have the advantage of slack-water periods. The method of closing was therefore different, although the materials were identical. A bottom protection had to be sunk, not at slack water but during flow conditions, although of course in a period of low river discharge. Sinking-up a sill is more complex due to the lack of variation is water level and flow. Therefore, the sill could not be made up to a high level and that made the construction of a willow-and-clay wall impossible, so another rather complex structure called "baardwerk" was made.

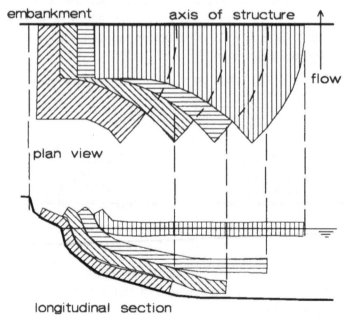

Figure A9-8 "Baardwerk"

In general terms, the technique consisted of the construction of a floating mattress, in situ, proceeding from the shores towards the centre. The grating was not rectangular and diagonal strengthening was provided to withstand the sideways shifting forces caused by the flow. Every time a part, (called beard because of its shape), was ready, it was ballasted and sunk, covering its predecessor. It was then sloping down from the water level at its root end, to the sill level at its outer end.

APPENDIX 10
Example of the determination
of a design storm

A10.1 Statistics of individual observations

In this computational example we will look for the design storm in front of the coast of the Netherlands. We will try to find a storm with an exceedance frequency of 1/225 per year. This storm yields 20% probability of failure in a lifetime of 50 years as has been explained in Chapter 3. It is stressed again that the use of these values does not constitute a recommendation on the part of the authors.

Figure A10-1 The Noordwijk measuring station

As described in Section 5.3.3, we start from a series of wave observations with a fixed interval, made at "Meetpost Noordwijk", off the Dutch coast (Figure A10-1). Each observation (e.g. with a duration of 20 minutes) results in a given value of the H_s. When an observation programme is continued during a long time, a time series is created of H_s-values. This time series is the basis of the statistical operations explained in this section.

In the example case we use, we have 20 years of data available from observations off the Dutch coast[1]. Every half hour a wave observation is made, but in the file used in

---

[1] The dataset is provided by the Netherlands Ministry of Public Works, the location Meetpost Noordwijk is at a waterdepth of approximately 18 m. Data from this, and other Dutch sites, can be downloaded from http://www.golfklimaat.nl

this example, these data are reduced to one observation every three hours. A one-month sample is plotted in Figure A10-2.

In Table A10-1 the clustered data are presented. The number of observations of each wave height bin is given both per bin as well as cumulatively. The probability of any wave height H'_s being equal or less than a specific wave height H_s is defined as:

$$P=P(H'_s \leq H_s) \tag{A10.1}$$

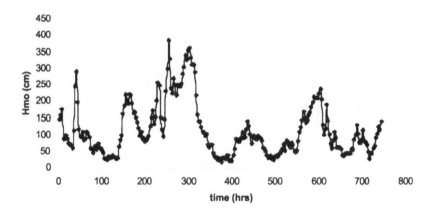

Figure A10-2 Data from Meetpost Noordwijk for one month (January 1979)

A probability of exceedance that H'_s is greater than a specific wave height H_s may also be defined as:

$$Q=Q(H'_s > H_s)=1-P \tag{A10.2}$$

Because a log relation is assumed, in the table also the value of $-\ln(Q)$ is given. Analysis shows that the correlation between H_s and $-\ln(Q)$ is 98.8 %. Plotting these data results in Figure A10-3. Although the correlation coefficient is quite large, it is clear that a log-distribution is not fully correct. The data are not on a straight line. For a small extrapolation this is not really a problem. The values from the regression line can be used to calculate the probability of exceedance of a given H_s. For example, a value of $H_s = 6$ m gives:

$$-\ln(Q)=1.51 \cdot 6 - 0.5 = 8.5$$
$$Q=2.035 \cdot 10^{-4} = 0.0002$$

Wave height class Hs (cm)		Number of observations		P	Q	−ln(Q)
		per bin	cumulative			
0	25	35	35	0.000599	0.999401	0.000599
25	50	8260	8295	0.141940	0.858060	0.153082
50	75	11424	19719	0.337423	0.662577	0.411618
75	100	10004	29723	0.508607	0.491393	0.710511
100	125	7649	37372	0.639493	0.360507	1.020245
125	150	5563	42935	0.734685	0.265315	1.326838
150	175	4389	47324	0.809788	0.190212	1.659615
175	200	3167	50491	0.863980	0.136020	1.994954
200	225	2360	52851	0.904363	0.095637	2.347200
225	250	1671	54522	0.932957	0.067043	2.702419
250	275	1234	55756	0.954073	0.045927	3.080692
275	300	851	56607	0.968634	0.031366	3.462047
300	325	556	57163	0.978149	0.021851	3.823487
325	350	392	57555	0.984856	0.015144	4.190168
350	375	276	57831	0.989579	0.010421	4.563938
375	400	206	58037	0.993104	0.006896	4.976819
400	425	136	58173	0.995431	0.004569	5.388507
425	450	82	58255	0.996834	0.003166	5.755400
450	475	66	58321	0.997964	0.002036	6.196632
475	500	38	58359	0.998614	0.001386	6.581307
500	525	30	58389	0.999127	0.000873	7.043930
525	550	20	58409	0.999470	0.000530	7.541769
550	575	22	58431	0.999846	0.000154	8.778531
575	600	9	58440	1.000000	0.000000	
Total		**58440**				

Table A10-1: Wave data from Meetpost Noordwijk

all data Noordwijk

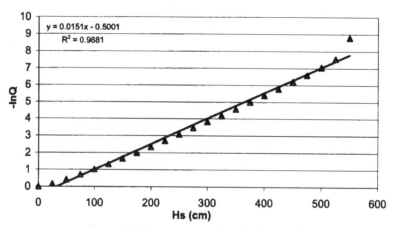

Figure A10-3 Cumulative data for Noordwijk

This means that a wave height $H_s > 6$ m will occur during 0.02 % of the time.

This result can also be obtained in a more direct manner. Plotting the values of Q on log-paper gives Figure A10-4. In this graph the exceedance of an observation larger than H_s' is given. The advantage of this plot is that the exceedance values can be read from the graph. (The statistical analysis and the plotting of data can easily be carried out using a spreadsheet software package. In this case, MS Excel has been used. It is beyond the scope of this book to outline the settings of the software). It is evident that the upper boundaries of the wave height bin should be used on the wave height axis (in this example 86% of the waves is higher than 50 cm, 99% is higher than 25 cm and 100% is higher than 0 cm).

Individual observations

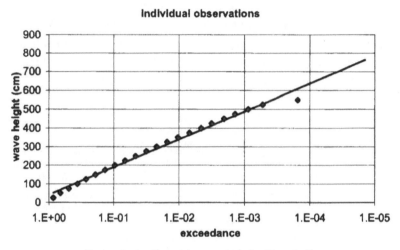

Figure A10-4 Exceedance graph for Noordwijk

However, for design purposes, these values have no real meaning. It means that, if one goes at a random moment to the sea, there is at that given moment a probability of 0.0002 (i.e. 0.02%) of observing a H_s of 6 m or more. But this has nothing to do with a design storm with a H_{ss} of 6 m.

This form of statistics can be applied for the assessment of workability. In fact the above example means that 0.02 % of the time a wave condition can be met in which H_s is larger than 6 m. Suppose that for a port in this area pilot service has to be suspended when $H_s > 3.5$ m, it can easily be read that during 1% of the time there is no pilot service.

For the design of structures, we need the exceedance of a design storm and not the exceedance of an individual wave condition.

A10.2 The Peak over Threshold method

In order to transform these individual observations into storms, we may use the fact that sequential wave height observations are not random (see also Figure A10.2).

When we measure at 12:00 hrs a H_s of 4 m, it is very unlikely that we measure at 15:00 hrs a H_s of 0.4 m. Usually the observations of 12:00 hrs and 15:00 hrs will belong to the same storm. So we will define a certain threshold, e.g. $H_t = 1.5$ m. And we will look in our record when the wave height exceeds 1.5m. The threshold-value of $H_t = 1.5$ m is arbitrarily selected. Later we will investigate the sensitivity of this choice. The month of the record in Figure A10-2 (Jan 1979) shows 9 storms, providing 9 data points.

The reason for introducing a threshold is to avoid that small variations in wave height during long, calm periods have significant influence on the final result. Basically one should assume the threshold as high as possible, as long as the base for statistics contains sufficient data for a reliable analysis. Later it will be shown that the final answer is not very sensitive to the choice of the threshold value.

When we process the whole data set of 20 years, we find 1746 storms in total with a H_{ss} higher than 1.5 m. The 1746 storms are classified in wave height bins, according to the maximum $H_{ss\,of}$ each storm which results in Table A10-2.

Wave height class H_{ss} (cm)	Number of storms per bin	cum.	P	Q	ln(Q)	−ln(H)	Qs	α=1.24 G	W
1.50 1.75	384	384	0.21993	0.78007	4.22098	-0.56	68.10	-0.415046	0.32522
1.75 2.00	381	765	0.43814	0.56186	3.89284	-0.69	49.05	0.192121	0.64136
2.00 2.25	266	1031	0.59049	0.40951	3.57655	-0.81	35.75	0.640938	0.91261
2.25 2.50	157	1188	0.68041	0.31959	3.32863	-0.92	27.90	0.954366	1.11202
2.50 2.75	148	1336	0.76518	0.23482	3.02042	-1.01	20.50	1.318085	1.34858
2.75 3.00	111	1447	0.82875	0.17125	2.70471	-1.1	14.95	1.672191	1.58094
3.00 3.25	81	1528	0.87514	0.12486	2.38876	-1.18	10.90	2.014645	1.80552
3.25 3.50	63	1591	0.91123	0.08877	2.04769	-1.25	7.75	2.375535	2.04065
3.50 3.75	31	1622	0.92898	0.07102	1.82455	-1.32	6.20	2.608194	2.19099
3.75 4.00	32	1654	0.94731	0.05269	1.52606	-1.39	4.60	2.916351	2.38832
4.00 4.25	23	1677	0.96048	0.03952	1.23837	-1.45	3.45	3.210883	2.57486
4.25 4.50	11	1688	0.96678	0.03322	1.06471	-1.5	2.90	3.387796	2.68590
4.50 4.75	20	1708	0.97824	0.02176	0.64185	-1.56	1.90	3.816515	2.95184
4.75 5.00	9	1717	0.98339	0.01661	0.37156	-1.61	1.45	4.089424	3.11883
5.00 5.25	7	1724	0.98740	0.01260	0.09531	-1.66	1.10	4.367707	3.28731
5.25 5.50	9	1733	0.99255	0.00745	-0.43078	-1.7	0.65	4.896399	3.60263
5.50 5.75	8	1741	0.99714	0.00286	-1.38629	-1.75	0.25	5.854211	4.15922
5.75 6.00	5	1746	1	0		-1.79	0.00		
	1746								

Table A10-2: Data from Noordwijk using PoT of 1.5 m

Using the same data, one can also determine the wave steepness. Without presenting the analysis, we found when looking at only those storms with a $H_T \geq 5$ m, a total of 29 storms in 20 years. The average duration of these heavy storms is 6.6 hrs, and the average wave steepness in these storms is 5.8 %, with a standard deviation of 0.6%.

These figures will be used in Section A10.4.3, where the probabilistic approach will be discussed.

The exponential distribution

Like was done before in equations (A10.1) and (A10.2) for H_s, the values of P and Q are computed for H_{ss}:

$$P = P(H'_{ss} \leq H_{ss}) \tag{A10.3}$$

$$Q = Q\left(H'_{ss} > H_{ss}\right) = 1 - P \tag{A10.4}$$

Plotting the values of Q logarithmically results in Figure A10-5, which gives the probability of exceedance of a storm. On can read this graph as: Given there is a storm (according to our definition $H_s > 1.5$ m), then the probability that that storm has an H_{ss} of more than 4 m is 0.05 (or 5%). [$4 = 0.78 \ln(x) + 1.63$]

threshold 1.5 m

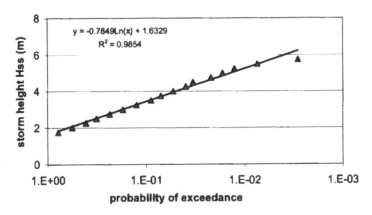

Figure A10-5 Wave exceedance using a threshold

Still we do not know the probability of a storm that occurs with a probability of for example 1/225 per year. In the analysis done so far we looked only at the probability of a single storm. We still need to transform our information to a probability per year. This can only be done when we know the number of storms per year. This number is known, since we have 1746 storms in 20 years, the average number of storms per year $N_s = 1746/20 = 87.3$. When we have 87.3 storms per year, a storm with a probability of exceedance of 1/87.3 has to be the "once-per-year" storm (in correct statistical terms: the storm with a probability of exceedance of once per year).

For transformation of the general probability of exceedance (Q) to the probability of exceedance of a storm in a year (Q_s), we multiply Q with the number of storms in a year.

Thus:

$$Q_s = N_s Q$$

The values of Q_s are also given in Table A10-2. Values of $Q_s > 1$ should not be called a "probability", because probabilities cannot be larger than 1. Physically however, these values do have a meaning. They represent the expected number of storms in a year.

The values of Q_s can again be plotted on log-paper. This results in Figure A10-6. In fact, Figure A10-6 is the design graph to be used. So we can calculate our 1/225 per year storm with:

$$H_{ss} = -0.785 \ln(Q_s)$$

$$= -0.785 \ln\left(\frac{1}{225}\right) + 5.141$$

$$= 9.39 \text{m}$$

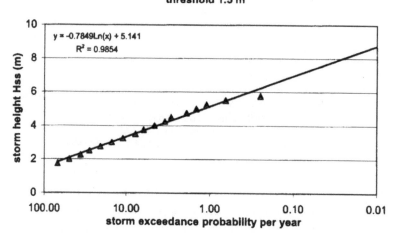

threshold 1.5 m

Figure A10-6 Storm exceedance using a threshold

The Gumbel distribution

A more detailed inspection of Figures A10-4 and A10-5 reveals that, like in Figures A10-2 and A10-3, the points are not exactly on a straight line. This is caused by the fact that a simple exponential relation has been used. Because we deal with extreme values, an extreme value distribution like Gumbel or Weibull may result in predictions that are more reliable.

The Gumbel distribution is given by:

$$P = \exp\left[-\exp\left(-\frac{H_{ss} - \gamma}{\beta}\right)\right] \tag{A10.5}$$

The coefficients β and γ can be found by regression analysis on the data. In order to do so, one has to reduce the equation of the Gumbel distribution to a linear equation of the type $y = Ax + B$. After that, standard regression will provide the values A and B, and subsequently values of β and γ. Taking two times the log, the Gumbel distribution can be reduced to:

$$\ln P = -\exp\left(-\frac{H_{ss} - \gamma}{\beta}\right)$$

$$\underbrace{-\ln(-\ln P)}_{G} = \frac{H_{ss} - \gamma}{\beta} = \underbrace{\frac{1}{\beta} H_{ss} - \frac{\gamma}{\beta}}_{AH_{ss} - B} \tag{A10.6}$$

The left-hand side of equation (A10.6) we call the reduced variate G:

$$G = -\ln\left(\ln\frac{1}{P}\right)$$

The values of G can be calculated for all P-values in Table A10-2. The values of G and H_{ss} are plotted in Figure A10-7. A linear regression leads to:

$$G = AH_{ss} + B$$

For the given dataset $A = 1.365$ and $B = -2.535$

So: $\beta = 1/A = 1/1.365 = 0.733$

$$\gamma = -\beta B = -0.733 \cdot -2.535 = 1.877$$

Like with the exponential distribution, the analysis given so far results in an absolute exceedance of a storm, not in a probability per year. As given before:

$$Q_s = N_s Q$$

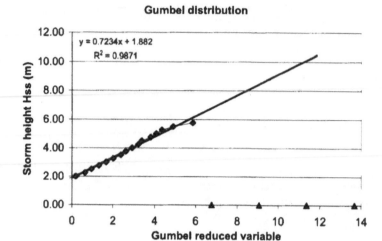

Figure A10-7 The Gumbel exceedance graph

From the Gumbel distribution it follows:

$$H_{ss} = \gamma - \beta \ln\left(\ln\frac{1}{P}\right)$$

$$= \gamma - \beta \ln\left(\ln\frac{1}{1-Q}\right)$$

$$= \gamma - \beta \ln\left(\ln\frac{1}{1-Q_s/N_s}\right) \qquad (A10.7)$$

$$= \gamma - \beta \ln\left(\ln\frac{N_s}{N_s - Q_s}\right)$$

So:

$$H_{s1/225} = 1.88 - 0.73 \ln\left(\ln\frac{87.3}{87.3 - 1/225}\right) = 9.10$$

A disadvantage of this approach is that in Figure A10-7 on the horizontal axis the reduced variable G is plotted, and not the probability of exceedance Q_s. Of course this can be transformed by:

$$G = -\ln\left(\ln\frac{N_s}{N_s - Q_s}\right) \qquad (A10.8)$$

So, for this given example (with $N_s = 87.3$ storms/year) this results in:

Q_t	G
1/10	6.77
1/100	9.07
1/1000	11.38
1/10000	13.68

Table A10-3

These values can be inserted in Figure A10-7. One should realize that when the number of storms per year (N_s) changes, also the units on the converted horizontal axis will have to be changed.

The Weibull distribution

Instead of a Gumbel distribution, we also may try a Weibull distribution. The Weibull distribution is given by

$$Q = \exp\left[-\left\{\frac{H_{ss} - \gamma}{\beta}\right\}^\alpha\right]$$

$$(A10.9)$$

Also here, in order to find the values α, β and γ from regression analysis, we have to reduce the equation.

$$-\ln Q = \left[\frac{H_{ss} - \gamma}{\beta}\right]^\alpha$$

$$(-\ln Q)^{\frac{1}{\alpha}} = \frac{H_{ss} - \gamma}{\beta}$$

$$\underbrace{(-\ln Q)^{\frac{1}{\alpha}}}_{W} = \underbrace{\frac{1}{\beta} H_{ss} - \frac{\gamma}{\beta}}_{A H_{ss} - B}$$

$$(A10.10)$$

So the reduced Weibull variate is

$$W = -(\ln Q)^{\frac{1}{\alpha}}$$

$$(A10.11)$$

The Weibull distribution has three variables (α, β and γ). Linear regression will provide only two constants, A and B (and subsequently β and γ). So the determination of the third coefficient (α) will require some trial and error. Assuming different values of α will change the curvature of the points in the plot of W vs. H_{ss}. The calculation can be carried out quite easily in a spreadsheet. Changing the α

immediately provides a new graph and a new value of the correlation coefficient. The value of α that provides the straightest line and the highest correlation coefficient is the best value for α. In our example this proves to be $\alpha = 1.24$, resulting in $\beta = 1.17$ and $\gamma = 1.22$. See also Table A10-2 and Figure A10-8. (Note: β and γ are not the values given in the regression line in Figure A10-8, because in this figure H_{ss} is plotted on the vertical axis; if one plots H_{ss} on the horizontal and W on the vertical, on gets the values of β and γ also directly).

Weibull distribution

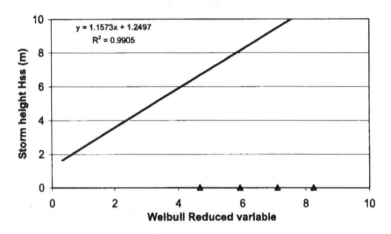

Figure A10-8 The Weibull exceedance graph

Also here the exceedance has to be transformed to a probability per year. From the Weibull distribution it follows:

$$H_{ss} = \gamma + \beta \{-\ln Q\}^{\frac{1}{\alpha}} \tag{A10.12}$$

Using $Q_s = H_s Q$ gives

$$H_{ss} = \gamma + \beta \left\{ -\ln\left(\frac{Q_s}{N_s}\right) \right\}^{\frac{1}{\alpha}} \tag{A10.13}$$

So:

$$H_{ss\frac{1}{225}} = 1.22 + 1.17 \left\{ -\ln\left(\frac{\frac{1}{225}}{87.3}\right) \right\}^{\frac{1}{1.24}}$$

$$= 1.22 + 1.17 \cdot 6.34$$

$$= 8.64\ m$$

Again, a manual plot has to be made of H_{ss} as a function of W (see also Figure A10-8). This can be transformed using

$$W = -\ln\left(\frac{Q_s}{N_s}\right)^{\frac{1}{\alpha}}$$

(A10.14)

So, for the given example (with $N_s = 87.3$ and $\alpha = 1.24$), Q_s and W are:

Q_s	W
1/10	4.68
1/100	5.92
1/1000	7.11
1/10000	8.24

Table A10-4

Summary

The coefficient γ theoretically has the meaning of the threshold value H_T as used in the PoT-analysis. So as a check, we can compare these values:

H_T threshold value	1.50	2.00	2.50	3.00	3.50	4.00
γ Exponential	1.63	1.65	1.65	1.65	1.65	1.65
γ Gumbel	1.86	2.37	2.84	3.44	3.98	4.52
γ Weibull	1.23	1.77	2.29	2.93	3.51	4.10

Table A10-5

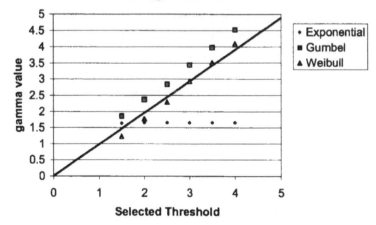

Figure A10-9: Relation between γ and the selected threshold

In this respect it is clear that the Gumbel and the Weibull distribution give a much better result. This becomes more relevant when the number of available storms in the database is lower.

Summarizing we find the following H_{ss} values for our 1/225 design storm for different threshold values:

H_T	1.50	2.00	2.50	3.00	3.50	4.00
N_s	87.3	59.6	38.9	19.4	10.5	5.3
$H_{ss\ 1/225}$ Exponential	9.39	9.30	9.00	8.85	8.58	8.31
$H_{ss\ 1/225}$ Gumbel	9.10	8.95	8.66	8.46	8.11	7.81
$H_{ss\ 1/225}$ Weibull	8.64	8.55	8.32	8.20	7.97	7.77

Table A10-6

In this comparison for the Weibull distribution a value $\alpha = 1.24$ has continuously been used. As explained before, α has to be determined by optimizing the correlation coefficient and visually on the fact that the points are on a straight line as much as possible. For easy reference we have continued to use $\alpha = 1.24$ as derived for a H_T of 1.5 m. For large values of H_T the number of wave height bins available for use in the analysis becomes lower. Consequently, the calculation becomes less reliable.

A10.3 What to do if only random data are available?

The above PoT-analysis can *only* be carried out in case a *sequential* database with observations is available. In case only the grouped statistics of observations are available (like Table A10-1) a PoT-analysis is not possible. The same problem occurs when data are collected by random observations, like the visual observations from the Global Wave Statistics[2]. Often the number of observations in each wave class bin is normalized in such a way that the total is 100% or 1000 ‰.

A typical sample of such a table is given in Table A10-7 (In fact the input data are identical to Table A10-1). In such a case, one can determine the exceedance frequency of a single H_s, as indicated in the section before. Because the data are not sequential, and because the number of storms per year (N_s) is not known, the PoT-analysis is not possible. Some of the data in the table may come from the same storm.

---

[2] The Global Wave Statistics [HOGBEN & LUMB, British Maritime Techonology, 1986] is a book containing visual observations collected by ships at given times, but random locations (i.e. the position of the ship at that moment). This makes that all observations in a given area are completely uncorrelated.

Hs-bin		Class Obs	Cumul	P	Q	LN(Q)	12 hrs s/y	storm duration ln()	α = 0.8 ln(H)	W
0	25	35	35	0.00060	0.99940	0.00	729.56	6.592	3.219	-10.564
25	50	8260	8295	0.14194	0.85806	0.15	626.38	6.440	3.912	-10.259
50	75	11424	19719	0.33742	0.66258	0.41	483.68	6.181	4.317	-9.747
75	100	10004	29723	0.50861	0.49139	0.71	358.72	5.883	4.605	-9.161
100	125	7649	37372	0.63949	0.36051	1.02	263.17	5.573	4.828	-8.562
125	150	5563	42935	0.73469	0.26531	1.33	193.68	5.266	5.011	-7.978
150	175	4389	47324	0.80979	0.19021	1.66	138.85	4.933	5.165	-7.353
175	200	3167	50491	0.86398	0.13602	1.99	99.29	4.598	5.298	-6.733
200	225	2360	52851	0.90436	0.09564	2.35	69.81	4.246	5.416	-6.095
225	250	1671	54522	0.93296	0.06704	2.70	48.94	3.891	5.521	-5.464
250	275	1234	55756	0.95407	0.04593	3.08	33.53	3.512	5.617	-4.808
275	300	851	56607	0.96863	0.03137	3.46	22.90	3.131	5.704	-4.165
300	325	556	57163	0.97815	0.02185	3.82	15.95	2.770	5.784	-3.573
325	350	392	57555	0.98486	0.01514	4.19	11.05	2.403	5.858	-2.992
350	375	276	57831	0.98958	0.01042	4.56	7.61	2.029	5.927	-2.422
375	400	206	58037	0.99310	0.00690	4.98	5.03	1.616	5.991	-1.822
400	425	136	58173	0.99543	0.00457	5.39	3.34	1.205	6.052	-1.262
425	450	82	58255	0.99683	0.00317	5.76	2.31	0.838	6.109	-0.801
450	475	66	58321	0.99796	0.00204	6.20	1.49	0.396	6.163	-0.315
475	500	38	58359	0.99861	0.00139	6.58	1.01	0.012	6.215	-0.004
500	525	30	58389	0.99913	0.00087	7.04	0.64	-0.451	6.263	0.369
525	550	20	58409	0.99947	0.00053	7.54	0.39	-0.949	6.310	0.936
550	575	22	58431	0.99985	0.00015	8.78	0.11	-2.185	6.354	2.657
575	600	9	58440	1	0	0	0	0	0	0

Table A10-7 Uncorrelated data (example from Noordwijk, but not taking into accout the fact that they are observed every three hours)

In order to solve this problem, we assume that we can divide the year in N_s periods of t_s hours, during which the wave height H_s does not vary. The basic idea behind this assumption is that, because of the persistence of winds, storm periods will have more-or-less the same duration. Therefore, the assumed time interval t_s is called "storm duration". Now, one can stipulate that each of the random observations describe an observation of one such storm.

This means that we usually have an unknown number of observations, however we know the percentage of exceedance of each observation. Because such an observation does in fact not represent the wave height of one (half-hour) sample (which we called H_s), but represents the average H_s during a t_s-hour storm, we will use the symbol H_{ss} for this observation.

Let us assume for the time being that we have a storm-duration of 12 hours. This means that we have $365 \cdot 24/12 = 730$ periods of 12 hours in a year ($N_s = 730$). This means that a "once-per-year" storm occurs in 1/730 times of the cases

$(=1.37 \cdot 10^{-3})$. And a "once-in-ten-year" storm occurs in $0.137 \cdot 10^{-3}$ times of the cases.

In the example one can see that observations with $H_{ss} > 4.75$ m occur in 0.001386 of the cases. It is clear that the "once-a-year" storm is approximately 4.75 m. The value $Q_{ss} = Q_s \cdot 730$ indicates the probability of a storm in a year.

It is obvious that from a statistical point of view values of $Q_s > 1$ have no meaning. However, physically this number indicates the number of storms per year.

The values of Q_{ss} can be plotted on log paper, which results directly in a design line, showing the probability of exceedance of a given design storm.

Statistically one can process the values of Q_s in the same way as has been done before with Q. Practically this computation can be done in a spreadsheet in a simple way. In fact, Table A10-3 is a print from a spreadsheet. A multiplier is computed with the value:

$$M = \frac{\text{number of observations}}{365 \cdot 24 \Big/ \text{storm duration}} = \frac{58440}{730} = 80$$

Now we calculate the total number of storms smaller than our bin-value by taking {total-(cumulative number)} and divide that by M. The result is given in the column with the heading s/y (storms per year). And this value can be plotted, resulting in Figure A10-10.

Also in this case, we calculate the slope and intercept of the regression line. In our example the slope (A) is 1.02, and intercept (B) equals 4.16. (In order to calculate these values, one should either make an extra column with ln Q_s or make a plot where Q_s is on the y axis and H_{ss} on the x-axis). For extrapolation one can use the equation:

$$Q = \exp\{-\beta(H_{ss} - \gamma)\} \tag{A10.15}$$

or in reversed order:

$$H_{ss} = \gamma - \frac{\ln Q}{A} \tag{A10.16}$$

in which A = slope of regression line and $\gamma = -B/A$ ($= 4.16/1.02 = 4.07$).
So, the 1/225 storm-height is

$$H_{ss \, 1/225} = 4.07 - \frac{\ln\left(\frac{1}{225}\right)}{1.02} = 9.39 \, m$$

storm exceedance

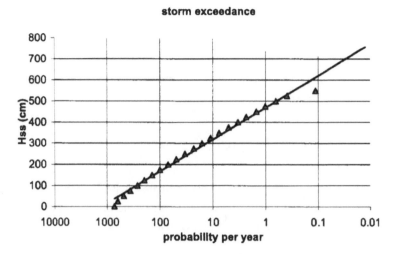

Figure A10-10 Storm exceedance graph using Random Observations

The value of 9.39 found in this way should be compared with the values between 8.31 and 9.39 found in the PoT analysis. This analysis too, can be improved by using the Weibull distribution instead of a simple log-distribution.

As was done with the PoT-analysis, the value of W_s is calculated using

$$W_s = \left(\ln \frac{1}{Q_s} \right)^{\frac{1}{\alpha}}$$

(A10.17)

See Table A10-3. At some place in the spreadsheet, a value of α is given and the data of the last column are calculated using that value of α. One can easily calculate slope, intercept and correlation. With trial and error, one can find that value of α which results in the highest correlation. In this example a value of $\alpha = 0.85$ (which is different than from the previous case) gives a correlation of 99.8 %. The parameters of β and γ follow from the slope and intercept of the regression line:

$$\beta = \frac{1}{A}$$

$$\gamma = -\frac{B}{A}$$

The values of α, β and γ can be used directly in the equation to calculate the probability:

$$W_s = \exp \left\{ -\left(\frac{H_{ss} - \gamma}{\beta} \right)^{\alpha} \right\}$$

(A10.18)

or in reversed order:

$$H_{ss} = \gamma + \beta\left(-\ln W_s\right)^{\frac{1}{\alpha}}$$
(A10.19)

So, the 1/225 storm height is:

$$H_{ss\,\frac{1}{225}} = 4.62 + 0.43\left\{-\ln\left(\frac{1}{225}\right)\right\}^{\frac{1}{0.85}} = 8.16\,m$$

One can make a plot of H_{ss} vs. W_s. This indicates that all points are nicely on a straight regression line. However, the value of W_s has no direct technical meaning. It becomes useful in case W_s is translated into Q_s using:

$$W_s = \left(\ln\frac{1}{Q_s}\right)^{\frac{1}{\alpha}}$$
(A10.20)

This means that one has to redefine the horizontal axis. Be aware that values of $W <$ 0 do not represent probabilities, but are only used to have a sufficient basis for extrapolation.

The resulting value of $H_{ss} = 8.16$ m should be compared with the values between 7.77 and 8.64 found in the PoT analysis. As stated before, the anticipated storm duration was 12 hours. Different durations result in different values for H_{ss}.

	3 hrs	6 hrs	9 hrs	12 hrs
Exponential	5.41	6.73	8.06	9.39
Weibull	8.48	8.32	8.23	8.16

Table A10-8

From this example it follows that in case a PoT-analysis is not possible, an analysis on the basis of tabulated (random) data can very well be done. The choice of the storm duration however remains a problem. Given the accuracy of the final answer, this choice is not extremely sensitive, at least when using the Weibull distribution.

A10.4 Computation of the armour units

A10.4.1 The classical computation

The classical way of computing the required block size is using the design formula, applying a design wave height with an exceedance of P_f during the lifetime of the structure. For example a lifetime of 50 years and a probability of failure of 20% during lifetime gives the following exceedance:

$$f=-\frac{1}{t_L}\ln(1-p)=-\frac{1}{50}\ln(1-0.2)=4.5\cdot10^{-3}=\frac{1}{225} \tag{A10.21}$$

Using the Weibull distribution from Section A10.2 for the example of Noordwijk, this results in a H_{ss} of 8.64 m. The design formula, as given by Van der Meer for cubes is:

$$\frac{H_{ss-design}}{\Delta d_n}=\left(6.7\frac{N_{od}^{0.4}}{N^{0.3}}+1.0\right)s_m^{-0.1} \tag{A10.22}$$

For N_{od} Van der Meer recommends a value of 0.5. The wave analysis in Chapter 5 has shown a wave steepness of 5.6%. There are approximately 4000 waves in a storm. For cubes with normal concrete one may use a concrete density of 2400 kg/m³, which results in a value of $\Delta = 1.75$.

Substituting these values in the above design formula yields a d_n of 2.4 m, or a block weight of 91 tons. In case one applies the Hudson formula with a K_D value of 5 (head) and a slope of 1:1.5, one gets a d_n of 3.3 m, and a weight of 83 tons (realise that the use of basalt split in the concrete may increase the density to 2800 kg/m³, which will give with Van der Meer a block weight of "only" 50 tons).

Because the depth in front of the Scheveningen breakwater is limited to 6 m below MSL (i.e. 9.5 m below design level), one may assume that waves never can become larger than 0.5 · 9.5 = 4.75 m (see Figure 5-26). Filling in these values results in a block weight of 15 tons.

However, one should realize that the number of occurrence of a storm with an H_{ss} of more than 4.75 is much more frequent. Using the Weibull distribution, one can find that a H_{ss} of 4.75 is exceeded once every 0.6 years. This means that during the design life (50 years) the breakwater will encounter 50/0.6 = 85 storms, with in total 400*85 = 33000 waves. Using this number of waves, we find a block size of 2.1 m, and a weight of 24 tons. Because the number of waves is not included in the Hudson formula, Hudson gives a block weight of only 14 tons.

During the design of the breakwater of Scheveningen, only the Hudson formula was available; the design of this breakwater however was not only based on this formula but also verified with model tests. The applied blocks in Scheveningen have a weight of 25 tons and a density of 2400 kg/m³.

A10.4.2 The method of Partial Coefficients

The method of partial coefficients is worked out in PIANC [1992]. In this method safety coefficients are added to the design formula. There are safety coefficients for load and for strength.

The partial safety coefficients for load

The probability of exceedance during service life of the design storm with a recurrence equal to the design life

$$P_{f\text{-lifetime}} = 1 - \left(1 - \frac{1}{t_L}\right)^{t_L} \tag{A10.23}$$

in which t_L is the design life. For a life time of 50 years, $P_{f\text{-lifetime}}$ is 0.64

The design storm to be applied has of course a much smaller probability, and consequently a longer return interval. The probability that a construction may fail during life time P_f is for example set to 20 %. The return period of the design storm then becomes

$$t_{Pf} = \left(1 - (1 - P_f)^{1/t_L}\right)^{-1} \tag{A10.24}$$

which gives 225 years according to our example.
To determine the partial safety coefficient one has to start with an extreme value distribution, for example the Weibull distribution. For this purpose the distribution is given as:

$$Q_{tL} = \left\{1 - \exp\left[-\left(\frac{H_{ss} - \gamma}{\beta}\right)^{\alpha}\right]\right\}^{N_s t_L} \tag{A10.25}$$

In this equation α and β are the parameters of the Weibull distribution, and γ is the threshold value. N_s is the number of observations per year.
This equation can be reworked to

$$H_{ss} = \gamma + \beta\left[-\ln\left\{1 - \exp\left(\frac{\ln Q_{tL}}{N_s T_L}\right)\right\}\right]^{1/\alpha} \tag{A10.26}$$

For Q_{tL} one should enter the non exceedance probability for $N_s t_L$ events during lifetime. This means that $Q_{tL} = P_{f\text{-lifetime}}$. For a design life of 50 years Q_{tL} is 0.364, for 100 years Q_{tL} is 0.366.
For a practical case this leads to the following values of H_{ss}:

Input from wave climate (from Meetpost Noordwijk):

α 1.24

β 1.17

γ 1.22

N_s 87.3

		lifetime 50 years	lifetime 100 years
H_{ss} for $t = t_L$ (50, 100)	$H_{ss}^{t_L}$	7.71	8.15
H_{ss} for $t = 3\ t_L$ (150, 300)	$H_{ss}^{3t_L}$	8.39	8.81
H_{ss} for $t = t_{20\%}$ (225, 450)	$H_{ss}^{t_{pf}}$	8.64	9.06

Table A10-9

The safety coefficient is given as:

$$\gamma_{H_{ss}} = \frac{H_{ss}^{t_{pf}}}{Q_{tL}} + \sigma'_{Q_L}\left(1 + \left(\frac{H_{ss}^{3tL}}{H_{ss}^{tL}} - 1\right)k_\beta P_f\right) + \frac{0.05}{\sqrt{P_f N}} \qquad (A10.27)$$

In this equation P_f is the allowable failure during lifetime (not to be mistaken with $P_{flifetime}$, which has been defined as the probability of exceedance of the "once in t_L-years storm" during the lifetime t_L.

The standard deviation $\sigma'_{Q_{tL}}$ is given in PIANC report [PIANC 1992] (copied in Table A10-10) as a function of the type of observations available. The data from Noordwijk are based on accurate observations, so a value of $\sigma'_{Q_{tL}} = 0.05$ is realistic. P_f was 20%, N is the number of "storms" in the PoT-analysis, for this example it is 1746. $k_\alpha = 0.027$ and $k_\beta = 38$ (see Table A10-2).

This leads to:

$$\gamma_{H_{ss}} = \frac{8.64}{7.71} + 0.2^{\left(1 + \left(\frac{8.39}{7.71} - 1\right)38 \cdot 0.2\right)} + \frac{0.05}{\sqrt{0.2 \cdot 1746}} = 1.13 \qquad (A10.28)$$

The safety coefficient consists of three parts. The first term gives the correct partial safety coefficient, provided no statistical uncertainty and measurement errors related to H_{ss} are present. The second term signifies the measurement errors and the short-term variability related to the wave data. The last term signifies the statistical uncertainty of the estimated extreme distribution of H_{ss}. The statistical uncertainty treated in this way depends on the total number of wave data N, but not on the length of the observation period.

Parameters	Method of determination	Typical value for σ'
Wave height		
Significant wave height offshore	Accelerometer buoy, pressure cell, vertical radar	0.05 – 0.1
	Horizontal radar	0.15
	Hindcast, numerical model	0.1 – 0.2
	Hindcast, SMB method	0.15-0.2
	Visual observation (Global wave statistics	0.2
H_{ss} nearshore determined from offshore H_{ss} taking into account typical nearshore effects (refraction, shoaling, breaking)	Numerical models	0.1-0.2
	Manual calculation	0.15-0.35
Other wave parameters		
Mean wave period offshore on condition of fixed H_{ss}	Accelerometer buoy	0.02-0.05
	Estimates from amplitude spectra	0.15
	Hindcast, numerical model	
Duration of sea state with H_{ss} exceeding a specific level	Direct measurements	0.1-0.2
	Hindcast, numerical model	0.02
Spectral peak frequency offshore	Measurements	0.05-0.1
	Hindcast, numerical models	0.05 0.15
Spectral peakedness offshore	Measurements and hindcast with numerical models	0.1-0.2
		0.4
Mean direction of wave propagation offshore	Pitch-roll buoy	
	Measurement of horizontal velocity components and waterlevel or pressure	5°
	Hindcast, numerical model	10°
		15-30°
Water level		
Astro tides	prediction from constants	0.001-0.07
storm surges	numerical models	0.1-0.25

Table A10-10 Typical variation coefficients for sea state parameters [from PIANC 1992]

If extreme wave statistics are not based on N wave data, but for example on estimates of H_{ss} from information about water level variations in shallow water, then the last term disappears and instead the value chosen for σ' must account for the inherent uncertainty.

In Table A10-11 the values of σ' and N have been changed to the values for simple manual calculations and a shorter dataset. It is clear from this table that the effect of the length of a dataset is less important than accurate observations.

	base example	use $\sigma' = 0.35$	use $N = 10$ storms	use $\sigma' = 0.35$ and $N = 10$
basic safety coefficient	100%	87%	99%	84%
measurement and short term errors	0%	13%	0%	13%
statistical uncertainty	0%	0%	1%	3%

Table A10-11

The partial safety coefficient for strength

The safety coefficient for the strength can be calculated using

$$\gamma_z = 1 - (k_\alpha \ln P_f) \tag{A10.29}$$

in which k_α and k_β are coefficients determined by optimisation and given in the PIANC manual [PIANC 1992]. These values are copied in Table A10-12. The value of P_f is the allowable probability of failure during lifetime.

Formula, type of construction	k_α	k_β
Hudson, rock	0.036	151
Van der Meer, rock, plunging waves	0.027	38
Van der Meer, rock, surging waves	0.031	38
Van der Meer, Tetrapods	0.026	38
Van der Meer, Cubes	0.026	38
Van der Meer, Accropodes	0.015	33
Van der Meer, rock, low crested	0.035	42
Van der meer, rock, berm	0.087	100

Table A10-12 Coefficients used to determine the partial safety factor γ_z

In the Coastal Engineering Manual the same approach is followed, however in that manual the values of γ_H and γ_z are directly given as a function of P_f and σ'.
For cubes, one can apply the Van der Meer equation:

$$\frac{1}{\gamma_z}\left(6.7\frac{N_{od}^{0.4}}{N^{0.30}} + 1.0\right)s_m^{-0.1}\Delta d_n \geq \gamma_{Hss}H_{ss}^{tL} \tag{A10.30}$$

For the harbour of Scheveningen, one may use the wave data of Noordwijk. Filling in values of $N_{od} = 1$, $N = 1500$, $\Delta = 1.75$, $s_m = 2.5\%$, this gives $d_n = 2.07$, or 25 tons.
Notice that in the above equation for the wave height the H_{ss} is used which has a probability of exceedance of once in the lifetime of the structure, i.e. the "once in 50 years storm" (7.71 m). This wave height is multiplied by γ_H (1.13), resulting in a total wave height of 8.71 m.
Traditionally one should use a wave height with a probability of 20% during the lifetime of 50 years. This wave has a yearly exceedance of $1/225 = 4.4 \cdot 10^{-3}$.
See also Chapter 3. The "once in 225 years wave" is 8.64 m (in the PIANC guidelines it is called H_{ss}^{tpf}) and compares quite well to the calculated value of 8.71 m.

Note that in the latter example, the limited water depth has not been taken into account.

A10.4.3 Probabilistic approach

Instead of the method with partial safety coefficients, one may apply a full probabilistic computation, either on level 2 or level 3. For level 2 one may apply the FORM method, for level 3 one may apply the Monte-Carlo method. In the examples below the computer program VaP from ETH-Zürich has been applied.

The first step is to rewrite the design equation as a Z-function. For cubes the Z-function is:

$$Z = \left(A \frac{N_{od}^{0.4}}{N^{0.30}} + 1.0 \right) s_m^{-0.1} \Delta d_n - H_{ss}^{tL} \tag{A10.31}$$

In this equation seven variables are used. In a probabilistic approach one has to determine the type of distribution for each parameter.

The constant 6.7 from the Van der Meer equation is replaced here by a coefficient A with a normal distribution. This coefficient has a mean of 6.7 and a standard deviation describing the accuracy of the equation itself. According to Van der Meer the standard deviation of the coefficient A is approximately 10% of its value.

Wave steepness is assumed to have a Normal distribution. The average steepness as well as the standard deviation of the steepness can be calculated from the dataset of Meetpost Noordwijk. If one considers only the heavier storms (i.e. storms with a threshold of e.g. 4.5 m), the average steepness is 0.058, with a standard deviation of 0.0025.

We then have 56 storms in 20 years (i.e. 2.8 storms per year) with an average duration of 7.2 hours. This means that the average period is 6.9 seconds, and that there are consequently 3700 waves in a storm. However, the duration of the storms varies quite a lot (it may go up to 20 hours, which means that we have 10000 waves. This means that N will have a large standard deviation. One can fit the calculated waves in each storm and fit this to a distribution. However, the effect of the number of waves is not that high so one may assume a Lognormal distribution (on cannot use a Normal distribution, because the number of waves may become negative when using high standard deviations).

Because the standard deviation in the block size (concrete cubes) is so small, this parameter can be considered as a deterministic value. The acceptable damage level is a target value, therefore this should also be a deterministic value.

H_{ss} has a Weibull or Gumbel distribution. For the wave-height one may enter for example a Weibull distribution, using the values of α, β and γ as determined before. This results in the following input table:

parameter	type	mean	standard deviation
A	Normal	6.7	0.67
N_{od}	Deterministic	1	-
N	Lognormal	3000	3000
S_m	Normal	0.058	.0025
Δ	Normal	1.75	0.05
d_n	Deterministic	2.4	-

Table A10-13

One can compute the probability of $(Z < 0)$, which is the probability of failure.

The target probability is $1/225 = 0.0044$ per year. However, VaP gives the probability per event. Because we have 87.3 storms per year, the target probability per event becomes $0.0044 / 87.3 = 50*10^{-6}$. The weight of a cube of 2.4 m is 37.5 tons.

For this example using the FORM method a probability of failure of $45*10^{-6}$ is found, which is quite near the target value. A Monte-Carlo computation gives a probability of failure of $70*10^{-6}$, so quite comparable.

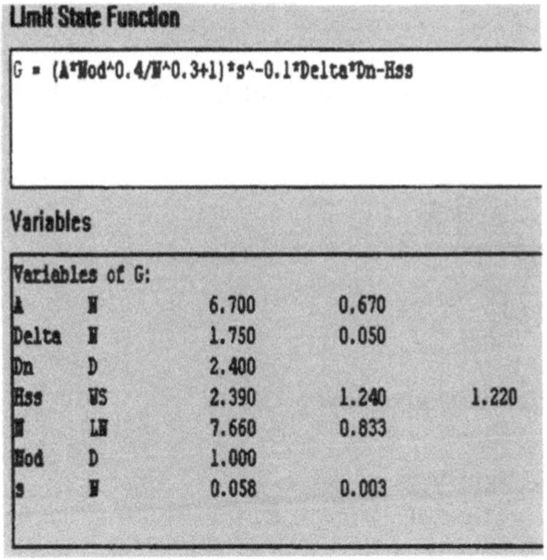

Limit State Function

$G = (A*Nod^0.4/N^0.3+1)*s^-0.1*Delta*Dn-Hss$

Variables

Variables of G:

A	N	6.700	0.670	
Delta	N	1.750	0.050	
Dn	D	2.400		
Hss	WS	2.390	1.240	1.220
N	LN	7.660	0.833	
Nod	D	1.000		
s	N	0.058	0.003	

Figure A10-11

This computation is based on the fact that we have defined $N_{od} = 1$ as start of failure. However, this is quite some damage. Lowering the value of N_{od} to 0.5 means that we have to increase the block size to 2.65 m (39 tons) in order to obtain the same probability of failure. Using a N_{od} of 0 means an increase to 3.80 m (44 tons)

In the above calculations the uncertainty in the determination of the parameters of the Weibull distribution was not taken into consideration. Although this is common practice, it is not fully correct. The mathematically correct way is to consider α, β

and γ as stochastic parameters with a mean and a standard deviation. One can determine these values directly from the dataset, but it is not possible to apply these values directly in a probabilistic computation. In practice this problem is solved by introducing an extra variable M. This variable has a mean value of 1 and a standard deviation which expresses the variation in the prediction of H_{ss} using a Weibull or Gumbel distribution. The Van der Meer formula then becomes:

$$Z=\left(A\frac{N_{od}^{0.4}}{N^{0.30}}+1.0\right)s_m^{-0.1}\Delta d_n - MH_{ss}^{tL} \tag{A10.32}$$

A problem is that the standard deviation of M depends on the value of H_{ss}. On can determine this value using the design value for H_{ss}.
The standard deviation is given as:

$$\sigma'_M = \frac{\sigma_M}{H_{ss-design}} \tag{A10.33}$$

For σ_M an expression has been derived by Goda [Goda, 2000]:

$$\sigma_M = \sigma_z\sigma_x \tag{A10.34}$$

in which σ_x is the standard deviation of all H_{ss} values in the basic dataset and σ_z is defined by:

$$\sigma_z = \frac{\left[1.0+a(y-c)^2\right]^{1/2}}{\sqrt{N}} \tag{A10.35}$$

$$a = a_1\exp\left[a_2 N^{-1.3}\right]$$

in which the coefficients are given by:

distribution	a_1	a_2	c
Gumbel	0.64	9.0	0
Weibull, $\alpha = 0.75$	1.65	11.4	0
Weibull, $\alpha = 1.0$	1.92	11.4	0.3
Weibull, $\alpha = 1.4$	2.05	11.4	0.4
Weibull, $\alpha = 2.0$	2.24	11.4	0.5

Table A10-14

y is the reduced variate for the design value, our design value is a 1/225 wave, so $y = [\ln(87.3*22)]^{1/1.24} = 6.34$.

In the example used, we have a Weibull with $\alpha = 1.24$; because this value is not in the table, an interpolation is needed. For the example the following values were being used: $a_1 = 2.0$, $a_2 = 11.4$ and $c = 0.35$.

This results in a value of $\sigma_z = 0.024$. Given a σ_x of 6.85 (follows from dataset), this means:

$$\sigma'_M = \frac{\sigma_M}{H_{ss-design}} = \frac{\sigma_x \sigma_z}{H_{ss-design}} = \frac{6.85 \cdot 0.024}{8.64} = 0.02 \tag{A10.36}$$

A recalculation with these figures gives a probability of failure of $157*10^{-6}$. In order to bring this back to the required $50*10^{-6}$ we have to increase the cube size to 2.42 m. So, this can be neglected.

However, if our dataset would have been considerably smaller (for example only 100 storms), this would change the value of σ_z to 0.175 and consequently σ_M' to 0.14. In order to get a probability of failure in the order of the required $50*10^{-6}$ we have to increase the cube size to 2.70 m (from 37.5 to 55 tons). Again this is for deep (18 m) water conditions.

This shows that the size of the dataset has a considerable impact on the required block size.

A10.4.4 Probabilistic calculation in case of a shallow foreshore

Statistical software does discard physical limitations. E.g., applying statistics in deep water circumstances will give very high waves with very low probability. But in shallow water these high waves cannot exist at all. In general the significant wave height in shallow water is limited by the depth. So:

$$H_{ss} = \gamma_b h \tag{A10.37}$$

in which the breaker index γ_b has a value in the order of 0.6. For individual waves γ_b may have values up to 0.78, but for the significant wave height this value is much lower, sometimes even down to 0.45.

For foreshores with a gentle slope one may assume that γ_b may have an average value of 0.55, with a standard deviation of 0.05. The water depth in this equation is the total depth, i.e. the depth below mean sea level + the rise of waterlevel due to tide and storm surge.

With this information, one can rewrite the Van der Meer equation for cubes to:

$$Z = \left(A \frac{N_{od}^{0.4}}{N^{0.30}} + 1.0 \right) s_m^{-0.1} \Delta d_n - \gamma_b (h_{surge} + h_{depth}) \tag{A10.38}$$

The water depth below mean sea level has a Normal distribution, but the standard deviation in this parameter is usually so low, that it can be considered as a deterministic value. Of course in case large bed fluctuations are to be expected, one may also enter this value as a stochastic parameter with a Normal distribution. The surge has an extreme value distribution. For this value one can apply a Gumbel distribution.

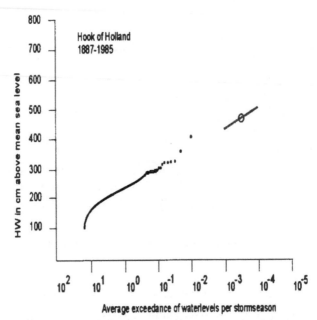

Figure A10-12 Extreme waterlevels in Hook of Holland

Again we look at the example of Scheveningen harbour. The water depth in front of the breakwater in Scheveningen is 6 m below mean sea level. Long term waterlevels are available from Hook of Holland, which are approximately identical to Scheveningen. The exceedance can be given by:

$$Q=1-\exp\left[-\exp\left(-\frac{h_{surge}-\gamma}{\beta}\right)\right]$$
(A10.39)

In this equation is γ the intercept at the 10^0-line, and β is the slope parameter. From the diagram one can derive that γ equals 2.3 and that the slope β is 0.30.

The equation is based on the maximum surges per year, so the exceedance is also per year, and not per storm. This implies that in this case the target probability of failure is $1/225 = 0.0044$.

This leads to $d_n = 1.85$ m, or a block weight of 17.7 tons.

In principle one can in this case also add the statistical uncertainty by adding a factor M in front of the surge height in the equation. However, because of the long dataset and the limited extrapolation, this uncertainty is very small and may be neglected.

Limit State Function

G = (A*Mod^0.4/N^0.3+1)*s^-0.1*Delta*Dn-gamma*(h+6)

Variables

Variables of G:

A	N	6.700	0.670
Delta	N	1.750	0.050
Dn	D	1.850	
N	LN	7.660	0.833
Mod	D	1.000	
gamma	N	0.550	0.050
h	GL	2.300	3.289
s	N	0.025	0.002

Figure A10-13

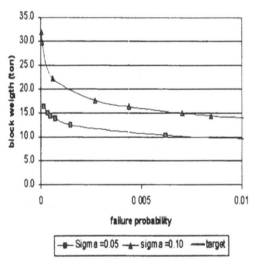

Figure A10-14 Effect of standard deviation of γ_b on block weight

Changing the standard deviation in γ from 0.05 to 0.1 has a considerable effect. To obtain the same target probability of failure, one has to increase the blocks to d_n = 2.2 m (30 tons).

A simplification made in this computation is that is assumed that in deep water, wave steepness remains the same after breaking. This is probably not the case. The higher waves will break (usually as spilling breakers) which decreases their height considerably, but usually not the period. As a consequence the wave steepness in broken waves is much less than the wave steepness at sea. Because in the Van der Meer formula for cubes a low steepness gives smaller cubes than a high steepness, neglecting the change in steepness is a conservative approach. Be aware that in case of natural rock the opposite is true.

References

Overview of useful lecture notes

Battjes, J.A. (1998), "Korte golven", Hydraulic Engineering Section, Faculty of Civil Engineering, Delft University of Technology, Delft, The Netherlands.

Battjes, J.A. (1998), "Stromingen in open Waterlopen", Hydraulic Engineering Section, Faculty of Civil Engineering, Delft University of Technology, Delft, The Netherlands.

d'Angremond, K. et al. (1998), "Introduction to Coastal Engineering", Hydraulic Engineering Section, Faculty of Civil Engineering, Delft University of Technology, Delft, The Netherlands.

De Ridder, H.A.J. (1999), "Organisatie van het Ontwerpproces", collegedictaat CTow4050, Sectie Civiele Bedrijfskunde, Faculteit Civiele Techniek en Geowetenschappen, Delft University of Technology, Delft, The Netherlands.

Schiereck, G.J. (1998), "Introduction to bed, bank and shore protection", Lecture notes Ctwa4310, Hydraulic Engineering Section, Faculty of Civil Engineering, Delft University of Technology, Delft, The Netherlands. These lecture notes have been included in Introduction to bed, bank and shore protection by G.J. Schiereck (2001), Delft University Press.

Various authors (1986), "Coastal engineering", pg. 90-92, edited by Massie, W.W., Coastal Engineering group, Faculty of Civil Engineering, Delft University of Technology, Delft, The Netherlands.

Velden, van der E.T.J.M. (1989), "Coastal Engineering", Coastal Engineering group, Faculty of Civil Engineering, Delft University of Technology, Delft, The Netherlands.

Vrijling, J.K. (1986), "Probabilistisch Ontwerpen in de Waterbouwkunde", Hydraulic Engineering Section, Faculty of Civil Engineering, Delft University of Technology, Delft, The Netherlands.

Other references

Angremond, K. d', et al. (1970), "Use of Asphalt in Breakwater Construction". Proc. 12th International Coastal Engineering Conference, ASCE, New York, USA.

Angremond, K.d', Meer, J.W. van der and Jong, R.J. de (1996), "Wave transmission at low-crested structures", Proc. 25th ICCE, Orlando. ASCE New York, USA.

Angremond, K.d' (1982), "Breakwater Design", International Institute for Hydraulic and Environmental Engineering, Delft, The Netherlands

Anonymous (1970), "Deltawerken, Driemaandelijks Bericht, nummer 54, November 1970", onder redactie van de Deltadienst, Staatsuitgeverij, 's-Gravenhage, The Netherlands.

Anonymous (1985), "Computer Aided Evaluation of the Reliability of a Breakwater Design", Final Report of the CIAD Project Group Breakwaters, ISBN 90-6818-019-3 CIAD, Zoetermeer, the Netherlands.

Anonymous, British Admiralty Tide Tables

Anonymous (1977), "Shore Protection Manual" Third Edition, U.S. Army Coastal Engineering Research Center. Fort Belvoir, Virginia, USA.

Anonymous (1984), "Shore Protection Manual", Fourth edition, U.S. Army Engineers Waterways Experiment Station, Coastal Engineering Research Center, Vicksburg, USA.

Anonymous, "Journal of Waterway, Port, Coastal and Ocean Engineering", ASCE, New York, USA

Anonymous, "Proceedings of various International Conferences on Coastal Engineering", ASCE

Anonymous, "An International Journal for Coastal, Harbour and Offshore Engineers", Elsevier Science Publications, Amsterdam, The Netherlands

Anonymous (1998), "Technisch Rapport Golfoploop en Golfoverslag", Technische Adviescommissie voor de Waterkeringen, Delft, The Netherlands.

Bagnold, R.A. (1939), "Interim Report on Wave Impact Research", Journal of the Institution of Civil Engineers, Vol. 12, London, UK.

Battjes, J.A., and Janssen, J.P.F.M. (1978), "Energy loss and set-up due to breaking of random waves", Proc.16th Int. Coastal Eng. Conf., ASCE, New York, USA.

Battjes, J.A. (1974), "Computation of set-up, longshore currents, run-up and overtopping due to wind-generated waves", Communications on Hydraulics, Delft University of Technology, Report 74-2, Delft, The Netherlands.

Blackman, D.J. (1982), "Ancient harbours in the Mediterranean", Int. Journal of Nautical Archaeology and Underwater Exploration, part 1 in 11.2 pp. 97-104; part 2 in 11.3 pp 185-211.

Bouws, E. (1978), "Wind and Wave Climate in the Netherlands Sector of the North Sea between 53° and 54° North Latitude", KNMI, De Bilt, The Netherlands.

Bradbury, A.P., Allsop, N.W.H., and Stephens, R.V. (1988), "Hydraulic performance of breakwater crown wall", Report SR146 Hydraulics Research, Wallingford, U.K.

Bregman, M. (1998), "Porositeit in geplaatste steenlagen, onderzoek naar factoren die de hoeveelheid geplaatst steenmateriaal beïnvloeden", Masters Thesis, Delft University of Technology, Delft and Royal Boskalis Westminster, Papendrecht, The Netherlands.

Browder, A.E., Dean, R.G.,Chen, R. (1996), " Performance of a submerged breakwater for shore protection", proceedings of 25th ICCE 1996 Florida, edited by ASCE, New York, USA.

Burcharth, H.F. and Frigaard, P. (1987), "On the stability of berm breakwater roundheads and trunk erosion in oblique waves", Proc. ASCE Seminar on unconventional rubble-mound breakwaters, Canada.

Burcharth, H.F. and Frigaard, P. (1988), "On 3-dimensional stability of reshaping berm breakwaters", Proc. 21st ICCE, ASCE New York. USA.

Burcharth, H.F. (1992), "Reliability Evaluation of Structures at Sea", Proc. Of the Short Course on Design and Reliability of Coastal Structures, Venice 1992, ed. Tecnoprint Bologna, Italy.

Burger, G. (1995), "Stability of low-crested breakwaters" Master's Thesis, Delft University of Technology, Delft, The Netherlands.

Burger, G. (1995), "Stabiliteit Golfbrekers met lage kruin" in Dutch, Master's thesis, Delft University of Technology, Faculty of Civil Engineering, Delft, The Netherlands.

CUR/CIRIA (1991), "Manual on the use of rock in coastal and shoreline engineering", CUR/CIRIA Report 154/83, CUR, Gouda, the Netherlands.

CUR/RWS (1995), "Manual on the Use of Rock in Hydraulic Engineering", CUR/RWS Publication 169, CUR, Gouda, The Netherlands.

Daemen, I.F.R. (1991), "Wave transmission at low-crested breakwaters", MSc Thesis Delft University of Technology, Faculty of Civil Engineering, Delft, The Netherlands.

Daemrich, K.F. and Kahle, W. (1985), "Schutzwirkung von Unterwasserwellenbrechern unter dem Einfluss unregelmassiger Seegangswellen", Internal report Heft 61 Franzius Institut fur Wasserbau und Kusteningenieurswesen, Germany.

Danel, P. (1953), "Tetrapods", Proc. Fourth Int. Conf. Coastal Eng., ASCE, Chicago, pp. 390 – 398, USA.

d'Angremond, K., Span H.J. Th., Weide, J. v.d. and Woestenenk, A.J. (1970), "Use of asphalt in breakwater construction", Proc. 12th cont.. Coastal Engineering, London, UK.

De la Pena, J.M., Prada, E.J.M., Redondo, M.C. (1994), "Mediterranean Ports in Ancient Times", PIANC Bulletin No. 83/84, Brussels, Belgium.

Dienst der Hydrografie (1994), "Stroomatlas HP15, Westerschelde-Oosterschelde, GW D09 M03", Koninklijke Marine, Dienst der Hydrografie, 's Gravenhage, The Netherlands.

Dienst der kusthydrografie Oostende (1971), "Stroom Atlas Vlaamse Banken Noordzee", Koninkrijk België, Ministerie van Openbare werken, Dienst der kusthydrografie Oostende, Oostende, Belgium

Docters van Leeuwen, L. , "Toe stability of Rubble Mouth Breakwaters", Thesis, Delft University of Technology, Delft, The Netherlands

Dorrestein (1967), R., "Wind and Wave Data Netherlands Light Vessels since 1949", De Staatsdrukkerij, 's Gravenhage, The Netherlands.

Escher, B.G. (1940) "Algemene Geologie" (in Dutch), Wereldbibliotheek, Amsterdam

Franco, L., Gerloni, M.de and Meer, J.W. van der (1994), " Wave overtopping on vertical and composite breakwaters", Proc. 24th ICCE Kobe, ASCE, New York, USA.

Franco, L. (1996), "History of Coastal Engineering in Italy", in: History and heritage of coastal engineering, ASCE, New York, USA.

Gerloni, M. de, Franco, L. and Passoni, G. (1991), "The safety of breakwaters against overtopping", Proc. ICE Conference on Breakwaters and Coastal Structures, Thomas Telford, London.

Gent, M.R.A. van, "Wave Interaction with Permeable Coastal Structures", Doctors Thesis, Delft University of Technology, Delft, The Netherlands

Goda, Y. (1985), "Random seas and Design of Maritime Structures", University of Tokyo Press, Tokyo, Japan.

Goda, Y. (1992), "The design of upright breakwaters", Proc. Of the sort course on Design and Reliability of Coastal Structures, Venice, Instituto di Idraulica Universita di Bologna, Bologna, Italy.

Graauw, A. de, Meulen, T. van der and Does de Bye, M. van der (1983) "Design Criteria for Granular Filters" Delft Hydraulics Publication no. 287, Delft, The Netherlands.

Groen, P. and Dorrestein, R. (1958), " Zeegolven", Deel 11: Opstellen op oceanografisch en maritiem meteorologisch gebied, KNMI, Staatsdrukkerij- en Uitgeversbedrijf, 's-Gravenhage, The Netherlands.

Groenendijk, H.W. (1998), "Shallow foreshore wave heights statistics", Master Thesis, Delft University of Technology, Delft, The Netherlands.

Groenendijk, H.W., and Van Gent, M.R.A. (1998), "Shallow foreshore Wave Height Statistics", W.L. Delft Hydraulics, Delft, The Netherlands.

Gumble, E.J. (1958),"Statistics of Extremes", Colombia Univ. Press.

Hogben, F., and F.E. Lumb (1967), "Ocean Wave Statistics", HSMO, London, United Kingdom

Hogben, N, Dacunha, N.M.C., and Oliver, G.F. (1986), "Global Wave Statistics", Edited for British Maritime Technology ltd. by Unwin Brothers Ltd, London, United Kingdom.

Holzhausen, A.H. and Zwamborn, J.A. (1992), "New Stability Formula for Dolosse", Proc. 23rd ICCE, pp 1231-1244, ASCE, New York, USA..

Hudson, R.Y. (1953), "Wave forces on breakwaters" Transactions ASCE 118, pp 653-674, ASCE New York, USA.

Hudson, R.Y. (1959), "Laboratory investigations of rubble mound breakwaters" Proc. ASCE 85 WW 3, ASCE, New York, USA.

Hudson, R.Y. (1961), "Wave forces on rubble mound breakwaters and jetties" U.S. Army Waterways Experiment Station, Misc. Paper 2-453, Vicksburg (Miss.) USA.

Huis in 't Veld, J.C, Stuip, J. ,Walther, A.W., van Westen, J.M. (1984), "The Closure of Tidal Basins", Closing of estuaries, Tidal inlets and Dike breaches, Delftse Universitaire Pers, Delft, The Netherlands

Huitema, T (1947), "Dijken, samenstelling, aanleg en onderhoud", Uitgeversmaatschappij Kosmos, Utrecht, The Netherlands.

International waves commission (1976), "Final report, second part", Brussels, Belgium (updated report in preparation)

Iribarren Cavanilles, R. (1938), "Una formula para el calculo de los diques de escollera", M. Bermejillo-Pasajes, Madrid, Spain.

Iribarren Cavanilles, R. and C. Nogales y Olano (1950), "Generalizacion de la formula para el calculo de los diques de escollera y comprobacion de sus coeficientes", Revista de Obras Publicas, Madrid, Spain.

Iribarren Cavanilles, R. and C. Nogales y Olano (1953), "Nueva confirmacion de la formula para el calculo de los diques de escollera", Revista de Obras Publicas, Madrid, Spain.

Iribarren Cavanilles, R. and C. Nogales y Olano (1954), "Otras comprobaciones de la formula para el calculo de los diques de escollera", Revista de Obras Publicas, Madrid, Spain.

Iribarren Cavanilles, R. and C. Nogales y Olano (1953), "Nouvelles conceptions sur les diques a parois verticales et sur les ouvrages a talus" XVIIIth Int. Navigation Congress, Rome, Italy

Iribarren Cavanilles, R. (1965), "Formule pour le calcul des diques en enrochements naturels ou elements artificiels" XXIst Int. Navigation Congress Stockholm, Sweden.

Jarlan, G.L.E (1961), "A perforated wall breakwater", The Dock and Harbour Authority 41, no. 486, London, UK.

Jong, R.J. de (1996), "Stability of Tetrapods at front, crest and rear of a low-crested breakwater" MSc thesis Delft University of Technology, Delft, The Netherlands

Jong, R.J. de (1996), "Wave transmission at low-crested structures", Part I MSc Thesis Delft University of Technology, Faculty of Civil Engineering, Delft, The Netherlands.

Kawakami, K. and Esashi, Y. (1961), "On Drainage Filter for Earth Structure" Abstracts of Papers 16th Annual Meeting, Jap. Society of Civil Engineers.

Klabbers, M, Van den Berge, A., Muttray, M. [2004] Hydraulic performance of Xbloc armour units; proc. 29th ICCE, Lisbon, Portugal

Kreeke, J. van de , and Paape, A. (1964), "On Optimum Breakwater Design", Proc. 9th ICCE, ASCE, New York, USA.

Latham, J.-P. et al. (1988), "The influence of armour stone shape and rounding on the stability of breakwater armour layers" Queen Mary College, University of London, U.K.

Le Mehaute, B. (1969), "An Introduction to Hydrodynamics and Water Waves," Water Wave Theories, Vol. II, TR ERL 118-POL-3-2, US Department of Commerce, ESSA, Washington D.C., USA.

Meer, J.W. van der (1988a), "Rock Slopes and Gravel Beaches under Wave Attack", Doctors thesis, Delft University of Technology, Delft The Netherlands

Meer, J.W. van der (1988b), "Stability of Cubes, Tetrapods and Accropode", Proc. Conference Breakwaters '88, Thomas Telford, London, United Kingdom.

Meer, J.W. van der, Angremond. K. d' and Gerding, E. (1995) "Toe structure stability of rubble mound breakwaters" Proc. Coastal Structures and Breakwaters, ICE, London U.K.

Meer, J.W. van der (1990), "Verification of Breakwat for berm breakwaters and low-crested structures.", Delft Hydraulics Report H986, Delft, The Netherlands.

Meer, J. W. van der, and Veldman, J.J. (1992), "Stability of the seaward slope of berm breakwaters", Journal of Coastal Engineering 16, pp 205-234, Elsevier, Amsterdam, The Netherlands.

Meer, J.W. van der, and Stam, C.J.M. (1992), "Wave run-up on smooth and rock slopes of coastal structures", ASCE, Journal of WPC & OE, Vol. 118, No. 5 pp 534-550, New York, USA

Meer, J.W. van der (1990), "Data on wave transmission due to overtopping", WL/Delft Hydraulics Internal Report No. H986, Delft, The Netherlands.

Meer, J.W. van der and Angremond, K.d' (1991), "Wave transmission at low-crested structures", Proc. ICE Conference Coastal Structures and Breakwaters, Thomas Telford, London, U.K.

Meer, J. van der (1993), "Conceptual Design of Rubble Mound Breakwaters", WL/Delft Hydraulics, Delft, The Netherlands.

Melby, J.A. & Turk, G.F. [1997] Core-Loc Concrete armour units, Technical Guidelines, US Army Corps of Engineers, Technical Report CHL 97-4

Meulen, T. van der, Schiereck, G.J. and Angremond, K. d' (1997), "Toe stability of rubble mound breakwaters", Proc. 25th ICCE, ASCE New York, USA.

Miche, R. (1944), "Mouvements ondulatoires de la mer en profondeur constante ou decroissante", Annales des Ponts et Chaussees, Vol 114, Paris, France.

Minikin, R.R. (1955), "Breaking Waves: A comment on the Genoa Breakwater", Dock and Harbour Authority, pp 164-165, London, UK.

Minikin, R.R. (1963), "Winds, waves and maritime structures: Studies in Harbor Making and in the Protection of Coasts", 2nd rev. ed., Griffin, London, UK

Oorschot, J.H. van, Discussion paper. Proc. symp. Research on Wave Action, Vol. III (1968), Delft, The Netherlands.

Owen, M.W., (1980), "Design of Seawalls allowing for wave overtopping", Report No. EX 924, Hydraulics Research, Wallingford, U.K.

PIANC (1992), "Safety of Rubble Mound Breakwaters", Report of PTC II Working Group 12, PIANC, Brussels, Belgium.

PIANC (1993), "Analysis of Rubblemound breakwaters", Report of PIANC working group no. 12 of permanent committee II, PIANC, Brussels, Belgium

PIANC (1997), "Approach Channels, a Guide for Design", Final Report of the joint working group PIANC and IAPH PTC II-30, PIANC, Brussels, Belgium.

PIANC (1995), "Criteria for movements of moored ships in harbours", a practical guide, Report of Working Group 24 of the Permanent Technical Committee II, PIANC, Brussels, Belgium.

PIANC (1976), "Final Report International Commission on the Study of Waves", Part 2, PIANC, Brussels, Belgium.

Postma, G.M. (1989), "Wave reflection from rock slopes under random wave attack", MSc thesis Delft University of Technology, Faculty of Civil Engineering, Delft, The Netherlands.

Powell, K.A. and Allsop, N.W.H. (1985), "Low-crest breakwaters, hydraulic performance and stability", Report No. SR 57 Hydraulics Research, Wallingford, UK.

Rundgren, L. (1958), "Water Wave Forces", Bulletin No. 54, Royal Institute of Technology, Division of Hydraulics, Stockholm, Sweden.

Sainflou, M. (1928), "Treatise on vertical breakwaters", Annals des Ponts et Chaussees, Paris, France.

Shaw, J. (1974), "Greek and Roman Harbour Works, from Navi e Civiltà", Milan.

Seelig, W.N. (1980), "Two-dimensional tests of wave transmission and reflection characteristics of laboratory breakwaters", CERC Technical Report No. 80-1, US Army Waterways Experiment Station, Vicksburg, USA.

Simons, Th.M.H.G., (1981) Contribution Themadag 3 Koninklijk Instituut van Ingenieurs, The Hague, the Netherlands

Tanimoto, K. and Takahashi, S. (1994), "Design and Construction of caisson breakwaters, the Japanese experience", Journal of Coastal engineering Vol 22 pp 55-77 Elsevier, Amsterdam, The Netherlands.

Takahashi, S. (1996), "Design of Vertical Breakwaters", Reference Document no.34, Port and Harbour Research Institute, Ministry of Transport, Yokosuka, Japan.

The Liverpool Thessaloniki network (1996)," European Coasts", an introductory survey, Hydraulic Engineering group, Department of Civil Engineering, Delft University of Technology, Delft, The Netherlands.

Thoresen, C.A., (1988), "Port Design, Guidelines and Recommendations", Tapir Publishers, Trondheim, Norway.

Velsink, H. (1987), "Principles of integrated port planning", 56th bulletin of PIANC, Brussels, Belgium

Vidal, C., Losada, M.A., Medina, R., Mansard, E.P.D. and Gomez-Pina, G. (1992), "A universal analysis for the Stability of both low-crested and submerged breakwaters", Proc. 23rd Int. Coastal Engrg. Conf., ASCE, New York, N.Y., USA.

Vierlingh, A (1570), "Tractaet van Dyckagie" (in old Dutch), reprint of an old manuscript. Martinus Nijhoff. Den Haag, 1973. Published by the Ned.Ver.Kust- en Oeverwerken, Rotterdam.

Vitruvius, M.L. (27 BC.), "De Architectura", vol II –6; vol. V –13 (in Latin). English translation by Morgan, M.H. (1914), in Cambridge or Loeb Classical Library , London, United Kingdom. Dutch translation (1998) published by Querido, Amsterdam, the Netherlands.

Waal, J.P. de, and Meer, J.W. van der (1992), "Wave run-up and overtopping at coastal structures", Proc. 23rd ICCE Venice, Italy, ASCE New York.

Young, I.R. and Holland, G.J. (1996), "Atlas of the Oceans, Wind and Wave Climate", Pergamon, Oxford, United Kingdom.

Zwamborn, J.A. (1980), "Dolosse", Oceans, 1980 no. 4.

Index

T - #0206 - 071024 - C0 - 246/174/21 - PB - 9780415332569 - Gloss Lamination